GIS/LIS
PROCEEDINGS

10-12 November, 1992
San Jose, California

VOLUME 2

Sponsored by:
American Congress on Surveying and Mapping
American Society for Photogrammetry and Remote Sensing
Association of American Geographers
Urban and Regional Information Systems Association
AM/FM International

ISBN 0-944426-90-5
ISBN 0-944426-92-1

Published by

American Society for Photogrammetry and Remote Sensing
American Congress on Surveying and Mapping
5410 Grosvenor Lane
Bethesda, Maryland 20814

Association of American Geographers
1710 16th Street, N.W.
Washington, D.C. 20009

Urban and Regional Information Systems Association
900 2nd Street, N.E.
Washington, D.C. 20002

AM/FM International
14456 East Evans Avenue
Aurora, Colorado 80014

Printed in the United States of America

TABLE OF CONTENTS
VOLUME 1

VOLUME 2

OPPORTUNITIES AND CHALLENGES OF PARALLEL PROCESSING IN SPATIAL DATA ANALYSIS: INITIAL EXPERIMENTS WITH DATA PARALLEL MAP ANALYSIS

Bin Li
Department of Geography
Syracuse University
Syracuse, NY13244

ABSTRACT

The rapid development of parallel computer technology introduces both opportunities and challenges to the field of spatial data analysis and modeling. There are some basic issues in applying parallel processing to solving geographic problems. We suggest that these issues can be studied by actual mapping of computational models in geography to parallel software models. This paper reports results of initial experiments on implementing the Map Analysis Model to the Data Parallel Model.

INTRODUCTION

Technological advances are gradually removing the technical and economic constraints to the accessibility of spatial data in various resolutions and scales. More and more sophisticated techniques are also developed to solve geographic problems using computers. This trend generates the needs for high performance computers in certain categories of spatial analysis.

The development path of high performance computers exhibits a trend towards adopting massive parallel architectures. The likely dominance of parallel computers in the future high performance computing environment presents both opportunities and challenges. On the one hand, many people believe that parallel processing may overcome the computational and conceptual bottlenecks of the Von Neumann sequential processing model (Langran, 1991). On the other hand, it is a general perception that parallel computers are difficult to use and difficult to program, and that high efficiency and speedup are not easy to achieve without tremendous overheads in developing efficient parallel algorithms.

Leading scholars in computational science suggest that the fundamental issue in solving problems on a parallel computer is whether we can efficiently map the problem architecture to a particular parallel architecture. Geoffrey Fox is among the few who attempted to establish a general framework for mapping "real-world" problems to a particular class of parallel computers. He suggested that the Theory of Complex Systems could be applied to the mapping problem (Fox, 1988). By treating "real-world" problems and parallel computers both as complex systems, Fox found that a subclass of the problem systems can be handled effectively by a subclass of parallel computers as long as such computers have a topology that contains, or is richer, than that of the problem systems (Fox, 1988). He further classified computational problems in major fields that utilize parallel computers into three major classes: synchronous problems, loosely synchronous problems, and asynchronous problems (Fox, 1991).

While Fox's work mainly addressed issues in low level parallel software development, we suggest that from a user's point of view, a practical approach to examining issues in solving problems with parallel computers is through actual mapping of existing computational models to a particular parallel software model. In this paper, we report our initial experiment with mapping the Map Analysis Model to the Data Parallel Software Model.

We chose these models because both play central roles in their respective fields. As we all know, the Map Analysis Model, or Cartographic Model, is at the core of spatial data analysis and modeling in current GIS. Spatial modeling functions provided by leading GIS packages, such as GRID, GRASS, and ERDAS, are all based on the Map Analysis Model, originally formalized by Dana Tomlin (Tomlin, 1990).

Data parallel model is one of the two major software models in parallel processing — data parallel and task parallel. The concept is promoted and implemented by Hillis at the Thinking Machine Corporation who produces the Connection Machine (Hillis, 1986, 1987). From the hardware perspective, data parallelism is achieved by distributing data to the processor array and having processors execute the same instructions on their respective data elements simultaneously. Theoretically, data parallelism offers the highest potential for concurrency, and therefore the greatest potential for gains in performance. Conceptually, data parallel processing provides more natural approaches to solving particular class of problems than sequential processing. Structurally, data parallel processing are most suitable for synchronize problems, i.e., problems with regular spatial and temporal structures, which seems to be a natural choice for raster-oriented map analysis. In reality, data parallel languages are more mature. Newly developed parallel languages, such as FORTRAN90, C*, *LISP, and MPL, are all based on the data parallel software model.

We conducted the experiment through three steps. First, basic operations in map analysis are classified according to their computational characteristics. Primitive data parallel operators are also generalized and grouped based on existing data parallel languages. Second, we study the data parallel implementations for basic map analysis operations in each class, examining the efficiency and potential performance improvements of particular data parallel algorithms. Third, we conduct actual performance comparisons on selective operations between sequential processing and parallel processing.

PRIMITIVE OPERATIONS IN MAP ANALYSIS

There have been several attempts to categorize primitives in map analysis since Tomlin (Tomlin, 1990; Burrough, 1991). They generally belong to two schemes of classifications — task oriented and operation oriented. The former characterizes what an operation does, while the latter is based on how the results are obtained. Since we are concerned about how particular operations can be implemented with data parallel instructions, operation oriented classification seem to be most appropriate.

Based on the spatial relations between individual data unit and its neighborhood, its region (zonal), and all other data units in the coverage in the operation, we can group map analysis primitives into four categories: local, focal, zonal, and global (figure 1).

  • In local operations, the resulting value is a function of all corresponding data units at the same location.

- **Local operations — Elementwise parallel operations**

    1. **Common operators:** same as sequential
    2. **Special operators** $\begin{cases} \text{Combinatorial: *} \\ \text{Logical: elementwise logical operation} \end{cases}$
    3. **Reclassification: WHERE, =**
    4. **Selection: WHERE**
    5. **Statistical** $\begin{cases} \text{Class I: inter-layer parallel operations} \\ \text{CalssII: corresponding data elements stored in local memory} \end{cases}$

- **Focal operations — Regular pattern inter-processor data movement**

    1. **Surface functions** Convolution
    2. **Statistical** $\begin{cases} \text{Class I: data movement around window with computation} \\ \text{CalssII: neighbor values must be stored in local memory} \end{cases}$

- **Zonal operations — Selection, Reduction, Scan, Permutation**

    1. **Contiguous** $\begin{cases} \text{Statistics} \begin{cases} \text{Class I: segmented scan} \\ \text{CalssII: parallel operations on each zone} \end{cases} \\ \text{Geometry: reduction, (thickness *)} \end{cases}$

    2. **Non-contiguous:** select each zone sequentially to perform parallel operations

- **Global — Scan functions**

    1. **Euclidean distance function:** 1. segment scan; 2. selection
    2. **Weighted distance function:** segment scan, (path *, corridor *)
    3. **Visibility** (Line of sight): segment scan
    4. **Grouping:** 1. segment scan; 2. distance doubling
    5. **Expand/shrink** : maximum distance neighborhood searching
    6. **Interpolation:** *
    7. **Raster-vector conversion:** segmented scan
    8. **Accumulative operation:** reduction

Figure 1. Primitive operations in Map Analysis and general strategies for data parallel implementation. The symbol * indicates the operation has not been implemented.

- In focal operations, the resulting value at each location is a function of data units in a specified neighborhood.
- In zonal operations, the resulting value at each location is a function of all data units in the zone it belongs to.

• In global operations, the resulting value at each location is a function of potentially all data units in the coverage (ESRI, 1991).

Within each group, map analysis operations also have different computational characteristics. In general, we can subdivide the mathematical/statistical operations into two groups. In the first group, computations do not require sorting or histogramming. These operations include *mean, sum, maximum, minimum, standard deviation, lessthan,* etc. Computations in the second group, such as *mode, median, majority,* and *variety,* must extract information from the sorted lists of all related values.

In zonal operations, depending upon if the zone is contiguous, computational characteristics can be different. In a contiguous zone, we can perform some mathematical/statistical operations such as *sum, max,* and *min* with repeated scan passes, while this is impossible if the zone is non-contiguous (this is only true in data parallel processing).

Although most global operations involve unique computations, we may be able to find similarities. For example, *connected component labeling* is a common procedure that *raster-vector conversion* and *"clump"* share; and many others, such as *visibility, buffering,* involve distance calculations. Nevertheless, since most global operations require complex procedures that can be carried out with varies approaches, we leave them each as a unique operation.

BASIC OPERATIONS IN DATA PARALLEL PROCESSING

Current data parallel languages are extensions of existing sequential languages with parallel operations added. Most of these data parallel operations are very similar to those in a vector processing language, such as APL and J. The primary objects in data parallel processing are parallel (plural) variables, as opposed to scalar (singular) variables in sequential processing. Parallel variables are equivalent to arrays in sequential processing. The difference is that data parallel software provides operations on the whole array (parallel variable) as a single processing unit. It is our concern how map analysis operations and other computational tasks in spatial analysis can be implemented with these parallel operators. Figure 2 gives a summary of the basic operations in data parallel processing (TMC, 1990, 1992; Maspar, 1991; Walter, et. al., 1990; Blelloch, 1991).

Elementwise operations, *mixed* operations, and *selections* are straight-forward and can be best explained by examples.

Suppose A, B, and C are both N by N raster coverage, n is an integer. A and B are treated as parallel variables, n as scalar variable.

- Elementwise operation: $C = A + B$;
- Mixed operation
    - *reduction*: $n \mathrel{+}= A$; (assign the sum of A to n)
    - *distribution*: $A = n$; (assign n to all elements in A)
    - *insertion*: $[i][j]A = n$; (assign n to location i, j in A)
    - *extraction*: $n = [i][j]A$; (assign value at location i, j in
        A to n)
- Selection
    where $(A == 1)$
        $n \mathrel{+}= A$;   (sums all the 1s in A and assign it to n)

Communication functions are the most important operations in data parallel programming because they provide mechanisms to manipulate the underneath

---

Basic operations in data parallel processing

1. Common operators in sequential languages

2. Elementwise operations

3. Mixed operations between parallel and scalar variables
     — reduction, distribution, insertion, extraction

4. Selection: WHERE, FOR ALL

5. Communication functions
     Simple data movement: **get, send**
     Data movement with combiner
     Permutation
     Scan functions:
         — **combined scan, rank, enumerate, reduce,**
            spread, copy

Figure 2. Basic operations in data parallel processing.

parallel hardware with massive number of processors connected in a particular network architecture.

Get and *send* are the two basic communication operations. In general communications, each data element in a parallel variable gets the value of data elements from any other locations. In regular communications, data movements follow a particular pattern, for instance, a NEWS grid, or an X network:

The *get* and *send* locations can be specified by two types of indexing. We can give the relative distance to the source along particular axes. For example, as a result of the following C* expression,

     $A = [.+1][.+1]A;$

each data element in A is replaced by its neighbor in the southeast. Or, we can use another parallel variable as the index for data communication (*permutation*). When multiple data values are sent to the same location, they can be combined in various manners, such as add, multiply, and logical operations. For example, assuming A is converted to a one dimensional parallel variable, we can calculate the histogram for A using *index send* with the *add* combiner. Here is the C* expression,

        $[A]HIS += ONE;$

where HIS and ONE are parallel variables of the same shape of A, HIS is initialize to 0, and ONE to 1, A serves as index. As the result of this combined send, HIS contains the histogram for A.

Another important class of communication functions is *scan*. It is an accumulative operation along a particular axis. Calculate the running total of a one dimensional array is a common example to illustrate the concept of *scan*.

data

0	1	2	3	4	5	6	7	8	9	10	11

add-scan

0	1	3	6	10	15	21	28	36	45	55	66

Combiners for *scan* could include all arithmetic and logical operators. *Maximum, minimum, multiply, or,* and *and,* are the commonly used combiners. Scan operations also allow efficient data movements and reorganizations. These functions include *rank, enumerate, spread,* and *copy.*

A parallel variable can be divided into segments to allow separate *scan* operations. Normally, a parallel variable is constructed as a mask to control segment scan. In C*, this mask is called the *sbit.*

Various combinations of these scan operations enable one to develop very efficient data parallel algorithms. In general, the complexity of a combined scan operation is $O(logN)$, with N as the length of the axis. Low level implementations of the scan operations could produce $O(1)$ algorithms (Blelloch, 1987; Ajit, et. al., 1989).

MAPPING

GENERAL GUIDELINES

For a given computational problem, there may be various ways to map it to a computer software model. The general criteria to evaluate a particular algorithm is the measurement of time complexity. The less the time complexity, the more efficient the algorithm. While the development of parallel algorithms shares similar evaluation criteria for sequential algorithms, there are additional considerations on interprocessor communications and efficiencies in machine utilization. Below are some simple rules to follow:

(1) The algorithm should minimize interprocessor communications by using regular communication patterns;

(2) Keep as many processors active as possible at each step of the process to achieve high efficiency in machine utilization;

(3) Choose a proper ratio between the number of virtual processors and the number of physical processors to maximize machine utilization and to reduce interprocessor communication.

Figure 1 gives a summary of the general strategies to implement map analysis operations with data parallel operators.

LOCAL OPERATIONS

• **Data Structure** There are two simple schemes to map a raster coverage to a parallel variable to facilitate local operations. For most operations, a direct mapping from a raster coverage to a two dimensional parallel variable is sufficient, with each location in the raster coverage corresponding to a virtual processor in the parallel computer. To accommodate the class II statistical measurements, however, we may use a parallel structure or a three dimensional parallel variable to store all raster coverages involved in the analysis. The following is a C* expression of a parallel structure:

```
struct layer{
        int     cov[k];
};
shape   [N][N]image;
struct layer:image   raster;
```

shape is the parallel data type in C*. Note that the sequential array *cov* is stored at each location (virtual processor). Each field in "coverage" contains the data value in a corresponding coverage.

• **Computation** For most local operations, we can directly use elementwise operators provided by a data parallel language. From the programming perspective, the only difference is the elimination of the double loop. The expressions are almost identical to the macro languages provided by major GIS packages such as GRID, GRASS (r.mapcalc), SPANS, and ERDAS.

The *selection* and *reclassification* operations are simply executed by the selection operator in a data parallel language. The *where* statement is a parallel expression of *if* in common sequential processing.

Algorithms use for statistical measurements are the same as those in sequential processing. The difference now is that each location (pixel, virtual processor) would carry out exactly the same computation simultaneously. For instance, a *localsum* may be computed by the following statements:

```
for(i=0;i<k;i++)
        localsum += raster.cov[i];
```

As indicated in previous discussion of basic operations in map analysis, class II statistical measurements require sorting or histogramming computation. With a parallel structure or a three dimensional parallel variable, we can easily perform a bubble sort on all data values at each location in parallel. The following is a portion of the C* codes,

```
for(i=0;i<k-1;i++) {
        j = i+1;
    while(j<k) {
        where(raster.cov[i] < raster.cov[j]) {
            raster.temp = raster.cov[j];
            raster.cov[j]=raster.cov[i];
            raster.cov[i]=raster.temp;
        }
        j++;
    }
}
```

• **Performance** Data parallel mapping of local operations is close to optimal. With each raster mapped to one processor, maximum parallelism is achieved. In sequential processing, the minimum steps for a local operation are equal to the size of the raster coverage times the number of coverages, or $N \times N \times k$; while in data parallel processing, the number of steps is reduced to k. Moreover, since we can map the raster coverages to a parallel structure or a three dimensional parallel variable, interprocessor communication is not required. All computations and data movements are performed within each local processor.

Actual running time is determined by the characteristics of the processors and the *V/P* ratio. As a comparison, performing elementwise sorting for a *1024* by *1024* coverage with *8* layers would take a SparcII approximately *35* seconds. It requires only *0.16* seconds to finish the task on the Connection Machine Model 2 with 16k processors employed.

FOCAL OPERATIONS

• **Data structure** Similar to local operations, we can map each raster coverage to a two dimensional parallel variable. This mapping is suitable for surface operations and class I statistical measurements. For class II, we need to use a particular parallel structure again. This time, each location would need to store all the neighborhood data values:

```
struct neighbor{
        int     win[w];
        int     majority;
};
shape   [N][N]image;
struct neighbor:image  raster;
```

win is the sequential array that holds all data values in a neighborhood of size *w*. Different from local operations, where memory requirement remains the same for both data mapping, the parallel structure proposed above increases total memory requirement by *w* times. Since the window size can increase quite dramatically from one level to another, this data mapping is only applicable for operations with small neighborhoods.

• **Computation** There are two classes of computations in focal operations. The first class includes calculations of *slope/aspect, flow, filter,* and class I statistical measurements. With the raster coverages mapped directly to two dimensional parallel variables, this class of operations can be implemented with simple regular data movements where each pixel distributes its value to the neighborhood locations. Various calculations are performed along with data distribution. The core of the operation is similar to a typical parallel convolution algorithm, as depicted below:

In this example, it takes 8 steps to distribute the center pixels. It can be reduced to 4 steps if we combine the results along a particular axis then distribute to another axis. The flowing is a C* implementation of *focalmax,*

```
focalmax = raster;
focalmax >?= [.+0][.+1]raster >? [.+0][.-1]raster;
focalmax >?= [.+1][.+0]raster >? [.-1][.+0]raster;
```

452

The left indexes specify the relative distances for the center pixel to travel along each axis. >? and >?= are both compound operators in C*.

As for the second class of operations, each location gets all the neighborhood values in its local memory then executes a sequential algorithm in parallel, as described earlier in local operations.

• **Performance** Focal operations in the first group are mapped efficiently with communication functions in a data parallel language. Total steps are only related to the window size and independent of the size of the raster coverage. Most of the operations are in the order of $2(\sqrt{w}-1)$ steps. Compared to $2N^2(\sqrt{w}-1)$ steps in sequential processing, this is clearly a dramatic speedup. The elapsed time to perform a 3 by 3 *focalmax* on a 1024 by 1024 raster coverage was 0.08 seconds on a 16k processor CM2, but 17 seconds on a SparcII.

The second group of focal operations requires sorting or histogramming. Current mapping strategy enables one to perform sequential tasks at all locations (pixel) simultaneously, which results in a speedup of N^2 steps over sequential processing. Actual elapsed time to perform a 3 by 3 *focaldiversity* on a 1024 by 1024 raster coverage was 0.5 seconds on a 16k CM2 but 57 seconds on a SparcII.

ZONAL OPERATIONS

• **Data structure** We can use two types of data mapping to facilitate the two groups of zonal operations. If the zone is contiguous and the operation belongs to class I statistics, we use two dimensional parallel variables to map the raster coverages. For class II statistics and non-contiguous zone operations, we can use one dimensional parallel variables to represent the raster coverages so that it is easier to perform parallel operations. We are able to do this because most zonal operations do not require neighborhood information.

• **Computation**

— **Group 1** Operations included in this group are *zonalmax, zonalmin, zonalsum,* and *zonalmean*. The core data parallel operation is *segment scan*. Since each zone is contiguous, we can use the zone boundaries as the segment bits to define the scan classes. Figure 3 depicts the basic algorithm for this group of operations.

The zonal boundaries can be generated by comparing each pixel with its neighbors. Note that the operation depicted in figure 3 is iterative. The number of steps taken to finish the operation varies. The more irregular the shape of the zone, the more iterations required. Also note that the above algorithm can be used in the global operation *clump*, or connected component labeling, and *raster-to-vector conversions*.

— **Group 2** When the zones are not contiguous, or when the operations require sorting, we can deal with only one zone each time. Again, we can divide zonal operations in non-contiguous zones into two subgroups. The first group includes common statistical measurements such as *zonalmean, zonalvariance,* and *zonalmax*. They can be easily implemented with the *reduction* operators in a data parallel language. The following C* codes shows a typical zonal operations for this group,

```
for(i=0;i<number_of_zones;i++) {
    where(coverage == i)
        zonalsum += coverage;
        coverage = zonalsum;
}
```

453

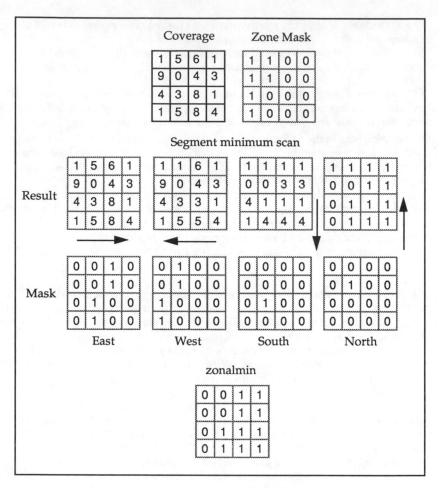

Figure 3. Using minimum scan to find zonal minimum.

Here *zonalsum* is a scalar variable. The sum reduction operator += calculates the total values in all active positions in *coverage* and assigns it to *zonalsum*. The assignment operators then distribute *zonalsum* back to all the positions in the zone.

For the second sub-group zonal operations, we found that most measurements can be easily extracted from a histogram. For example, *zonalvariety* is simply the number of non-zero entries in the histogram, *zonalmajority* is the position in the histogram that has the largest value (highest frequency), *zonalproportion* can be obtained by dividing each entry in the histogram by the total number of pixels in the zone. As described earlier, a histogram can be generated by a simple *combined send*. The following is a portion of the C* codes to implement zonal operations.

```
where(histogram > 0) {
```

```
        zonalvariety += one;    /* one = 1 */
        maj_tmp = >?= histogram;
        min_tmp = <?= histogram;
        where(histogram == maj_tmp)
              zonalmaj = (int) histogram;
        where(histogram == min_tmp)
              zonalminority = (int) histogram;
}
```

• **Performance** The *scan* functions used for the first group of zonal operations are very efficient data parallel operations. Each *scan* operation requires about $logN$ steps, with N being the number of columns or rows. The number of iterations is also quite small compared to the size of the coverage.

However, for operations in the second group, current mapping is not efficient. Although the calculations can be directly implemented with efficient data parallel operators, the operations are limited to one zone a time, leaving other zones in the coverage inactive. Nevertheless, the actual performance measurements are still quite impressive. For a 1024 by 1024 coverage with 10 zones, it took only 0.03 seconds to calculate *zonalmax*. When the number of zones increased to 100, the elapsed time only increased to 0.7 seconds. The increase rate is about 0.06 seconds/10 zones. Calculating the histogram is relatively slow because the *send combine* operation uses the general communication function that is much slower than the regular one. We used a one dimensional parallel integer to hold the histogram. It took the 16k CM2 about 2 seconds to generate the histogram for a 1024 by 1024 raster coverage. The rest of the reduction operations on the histogram are extremely fast; for example, it took only 0.003 seconds to calculate the *zonalmajority* for a zone.

GLOBAL OPERATIONS

• **Data structure** Most global operations are likely to be implemented by the *scan* functions. Data mapping should be flexible among each class of operations to facilitate particular scan operations. In general, we can map a raster coverage to a two dimensional parallel integer.

• **Computation** Most of the global operations is quite complex and efficient mapping to a data parallel language is not as intuitive as those in other classes of operations. However, many of these global operations are related to well-known computational problems, such as the *line-of-sight* problem and the *component labeling* problem. Much effort has been devoted to develop efficient parallel algorithms for these problems by people in digital image processing and computer graphics. Our strategy, therefore, is to transfer the knowledge to the implementations of global operations in map analysis.

For example, the core of *visibility analysis* is the *line-of-sight* problem in computational geometry and computer vision. Among various algorithms, the vector algorithm proposed by Blelloch can be directly implemented for data parallel processing (Blelloch, 1991). The core of the algorithm is to generate a segment vector (one dimensional parallel integer) that contains the rays from the view point to all boundary pixels. *Arctan* value for each pixel is calculated by distributing the x, y coordinate and elevations from the view point. Then these values are assigned the corresponding positions in the segment vector. A maximum scan along this vector finally determines if each location is visible from the previous one. A similar algorithm has been implemented with C* by Kim Mills and the performance was quite satisfactory (Kim, 1992).

Connected component labeling problem is always at the center of attention in image processing, computer vision, and other related fields. Many algorithms have been developed to solve this problem (Apostolakis, 1992). The algorithm described in the zonal operations is one of those many approaches. A much faster data parallel algorithm was designed and implemented by Willie Lim at the MIT Artificial Intelligence Lab (Willie, et. al., 1987). The time complexity of their algorithm is *O(logN)*.

We implemented the *clump* operation and a simple raster-to-vector conversion program with the minimum scan algorithm using C*. For a *1024* by *1024* randomly generated raster coverage, it took the 16k processor CM2 *5.7* seconds to finish *clump*, compared to *198* seconds on the SparcII. For a *256* by *256* random coverage, the elapsed time for the raster-to-vector conversion operation was *0.98* seconds on the CM2 and *64* seconds on the SparcII, respectively.

• **Distance functions** The sequential algorithms for both categories of distance functions, Euclidean distance and weighted distance, are relatively simple. Data parallel implementations of these functions seem to be straight-forward. For instance, to generate the shortest distance surface to source points, we can simply distribute the x, y coordinates of each source point to all locations in the coverage and calculate the distance simultaneously. Time complexity for single source point calculations is $O(1)$ and there is little interprocessor communication involved. It is also independent of the size of the raster coverage. The only problem is when there are many source points, the algorithm has to loop through each of them. To overcome this shortcoming, more general data parallel algorithms need to be developed.

The following describe the major steps for a data parallel algorithm to generate a shortest distance surface.

(1) Calculate the distances in the horizontal direction, using two pass *add-scan* with source points as segment masks (start bits). Use a sufficiently large integer to mask cells that have the same values as their coordinates. The results are compared elementwise. The smaller value is the shortest distance in the horizontal direction.

(2) Generate a segment mask with positions that share the same row with the source points become start bits.

(3) Calculate upper diagonal shortest distances.

• Distribute the shortest horizontal distances to each location by performing an downward (from bottom to top) minimum scan on the result of (1);

• Calculate the vertical distance at the upper level by an downward *add-scan* (from bottom to top) on the source coverage with source points as start bits;

• Calculate and distribute the shortest distances at the upper level by a two way minimum scan along horizontal direction;

• Use the shortest vertical distances and horizontal distances to calculate the upper diagonal distances.

(4) Calculate lower diagonal shortest distances with similar procedures in (3);

(5) Compare distances from horizontal, vertical, upper diagonal, and lower diagonal elementwise. The smallest value is the distance to the nearest source point.

In this algorithm, we apply segment scan repeatedly to find the solution. The algorithm requires totally eight scan operations, each with a time complexity $O(logN)$. The time complexity does not change with the number of source points. For a 1024 by 1024 coverage, it took the CM2 1.7 seconds to generate the distance surface. Using the first algorithm, the elapsed time was 1.16 seconds for 4 source points and 1.78 seconds for 8 source points.

CONCLUSION

As a result of the above experiment, several points are worth noting.

First, from both data structure and computation perspectives, mapping the raster-based Map Analysis Model to the Data Parallel software model can be achieved quite efficiently. Raster coverage is naturally mapped to parallel variables; or, from the hardware perspective, each pixel in the raster is mapped to a processor in a massive parallel computer. For local, zonal, and focal operations, most of the computations can be directly implemented with data parallel primitives. For global operations, algorithm design and implementation are not as straight-forward. However, many data parallel algorithms related to global operations exist and can be adopted to map analysis.

Second, for most operations, the speedup over sequential processing is dramatic. Parallel implementations of many operations are at least 100 times faster than sequential processing. Since most operations can be done in constant steps or in a complexity $O(logN)$, the advantage of data parallel processing is more obvious with size of the problem becomes larger and larger. The significant speedup in these primitive operations provides a promising foundation for spatial modeling in geography, ecology, and other related fields.

Third, although more efficient algorithms and data structures need to be developed for particular categories of operations, such as the class II statistical problems and the iterative problems in non-contiguous zones, software interface to data parallel computers is not a major obstacle in solving geographic problems with raster-based map analysis model. From this perspective, the challenge introduced by parallel computers lies in the conceptual arena — the development of more powerful spatial models to explore and understand real world problems.

REFERENCES

Ajit Agrawal, et. al., 1989. "Four Vector-matrix Primitives," *Proc. Symposium on Parallel Algorithms and Architectures*, June, 1989, pp. 292-302.

Blelloch Guy, 1987. "Scan as Primitive Parallel Operators," *International Conference on Parallel Processing*, August, 1987, Sartaj K. Sahni, ed., Penn State Univ Press, pp. 355-362.

Blelloch Guy, 1991. *Vector Models for Data-Parallel Computing*, The MIT Press.

Burrough P. A., 1986. Chapter 5, "Methods of Data Analysis and Spatial Modeling," *Principles of Geographical Information Systems for Land Resources Assessment*.

ESRI, 1991. *Cell-based Modeling with GRID, Analysis, Display, and Management*. Environmental Systems Research Institute, Inc.

Fox Geoffrey, etc., 1988. *Solving Problems in Concurrent Computers, General Techniques and Regular Problems*, Vol. 1, Prentice Hall.

Fox Geoffrey, 1991. "The Architecture of Problems and Portable Parallel Software Systems," *SCCS-134*, Syracuse University.

Hillis W.D. and Steele G.L., Jr., 1986. "Data Parallel Algorithms," *Communications of ACM*, Dec. 1986, pp. 1170-1183.

Hillis W.D., 1987. "The Connection Machine," *Scientific America*, Vol. 256, No. 6, pp. 108-115.

Langran Gail, 1991. "Generalization and Parallel Computation", in *Map Generalization: Making Rules for Knowledge Representation*, edited by Barbara Buttenfield and Robert McMaster, pp.204-216, Longman Scientific Technical.

Mills Kim, Geoffrey Fox, Roy Heimbach. 1992. "Implementing Spatial Environmental Models on Parallel Computing Systems," SCCS-150, Syracuse University. submitted to *Computers & Geosciences*.

Thinking Machines Corporation, 1990. *C* Programming Guide*. TMC, Cambridge, Massachusetts.

Thinking Machines Corporation, 1992. *CM FORTRAN User's Guide for the CM-5*, TMC, Cambridge, Massachusetts.

Maspar Computer Corporation, 1991. *Maspar Parallel Application Language (MPL) User Guide*, Maspar Computer Corporation, Sunnyvale, California.

Tomlin Dana, 1990. *Geographic Information Systems and Cartographic Modeling*, Prentice Hall.

Water Brainerd, Charles Golberg, Jeanne Adams, 1990. *Programmer's Guide to Fortran 90*, McGraw-Hill Book Company, New York.

Willie Lim, Ajit Agrawal, Lena Nekludova, 1987. "A Fast Parallel Algorithm for Labeling Connected Components in Image Arrays," Technical Report, Thinking Machine Corporation, 1987.

USING TRANSCAD DATA TO ESTABLISH AN ARC/INFO TRANSPORTATION GEOGRAPHIC INFORMATION SYSTEM FOR THE CHARLOTTE METRO REGION

Yuanjun Li
Department of Geography and Earth Sciences
University of North Carolina at Charlotte
Charlotte, NC 28223

ABSTRACT

The purpose of this project was to establish a transportation geographic information system for the Charlotte Metro Region using ARC/INFO. The development of a transportation GIS requires a large amount of spatial and non-spatial data input. Rather than have to digitize and enter manually, the existing TransCAD data were utilized directly. The rich geographic data and attribute data, especially the transportation information and analysis results in TransCAD could be converted or imported into ARC/INFO to generate information layers or attributes. The computer programs to perform the data conversion have been developed by the author. Five layers—States, Counties, Census Tracts, Zip-Code Zones and Roads— were generated. By means of the Arc Macro Language (AML), a spatial information query system was created. This paper describes data organization, data conversion and processing, and the procedures to create ARC/INFO coverages. The potential to integrate the transportation information with other kinds of information layers to make comprehensive spatial analysis is briefly discussed.

INTRODUCTION

The Charlotte Metro Region includes 13 counties which belong to either the State of North Carolina or South Carolina. The fast growth of economy in the region will largely depend on the development of transportation. The transportation systems have been well developed and are going to be further developed in this region. In order to meet the goals, a variety of information is required. It is increasingly realized by the transportation planners that a comprehensive geographic information system can help them analyze the present transportation system, forecast the future plan and make decisions from the view of the whole metro region. Therefore, a transportation GIS for Charlotte Metro Region may play an important role in the transportation development of this region. The full potential of a GIS can be realized when it brings some benefits for its users.

As it is well known, geographic information system (GIS) technology has been proved to be an effective tool to provide powerful data management and spatial analysis in various applications. The flexible capability of a GIS allows users to better use their own data as well as to share the information from other sources. ARC/INFO is one of the major GIS software packages widely used now. It has the functions of digital data input, data management, analysis, display, and conversion. TransCAD is a GIS package designed specially for transportation systems. It works with PC DOS environment. TransCAD can be used in highway planning, transit planning and operations, accident reporting and analysis, demand modeling and forecasting, and it is also developed for some other applications. It is surely significant to take advantage of both the rich transportation related data in TransCAD and plenty information about other features such as soil, water, land use, etc. as well as the multiple analysis functions and flexible display capability in ARC/INFO. It is obvious that utilizing the existing digitized data can save a lot of time and money.

In this project, It was attempted to establish a simple transportation GIS for the thirteen counties in the Charlotte Metro Region by converting the data from TransCAD into ARC/INFO installed on SUN/4 workstation. The original georeferenced data used in

459

this project were exported through TCBuild from TransCAD database which was developed in the Interdicipelin Transportation Studies Center at University of North Carolina at Charlotte. TCBuild is the database builder of TransCAD. The contents of an existing TransCAD database can be exported into comma-delimited ASCII data files, then, after reformat, imported into another GIS application. 1990 U.S. Census documents and County And City Data Book 1988 were also used as the reference data to build the attribute table for the county coverage in INFO. Based on the county coverage and road coverage, an information query system was developed by using ARC/INFO programming language— AML. The query can be performed either at county level or region level. Different graphic outputs can be obtained by means of Arcplot software and a printer or plotter.

PROCEDURES

Preparing Data

Three layers were designed in the regional transportation information system. They are state layer which covers North and South Carolinas, county layer, census tract layer, zip-code zone layer, and the road layer which includes the major link network in the Metrolina region. The geographic data needed for generating ARC/INFO layers were directly exported from TransCAD database rather than digitized manually. The data on a disk, then, could be transferred into SUN/UNIX workstation on which the ARC/INFO is hosted.

Since certain data format is required to generate an ARC/INFO coverage, the computer programs were developed using Pascal programming language to reformat the data. Also, the attribute data which describe the links, tracts, or counties were exported and reformated. Some other attribute data which were not in the original TransCAD database were collected from other sources. For example, FIPS codes, 1990 census population data, employment, farm, and retail trade data for counties, etc. could be obtained from the County City Data Book and 1990 U.S. Census files.

Generating ARC/INFO coverage

ARC/INFO coverages could be obtained by using GENERATE command and formated data. Several coverages in this transportation information system were generated. They are:

(1) State coverage — state of North Carolina and South Carolina;
(2) County coverage — all counties in both North and South Carolinas, and Metrolina county coverage — thirteen counties in the Charlotte Metro Region;
(3) Census Tract coverage - the Census tracts of the related counties;
(4) Zip-Code Zone coverage - the Zip-Code Zones in the related counties; and
(5) Metrolina road network coverage — Major roads in the Charlotte Metro Region.

Federal Information Processing Standard (FIPS) code were added as the labels of county polygons. In order to use the Metrolina coverages with other information layers in the future, the Metrolina county coverage and road coverage were projected to State Plane Coordinate System which is commonly used.

Adding Attribute Data to INFO Database

On completion of the spatial domain of the geographic information system, the attribute data should be added into the INFO database. The data for the road coverage were directly selected and exported from TransCAD link layer. The attribute data include length, route number/street name, speed, number of lanes, link type, county code, travel time, capacity, as well as the traffic measure data such as average daily travel volume, vehicle miles travel, and vehicle hours travel, etc. The data for the Metrolina county coverage were obtained from TransCAD database and census files, respectively.

Creating a Spatial Information Query System

The programming language Arc Macro Language (AML) was used to create a simple spatial query system. This query system performs two functions useful to users. These include map displays for different layers, either separately or simultaneously, and query of information on specific locations interesting to users. The query system scheme chart is shown in Figure 1.

Figure 1. The scheme of the spatial information query system

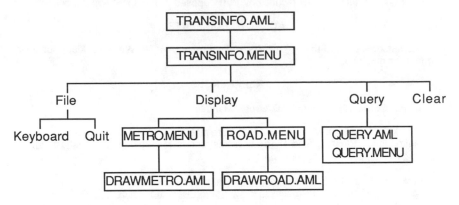

PROBLEMS AND FUTURE WORK

The accuracy of the attribute data from different sources may be inconsistent. How to evaluate the accuracy of a GIS is still a problem needed to be studied.

Since this is only a demonstration project. Much work still needs to be done, for example, more layers can be converted into this system. In addition, This system is not a professional transportation GIS because it does not have the function of transportation planning modeling, i.e., lack of vehicle routing and traffic assignment procedures. However, the TransCAD analysis results such as traffic forecast can be easily exported into the ARC/INFO database. The goal of this transportation geographic information system is to integrate the transportation information of the Charlotte Metro Region with other information to support regional growth analysis, regional planning, and decision making. For example, satellite imagery and aerial photos may be used to detect the changes such as new roads, interchanges or bridges, to evaluate the environmental impact on a highway corridor, and to classify land cover/land use in the highway corridor. It is also possible to use existing data layers such as land use, soil, and water to access the feasibility of a highway system. TIGER and Census files can be inexpensive data sources for a transportation GIS. However, to convert the TIGER line data file usually takes a large amount of computer memory.

On the other hand, this system can provided other researches concerned the Metrolina Region with transportation information. For example, road network data may be needed to analyze the interaction between zones, to solve retail location problem, to evaluate the economic growth in the region, and so on. The functions such as buffer and overlay offered by ARC/INFO will be very helpful to perform these spatial analyses.

CONCLUSIONS

It is obvious that to take full advantage of existing data and research results can be an economic way to establish or update a GIS application. Sharing information and data among different research groups is a good way to better deal with research problems. This paper explains the use of TransCAD data to generate ARC/INFO coverages for the Charlotte Metro Region which is developing dramatically. A transportation GIS including the Metrolina road network and census data was build in ARC/INFO environment. A simple information query system has been developed by using AML. It may be used for road map display and information query. Although this information system can not be used for transportation modeling, more traffic analysis results can be easily exported from TransCAD models into the ARC/INFO transportation GIS. Also, it is possible to integrate this transportation GIS with other information such as land use, soil, water, and satellite imagery to support the Metrolina regional planning. It is believed that there will be a bright future for GIS technology to be applied in transportation aspect.

ACKNOWLEDGMENTS

The author wishes to thank Prof. David T. Hartgen and Dr. Wei-Ning Xiang for their suggestions and help.

REFERENCES

Caliper Corp., 1990. TransCAD User's Manual.

Petzold R. G. and D. M. Freund, 1990. Potential for Geographic Information Systems in Transportation Planning and Highway Infrastructure Management, Transportation Research Record,TRB, No.1261, pp.1-9.

Simkowitz S. J., 1990. Using Geographic Information System Technology To Enhance the Pavement Management Process, Transportation Research Record, TRB, No.1261, pp.10-19.

Gallimore W. P., D. T. Hartgen, and Y. Li, 1991. Applications of GIS-Transportation Analysis Packages in Super-Regional Transportation Modeling, paper in press, Transportation Research Record, TRB.

A DYNAMIC WORK ORDER PROCESSING SYSTEM FOR AN ELECTRICAL DISTRIBUTION NETWORK

Stephen Lindsey
Computervision GIS
45 Vogell Road, 7th Floor
Richmond Hill, Ontario
CANADA, L4B 3P6

ABSTRACT

Electrical Distribution Networks contain many elements of vastly different scale. Organising access to these elements for work order processing in a multi-user environment presents peculiar challenges in allowing flexible selection approaches, while ensuring data consistency in a composite data model. The data model for an Electrical Distribution Network must be sufficiently sophisticated to support many, varied electrical analysis functions.

This paper will describe work carried out in implementing the work order processing component of a GIS-based data management system supplied to an Electrical Utility Company.

The data model necessary to support analysis functions on a full range of electrical network components will be outlined. The mechanisms to allow selection of data for work order processing in a flexible fashion, consistent with existing network maintenance practices will be discussed. This includes search techniques by which all associated network components are selected, providing a complete sub-network in the workspace for the user. The methods for supporting the integrity checks and constraints required, both to maintain consistency with the data model, and to allow the multi-user functionality required by the utility, will be discussed. Finally the elements of the user interface designed to allow changes to the network to be easily identified and checked before final commit will be discussed.

The application described in this paper was designed and implemented for Union Electric Fenosa, Spain by a SYSTEM 9 development team.

INTRODUCTION

The quantity of data required to fully describe an electrical distribution network,

463

covering an area of approximately 500 square km, is very large. In this case the data will occupy several gigabytes.

The uses a utility has for this data when stored in a spatial database, are many. They include:

— i) Retrieval of low level spatially related details to allow work crews to have accurate information for field repairs.

— ii) Analysis of electrical networks and sub-networks. For example when calculating transmission line losses, which depend on knowledge of line lengths and connectivity.

— iii) On-line display and control of the status of electrical components.

— iv) Maintenance of data allowing network modifications to be entered immediately, keeping the master database up-to-date.

As the requirements of different applications can be very different, great care must be taken in arranging the data model. Important considerations are:

— The connectivity of network components.

— The relation of these network components to topographic features.

— Allowing the use of both schematic and graphic representations of network components.

— Allowing data to be worked on at varying scales. (A user changing the position of a cable within a conduit requires a much different view of the data than a user analyzing the routes for high level voltage lines within a region).

An important feature of a network system which allows the macro and micro approaches to coexist, is that it should provide a seamless database, with even the largest features being monolithic.

This paper details the critical sections of the work done to implement the work order processing component of a system supplied to a Spanish Electrical Utility (Union Fenosa) supplying Madrid and surrounding areas. The database is being used for a full range of applications by the utility. These applications are beyond the scope of this paper, but their needs have a profound effect on the data model that was implemented.

The work order processing application supports the mechanism whereby modifications performed in the field are entered as updates to the database. This work is done in isolation from the master database.

DATA MODEL

The uses for the data representation of an Electrical distribution network are many, and the data model must be designed to allow these separate applications maximum flexibility with minimum interference.

The existing data was split into four types, or views; Orthogonal, Positional, Detailed and Urban.

464

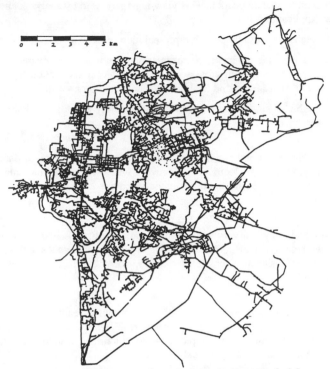

Medium Tension Network For Madrid

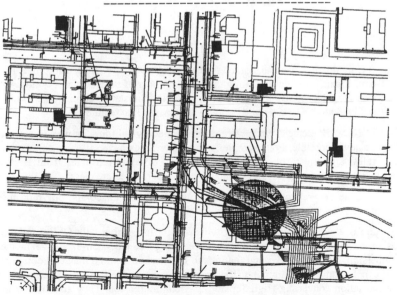

Detail of Electrical Network Viewed Against Urban Features

Orthogonal data is a schematic representation. This representation has only the vaguest relation to spatial location. This view is largely used by analysis, and control and display applications.

Positional Data is the spatially correct representation of network components.

Detailed data covers the schematic representation of individual elements in the electrical network such as conduit cross-sections and internal sub-station layouts. This data is related to a specific spatial location (e.g. the conduit or substation).

Urban data is the background topographical data (mainly buildings and streets) used for spatial reference.

In addition to the views outlined above, a further critical subdivision of data was by tension (voltage) level. An electrical sub-network will always contain elements all of the same tension level. Many applications only require information on a particular tension level.

A composite hierarchy is a very suitable representation for a network. Connectivity can be shown between elements if they belong to a common composite feature. Each level of the composite hierarchy will then represent a diffent level of network decomposition.

A spatial database should allow the formation of composite features, which can have a specified set of allowed classes of child features. These child features can themselves be composite features, or simple features which have a unique spatial representation. There should be no practical limit to the depth or breadth of a composite hierarchy.

The following general rules were followed in designing the data model.

— i) Common data was abstracted to the highest level. e.g. All tension lines must have a data attribute indicating the serial number of the Substation(s) to which they are attached. By including this data in the Substation composite class, it is easily available to any queries involving tension line features.

— ii) Network element types were differentiated by splitting features to the lowest level possible. For example Low tension lines have different connectivity rules, different display representations and different attributes from High or Medium tension lines, and so they form a separate feature class.

Both of these general rules are in keeping with the inheritance and polymorphism precepts of Object Oriented Design.

A feature of the composite hierarchy which is not in keeping with Object Oriented Design concepts, is that composite feature classes can be non-exclusive in their membership, allowing individual features to be attached to more than one composite class. An example is Medium Tension lines which belong to the sub-station network of a substation converting High Tension to Medium Tension. They also belong to a transformer centre network, centred on the transformer centre which converts

Medium tension to Low tension.

The attached diagram shows a section of the data model which has been simplified for ease of understanding. (The complete data model has 28 composite feature classes and 153 simple feature classes including everything from individual support poles, to many varied forms of switching gear).

The diagram shows how a substation network contains the substation feature and the tension lines attached to it. These tension lines also belong to a composite feature class, which represents everything contained in, and attached to, a given electrical conduit. This includes the conduit itself, cross-sections showing cable placings and the cables themselves. Many of the features have separate representations in the Positional, Orthogonal and Detailed views, and are split into separate feature classes for each. Only the Medium tension cable feature has been shown split to this level.

Some of the details omitted directly from this diagram are other tension levels (there are actually four tension levels), the multitude of separate feature classes for different types of switches, and many of the elements contained within the substation itself.

With the data model as described in place, we can display (or query) all elements connected to a Substation by displaying (or querying) the single Substation composite feature. At the micro level, the association between individual cables and the transmission lines to which they belong, is easily available, as it is embedded in the data model.

DATA ORGANISATION

The data to be worked on in the process of fulfilling a work order, is copied from the master database to a temporary data work area. Separate work areas have many advantages. The master copy is protected from uncontrolled editing and can be verified before being committed to the master database. Some applications perform faster on a data subset as the target ranges for queries are reduced without adding spatial qualifiers. Additionally, for this customer, work areas sometimes need to be taken to sites completely remote from the central database, hence these work areas must be self contained and provide all of the data a user might require for a particular task.

SELECTING DATA INTO WORK AREAS

As mentioned above work areas are self contained. As work orders refer to changes which may be very minor, it was very important to the utility that the creation of data sub-sets and the integration of modifications be very fast, to maximise the turnover of work.

The user interface and the underlying structure of work area selection, were carefully tailored to the existing user methodology, which in turn is based on the types of work order commonly processed. The system allows the user to define the area for work to

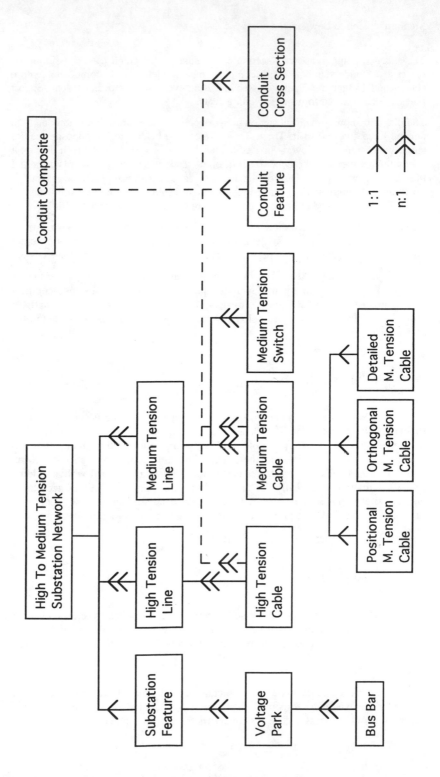

be carried out, in the following manners:

—   i) By a Substation identifier and radius.

—   ii) By a Transformer Centre identifier and radius.

—   iii) By a particular Tension line identifier and distance.

—   iv) By a mapsheet number (The old system used a set of mapsheet identifiers which were referenced in our data model for backwards compatibility).

—   v) By specifying two street identifiers, forming the corners of an area.

—   vi) By specifiying a Village name and radius.

—   vii) By specifying any individual network element.

—   viii) By a general set of coordinates.

Of vital importance to the data contained in a work area, is that all network references be intact, ie that the data form a sub-network, with splitting being done only at the highest specified level. By default this highest level is the top of the data model, in our case the Substation classes. Where a type of work only requires that data consistant from, say the conduit level down, is required (e.g. where a work order involves only localised changes, such as repositioning the cables within a single conduit), then the level of network focus can be specified. (In SYSTEM 9 this is achieved with a theme specification, which specifies which feature classes are to be included in a work area). The composite data model allows us to isolate these sub-networks with simple and fast logic. Without this form of data structure, this would be an arduous task.

Owing to the speed critical nature of data extraction, the algorithms for network tracing were tailored to carefully exclude unnecessary data. The "theme" control prevents tracing elements through levels that are of a higher network level than is required.

The seed network elements are determined either by being directly specified by the user (items (i),(ii),(iii),(vii) from above list), or as a result of a spatially qualified query on all feature classes included in the theme. Tracing through the data hierarchy is done downwards first, to extract the lowest level features. Tracing then is then done upwards, following all paths allowed by the theme, and extracting all composite features along these paths. Tracing is now reversed downwards again, using the composite features reached at the top of the upwards traversal as new seed features. All features extracted on any path down are now included in the final work area. Care is taken to ensure efficiency, by eliminating duplicate references at appropriate points.

Where a spatial select is used in the selecting of the seed data items, the use of spatial indexing is critical for speed.

Data integrity is of vital importance to a power distribution network representation. It is not acceptable for decisions controlling the infrastructure on which millions depend for power, to be based on data which can easily be corrupted.

There are many levels at which integrity is controlled in the final system implementation. We have already seen how we ensure that all work areas selected from the master database are integrally correct, particularly in terms of network connectivity. Fundamentally the sense and connectivity of a network is enforced by the data model. All composite classes have a defined set of allowed sub-classes. This prevents, for example, a medium tension cable being added to a high-tension line. The software layer controlling inserts and removals from the composite hierarchy, flags to the user any editing which would leave a sub-network with any breaks in it. A common example of this is checking whether a change would leave a composite with no children. This is never valid unless the user is deleting or completely redefining a full sub-network.

Most of the attribute data associated with the network items, is mandatory (e.g. all lines must include their voltage rating). This can be controlled at the database level with relational database constraint tools, and is also specifically enforced by the application interface, which does not allow a change to be committed, unless all appropriate data is present.

Multiple users processing work orders where they are required to work on private copies of the data, presents additional problems for ensuring data integrity.

Generally the separate work area technique leads to a control mechanism where the last change committed to the master database becomes the up-to-date copy. This is adequate for individual network elements as the management of assigning work orders should prevent the case where multiple users are simultaneously working on the same elements at the micro level. However in a network application where logical networks can cover large areas, there are problems in ensuring that concurrent changes to the connectivity arrangement of a network by separate users can be successfully integrated. A prime example of such a problem would be separate users simultaneously submitting work areas, where one had added a tension line to a substation network, and another had deleted a tension line from the same network. As these operations could be in spatially very different regions, this would be a hard case to detect from a purely manual organisational control.

Specialised logic is used to handle this integration of multi-user changes to network connectivity from separate work-areas. This logic for detecting multiple changes to the same sub-network, relies on comparing different copies of the same network tree to determine the differences between the trees. It then examines these particular areas, differentiating individual features as new, modified or deleted, to allow a correctly merged network tree to be finalised.

INTERFACE FOR CHECKING WORK ORDER CHANGES

Although the dbms handles the enforcement of data integrity as described, the prescribed customer workflow was that all operator work had to be checked by a supervisor before being committed to the master database. This approval checking provides a level of security whose rules cannot be expressed in database integrity terms. For example, to check if the route of a tension line is appropriate, or if a work order has been misinterpreted and the wrong conduit edited.

The important requirements for the interface used by a supervisor to check the work of an operator before allowing it to be committed to the master database, were that the process of viewing changes be simple, very fast and informative.

The SYSTEM 9 database model already allows for the simple extraction and differentiation of new, modified and deleted data. A graphical user interface was constructed that allows features in these categories to be highlighted in different colours against a background showing the rest of the unaltered data. This simple and elegant solution was a highlight of the system for the customer.

CONCLUSION

The following items were highlighted, during the design process, as being of particular importance for this type of application.

— For a network application the correct data model must be the basis for all work. The correct model can greatly facilitate application flexibility. In our case the relatively easy handling of different tension components was gained directly from their support in the data model.

— Data volume can be a critical factor for network applications. Algorithms and applications must be optimised wherever possible, while still maintaining the seamless integrity of the model.

— For a spatial network, it is necessary to make selection of data flexible enough to allow selection by a combination of spatial and network connectivity rules, that allow the end-user to use the selection technique most suited for their end purpose.

— The proper data model can be used to enforce data integrity rules. Connectivity rules are particularly important to a network application and are well served by a hierarchical data model.

— Special care is needed for multi-user network applications, in merging parallel changes to the network.

— Any work order system should take careful note of existing work practices, terminology and user roles and responsibilities. A system that mimics accepted work flows will find quicker and more ready acceptance from end-users.

ACKNOWLEDGEMENTS

This paper was part of a very large project, many people were involved from the Toronto and Zurich offices of Computervision GIS. Space prohibits listing all these people here. However special thanks is due to those who worked most directly on the elements described in this paper. They are Lars Ole Pedersen, who worked with the users in translating their needs required to build the data model, Caroline Williams who implemented the Maintenance application, and Ed Rozentals who helped implement the data selection work.

REFERENCES

— A) Berrill, A. R. and Mclasen R. A., Feb 1984. "Photogrammetric Data Acquisition : The Intelligent Approach", 15th Congress of the International Society for Photogrammetry.

— B) Cox, B. 1986 "Object Oriented Programming, An Evolutionary Approach", Addison-Wesly, Reading, Masachusetts.

— C) Date, C.J. 1977, "An Introduction To Database Systems", Addison-Wesley, Reading, Massachusetts.

— D) Guttman, A. "R-Trees: A Dynamic Index Structure For Spatial Searching", Proc ACM SIGMOD Int. Conf. on Management of Data, p47-57, 1984.

— E) Stevenson, W. D. "Elements of Power Systems Analysis". NG100.

ACCURACY OF A GIS-BASED SMALL-AREA POPULATION PROJECTION METHOD USED IN SPATIAL DECISION SUPPORT SYSTEMS

Panagiotis Lolonis
Marc P. Armstrong
Claire E. Pavlik
Sidan Lin

Department of Geography
The University of Iowa
Iowa City, IA 52242

ABSTRACT

This paper describes a recently developed GIS-based population projection method for small spatial units, known as the Modifiable Spatial Filter (MSF) method, and evaluates the accuracy of its projections with respect to a parameter that controls the sample size of the data that are used to estimate projection parameters. This assessment is made with respect to average absolute error, average relative error, distribution of absolute error, and absence of bias for subgroups. Results show that MSF projections appear to be stable when the parameter is greater than fifteen.

INTRODUCTION

Future populations are projected using a wide variety of techniques that differ in their underlying assumptions, the real-world processes modeled, and the input data required (Isserman 1977; Lee and Goldsmith, 1982). Regression based projection techniques (Ericksen, 1974; O'Hare, 1976), for example, associate past population changes with environmental, socioeconomic and political changes that have occurred in a study area. Cohort component techniques (Isserman, 1977; Isserman, 1984), on the other hand, project populations by progressing the population cohorts of a given base year into the future. These traditional population projection methods are most often used to project populations for spatial units with relatively large populations, such as metropolitan areas or states. Many locational planning problems, however, require population projections for much smaller spatial units, such as city blocks or block groups, that contain comparatively small populations. For these problems, traditional methods are inadequate; either the input data required for such methods are unavailable at this scale or the resulting projections are unacceptably inaccurate (National Research Council, 1980; Tayman, 1992).

Recent advances in GIS software and geo-coded databases now enable the development of methods that project populations for arbitrarily small areas. This paper presents one such method, the Modifiable Spatial Filter (MSF) method (Rushton et al., 1992), and assesses the accuracy of its projections against observed values. Previous assessments indicate that when applied at the census block level, the MSF method predicts populations more accurately than traditional cohort component methods (Rushton et al., 1992). In these evaluations of predictive accuracy, however, the minimum population threshold parameter, which controls the size of population sample that is used in the estimation of the projection parameters, was fixed and consequently its effect on projection accuracy is unknown. In this paper, we extend the assessment of the MSF method by evaluating the effects of variation of the minimum population threshold on the accuracy of one-year predictions.

THE MSF METHOD

The MSF method, an adaptation of the cohort component method, uses disaggregated geo-demographic data to estimate projection parameters and project populations for small areas. The MSF method estimates projection parameters for each spatial unit using data not only from that unit but also from neighboring units. The assumption underlying this method is that neighboring spatial units are characterized by similar demographic processes and, therefore, have similar projection parameters. Thus, the variance of cohort component parameters for spatial units with small population cohorts can be reduced by estimating progression rates using pooled data from neighboring areas.

The MSF method consists of five steps. First, the residence locations of individuals living in a study area are determined by linking population data to a digital geographical database using the address matching capabilities of GIS software. Second, the address matched population data are aggregated to spatial units appropriate for the study (e.g. city blocks). This step is performed using the spatial aggregation capabilities of GIS software. Third, the "neighborhood" of each spatial unit, which consists of all units (including itself) that are within a user-specified maximum distance from the unit, is determined using geometrical data stored in the GIS database and special purpose software (e.g. Lolonis, 1992). Fourth, progression rates for each spatial unit and age group are estimated using the population and neighborhood data from the two previous steps. This estimation is made by sequentially accumulating data from the center of the unit's neighborhood to its periphery, until a minimum population level is reached. If the required minimum population level is not reached within the unit's neighborhood, then progression rates for the study area as a whole are assigned to the unit. Currently, this estimation requires special purpose software that implements the MSF method (Lolonis, 1992). Finally, projections are computed using the populations of the "base year", the estimated progression rates, and projections of each year's initial cohort (e.g. projected births).

The estimation of the progression rates in the MSF method is made using the equation:

$$r_i(g, g+1) = \frac{\sum\limits_{j \in Fi(g,g+1)} \sum\limits_{t=T_0+1}^{T_1} n_{jt}(g+1)}{\sum\limits_{j \in Fi(g,g+1)} \sum\limits_{t=T_0}^{T_1-1} n_{jt}(g)} \qquad (1)$$

where:

$r_i(g,g+1)$: the progression rate of spatial unit i and age groups g and $g+1$;
i : basic spatial unit index;
g : age group index;
T_0: initial year of population data;
T_1: final year of population data;
t: index of year in the sequence T_0 through T_1;
j: index for the spatial units within the MSF of the ith spatial unit;
$F_i(g,g+1)$: set of spatial units in the filter area of spatial unit i, for age
   group g; and
$n_{jt}(g)$: number of individuals at spatial unit j, year t, and age group g.

The *filter area* of unit i for the gth age group is defined as the k closest spatial units in its neighborhood (including itself) that contain an average of at least M students for age groups g and $g+1$ over the observed years. If this minimum population threshold is not reached within the neighborhood defined for unit i, then the filter area is assumed to be the entire study area. Mathematically, the filter area is defined as:

474

$$F_i(g, g+1) = \{s_i(1), \ldots, s_i(k), k \leq p_i\} \tag{2}$$

$$\text{if} \qquad \frac{\sum_{j=1}^{k_i-1} \sum_{t=T_0}^{T_1} [n_{s_i(j)t}(g) + n_{s_i(j)t}(g+1)]}{2*(T_1-T_0)} < M \tag{3}$$

$$\text{and} \qquad \frac{\sum_{j=1}^{k_i} \sum_{t=T_0}^{T_1} [n_{s_i(j)t}(g) + n_{s_i(j)t}(g+1)]}{2*(T_1-T_0)} \geq M \tag{4}$$

$$\text{otherwise} \qquad F_i(g, g+1) = \{ \text{ all units in the area} \} \tag{5}$$

in which the additional variables are defined as:

M : minimum population threshold in the filter area;

p_i: number of spatial units in the neighborhood of i;

k: number of spatial units in the filter area of i;

j: index indicating the order of a spatial unit in the filter area;

$s_i(j)$: jth spatial unit in the neighborhood of unit i. The elements of that set are ordered in ascending order with respect to distance from spatial unit i;

$n_{s_i(j)t}(g)$: population for spatial unit $s_i(j)$ at year t, and age group g.

Population projections for spatial unit i, age group $g+1$, and time period T+1 are computed as:

$$n_{iT+1}(g+1) = r_i(g, g+1) * n_{iT}(g) \tag{6}$$

where:

$n_{iT+1}(g+1)$: population of unit i and age group $g+1$ at time T+1;

$r_i(g, g+1)$: progression rate for spatial unit i and age groups $(g, g+1)$; and

$n_{iT}(g)$: population of unit i and age group g at time T.

In summary, equation (1) evaluates the proportion of the cohort in age group g, residing in the filter area of spatial unit i, that advanced to the next age group $(g+1)$ in the next time period (T+1) over the observed time interval. Equations (3), (4) and (5) define the filter area for each spatial unit. Finally, equation (6) computes the projected population using the progression rates, the past population estimates, and projections of the youngest cohort.

The MSF method addresses the apparent inability of cohort component methods to model the small populations at the block level through pooling the demographic data of proximal areas. The pooling done for a projection is dependent on the value of M, the minimum population threshold, which determines the size of the filter for each projection unit given the spatial distribution of observed cohort component populations and the maximum distance constraint.

Two factors affect the accuracy of the estimates generated using the MSF method. If M is small, then the size of the spatial filter tends to be small and data are pooled over an area that is likely to be characterized by similar demographic processes. Projections computed using small M values, however, tend to be subject to the same problems observed in traditional cohort component models applied to small areas: when population counts within the filter area are low, even single count changes in the numerator or denominator of equation (1) strongly affect the resulting grade progression rate estimate. This results in cohort population projections that may deviate substantially from actual values because of the sensitivity of parameter estimates to small differences in observed data. On the other hand, if M is large, the problem of parameter sensitivity to small changes is reduced by increasing the number observations included in the estimate. The result, however, is that the size of the spatial filters also tends to increase, so that filter areas probably will contain spatial units with dissimilar demographic characteristics. With large M, therefore, the filtered progression rate estimates for a particular unit may include observations from units that are substantially different demographically, thus increasing the likelihood of inaccurate projections.

Previous results (Rushton *et al.*, 1992) demonstrate that with M set to 30 and a distance constraint of one mile, the MSF method provides more accurate one year projections than the traditional cohort component method. While these results indicate the general utility of the method in small area population prediction, they must be extended to include the influence of the minimum population threshold parameter, M, on predictive accuracy. Through such an extension, the tradeoff between parameter sensitivity and spatial homogeneity can be evaluated, leading to recommended M values for prediction. Here, we examine the impact on predictive accuracy of systematic variation in M, comparing observed and estimated data values for block level cohort components. The evaluation of the accuracy of the predictions is based on criteria proposed by the Panel on Small-Area Estimates of Population and Income (National Research Council, 1980).

THE STUDY AREA

The assessment of predictive accuracy of the MSF method presented here is based on the elementary school population of the Iowa City, Iowa, Community School District. In this assessment, census blocks are the spatial units used, while population cohorts are defined as the students of the 7 elementary school grades (K through 6). The district contains 1,187 census blocks and served 5,587 elementary students in 15 elementary schools during the 1991-92 school year. For the analysis, 896 of the 1,187 census blocks, which extend over areas with a well established mailing address system, were used. The remaining 291 census blocks, which are primarily rural and without an established street address system, were excluded from the analysis because the residential locations of students living in those areas could not be determined through address matching.

Official school enrollment records for the five year period 1987-91 containing each student's name, mailing address, and grade were used as the population data source. Enrollment data for the first four school years were used to estimate MSF projection parameters; enrollment data for the final school year, 1991-92, were used to assess the accuracy of projections based on the estimated parameters. In the comparison of projected and actual enrollments, projections for the youngest cohort, kindergarten students, were obtained with a regression-based, trend extrapolation technique (see Lolonis, 1992). Students were assigned to census blocks by address matching using a commercially available GIS software package (Caliper, 1990). In the address matching stage, approximately 90% of student addresses were successfully matched; unmatched student records were not included in the analysis.

ASSESSMENT OF THE ACCURACY OF MSF PROJECTIONS

The accuracy of the cohort projections for minimum threshold parameter values ranging from 0, which implements the traditional cohort component model, to 30 were evaluated by comparing actual and predicted student values by block and grade for elementary school students in the Iowa City Community School District during the 1991-92 school year. Examining these differences district-wide for total enrollments (Table 1), we see that while the traditional cohort component model over-predicts total enrollments by 510, this is approximately halved with the minimum population threshold parameter, M, set to 2, and decreases to 69 as M increases to 30. Thus, at the district level of aggregation the improvement in predictive accuracy of cohort component models due to spatial filtering is clear. In order to assess the accuracy of the method in more detail, we evaluated the differences between actual and predicted values with respect to average absolute error, average relative error, shape of the error distribution, and absence of bias for population subgroups.

Table 1. District-wide Actual and Projected Elementary School Enrollments by Grade, 1991

M	0	1	2	3	4	5	6	Total
0	791	837	800	801	817	809	720	5575
	50	*71*	*62*	*79*	*94*	*103*	*51*	*510*
2	762	804	781	774	775	748	688	5332
	21	*38*	*43*	*52*	*52*	*42*	*19*	*267*
5	750	798	762	759	762	742	680	5253
	9	*32*	*24*	*37*	*39*	*36*	*11*	*188*
7	749	789	755	757	758	731	679	5218
	8	*23*	*17*	*35*	*35*	*25*	*10*	*153*
10	748	786	749	750	750	730	675	5188
	7	*20*	*11*	*28*	*27*	*24*	*6*	*123*
15	748	781	748	746	741	724	673	5161
	7	*15*	*10*	*24*	*18*	*18*	*4*	*96*
20	749	781	747	742	743	721	669	5152
	8	*15*	*9*	*20*	*20*	*15*	*0*	*87*
25	748	780	745	739	739	720	667	5138
	7	*14*	*7*	*17*	*16*	*14*	*-2*	*73*
30	747	777	742	741	738	721	668	5134
	6	*11*	*4*	*19*	*15*	*15*	*-1*	*69*
Actual	741	766	738	722	723	706	669	5065

Note:
Numbers in *italics* show the differences between projected and actual enrollments.

Average Absolute Error
Average absolute error indicates how different, on average, a projected value is from its corresponding observation. Here, average absolute value was computed by averaging the absolute value of the difference between observed and projected grade population values for each block over the 896 blocks and 7 elementary grades in the study (6,272 observations). Average absolute error was computed for each of the minimum population threshold parameter values using:

$$e_M = \frac{\sum_{i=1}^{N} \sum_{g=g_i}^{g_f} |n_i(g) - \hat{n}_{Mi}(g)|}{N * (g_f - g_i + 1)}$$
(7)

where:

    e_M: average error for minimum population threshold value M;
    i : basic spatial unit index;
    N : total number of spatial units in the study area;
    g : age group index;
    g_i: first age group;

g_f: last age group;

$n_i(g)$: actual population of unit i and age group g in 1991-92 academic year;

$\hat{n}_{Mi}(g)$: projected population of unit i and age group g in 1991-92.

Table 2. Mean Absolute Error as a Function of the Minimum Population Threshold (M)

Minimum Population Threshold	Number of Observations	Average Absolute Error	Sample Standard Deviation	Standard Deviation of the Mean
0	6272	.445	.986	.012
2	6272	.403	.824	.010
5	6272	.378	.762	.010
7	6272	.371	.751	.010
10	6272	.368	.746	.009
15	6272	.362	.736	.009
20	6272	.359	.734	.009
25	6272	.359	.732	.009
30	6272	.358	.727	.009

For the Iowa City study area, as M increases to approximately 15 (Table 2) the average absolute error decreases, indicating convergence between observed and expected values. For M values from 15 to 30, however, the average absolute error remained approximately constant.

Average Relative Error
While average absolute error indicates how far predicted values are from observed values in the units of measurement used, the average of prediction errors relative to the observed values themselves is an additional criterion of predictive accuracy (Isserman, 1977). If, for example, we find in evaluating a set of predictions that the average absolute error is small but relative errors are large, we may not be satisfied with the accuracy of the projections. Average relative errors for each of the minimum population threshold parameter values was computed as:

$$\text{Re}_M = \frac{\sum_{i=1}^{N} \sum_{g=g_i}^{g_f} \left(\left| n_i(g) - \hat{n}_{Mi}(g) \right| / n_i(g) \right)}{N * (g_f - g_i + 1)} \qquad (8)$$

where:

Re_M: average relative error for threshold value M.

The remaining variables are defined as in equation (7).

The average relative error values indicate a problem symptomatic of small area projections: because of the low counts involved and the fact that only whole numbers are allowed as actual values, relative errors tend to be high (Table 3). On the whole, we observe that the average relative error decreases as M increases to 15 and stabilizes as M increases from 15 to 30.

Distribution of Absolute Errors

The distribution of absolute errors is used to examine aspects of accuracy, such as the dispersion of errors and the presence of an unacceptably large number of outliers, that are not captured by average absolute error or average relative error. Ideally, the distribution of absolute errors should be highly skewed to the left with few outliers. The presence of a large number of outliers would indicate a weakness in the predictive accuracy of the projection method and would decrease its reliability in applications (Isserman, 1977). Also, lack of clustering of error values close to zero may indicate bias in the projection method.

The distribution of the absolute errors of the MSF projections is heavily skewed to the left with few outliers (Table 4). Indeed, for all values of M, approximately 97-98% of the projected values differ by less than three students per census block and grade from the observed values. For each M, most observations fall within the interval [0,1], and their frequency decreases gradually towards zero. The most prominent changes in the distribution occur when M increases from zero (traditional cohort component) to non-zero values (filter). For M greater than two, the changes in the distribution are negligible.

Table 3. Average Relative Error as a Function of the Minimum Population Threshold (M)

Minimum Enrollment Threshold	Number of Observations	Average Relative Error	Sample Standard Deviation	Standard Deviation of Mean
0	2221	0.550	0.620	0.013
2	2221	0.492	0.521	0.011
5	2221	0.451	0.488	0.010
7	2221	0.441	0.479	0.010
10	2221	0.435	0.481	0.010
15	2221	0.428	0.486	0.010
20	2221	0.425	0.488	0.010
25	2221	0.422	0.488	0.010
30	2221	0.420	0.487	0.010

Note:
To avoid divisions by zero in computing average relative errors, we used only observations with non-zero actual enrollments in creating this table.

Table 4. Frequency and Cumulative Percentage of Absolute Errors as a Function of the Minimum Population Threshold

M	[0]	(0-1)	[1-2)	[2-3)	[3-4)	[4-5)	[5-6)	[6-7)	>=7
0	4050	868	935	248	83	41	16	13	18
	65	78	93	97	99	99	100	100	100
2	3816	1265	884	173	74	30	12	8	10
	61	81	95	98	99	100	100	100	100
5	3757	1387	852	174	57	23	11	4	7
	60	82	96	98	99	100	100	100	100
7	3719	1419	872	160	57	24	11	3	7
	59	82	96	98	99	100	100	100	100
10	3712	1443	852	162	58	24	12	3	6
	59	82	96	98	99	100	100	100	100
15	3670	1500	839	168	54	23	9	2	7
	59	82	96	99	99	100	100	100	100
20	3645	1526	847	156	62	18	9	3	6
	58	82	96	98	99	100	100	100	100
25	3626	1552	832	169	57	18	11	1	6
	58	83	96	98	100	100	100	100	100
30	3625	1553	831	171	58	16	10	2	6
	58	83	96	99	99	100	100	100	100

Notes:
Numbers in *italics* show the cumulative percentages.

Absence of Bias in Population Subgroups

While the first three criteria summarize the accuracy of the MSF projections over the entire set of observations, they do not assess systematic biases in the projections of population subgroups. Such biases may occur because of differences in the demographic processes of particular subgroups. Families, for example, may be concerned about the negative effects of migration on the social life and educational progress of their children and may decide not to migrate when children are in certain grades. Thus, if the accuracy of the projection method depends on migration, it may vary from grade to grade. Similar biases in the accuracy of the projections will be observed if migration patterns vary spatially through time. For example, if new housing is built in certain sub-regions of a study area, thus increasing the in-migration rates for those sub-regions, then projections based on progression rates estimated using population data that describe the demographic processes before the growth occurred will systematically underestimate the populations for those areas.

Analyses for subgroup biases were performed by grouping deviations of observed and projected grade by block values with respect to student grades and elementary school attendance areas. When a grouping is made with respect to grades, a decreasing trend in the values of average absolute error is observed for grades two through six (Table 5). The error for kindergarten projections is larger than that for the rest of the grades. This difference, however, could be caused not only by the underlying demographic processes but also by the method that was used to obtain the kindergarten projection; that method (Lolonis, 1992) is based on trend extrapolation and is not part of the MSF method.

Table 5. Average Absolute Error and Sample Standard Deviation by Grade as a Function of the Minimum Population Threshold (M).

M	0	1	2	3	4	5	6
			G r a d e				
0	.65	.41	.44	.43	.42	.41	.35
	.95	.99	.97	.93	.93	1.25	.79
2	.62	.37	.42	.39	.37	.34	.31
	.92	.88	.93	.79	.76	.74	.67
5	.61	.35	.39	.37	.34	.31	.28
	.89	.76	.88	.73	.68	.68	.63
7	.61	.34	.37	.36	.34	.30	.27
	.89	.75	.86	.73	.67	.64	.63
10	.61	.34	.36	.36	.33	.30	.27
	.90	.75	.83	.72	.66	.65	.62
15	.60	.34	.37	.35	.33	.29	.26
	.89	.73	.83	.70	.65	.63	.61
20	.61	.33	.36	.34	.33	.29	.26
	.89	.72	.82	.70	.66	.63	.62
25	.61	.33	.36	.34	.32	.29	.26
	.89	.73	.82	.69	.64	.64	.62
30	.60	.33	.36	.34	.32	.29	.26
	.88	.74	.81	.70	.63	.63	.62

Notes:
Numbers in *italics* are sample standard deviations.
The number of observations per grade is 896.

The error trends observed above are also evident here. Indeed, with the exception of kindergarten enrollments, the average absolute error decreases as M increases. Again, the most prominent changes occur when M ranges between zero and fifteen.

When subgroups are defined with respect to existing school attendance areas, the projection accuracy exhibits two distinct characteristics. First, the average errors appear to stabilize at a smaller value of M (Table 6) and second, their magnitude and rate of change vary substantially over space. The most accurate projections are obtained for schools serving primarily urban areas with low growth rates, such as Lincoln (1) and Longfellow, while the least accurate projections are obtained for schools, such as Lemme, Lucas, and Wood, that serve areas which have experienced growth in the recent past. The most evident improvements in the accuracy of projections as a function of M are observed in areas around the periphery of the metropolitan area that have experienced growth; the average absolute error for Lincoln (2), Roosevelt (2) and Shimek decreases by 29%, 33% and 38% respectively, as M increases from zero to thirty. For those three schools, the most rapid decrease in the error values occurs when M increases from zero to seven.

Table 6. Average Errors and Sample Standard Deviations by School Attendance Area as a Function of the Minimum Population Threshold

School	0	2	5	7	10	15	20	25	30
C. Central	0.57	0.50	0.47	0.46	0.46	0.45	0.45	0.44	0.44
(64; 441)	*0.96*	*0.84*	*0.76*	*0.75*	*0.75*	*0.77*	*0.76*	*0.73*	*0.72*
Hills	0.65	0.62	0.63	0.63	0.63	0.61	0.60	0.60	0.60
(17; 179)	*1.31*	*1.30*	*1.31*	*1.33*	*1.28*	*1.19*	*1.19*	*1.18*	*1.19*
Hoover	0.46	0.37	0.35	0.34	0.34	0.34	0.33	0.33	0.33
(72; 356)	*0.85*	*0.58*	*0.53*	*0.52*	*0.51*	*0.51*	*0.51*	*0.51*	*0.51*
Horn	0.55	0.50	0.46	0.46	0.46	0.44	0.43	0.43	0.43
(41; 333)	*0.83*	*0.72*	*0.68*	*0.68*	*0.69*	*0.67*	*0.67*	*0.67*	*0.66*
Kirkwood	0.86	0.83	0.77	0.76	0.76	0.76	0.76	0.76	0.76
(34; 417)	*1.28*	*1.23*	*1.11*	*1.11*	*1.13*	*1.13*	*1.11*	*1.10*	*1.10*
Lemme	0.73	0.72	0.69	0.66	0.67	0.66	0.66	0.66	0.66
(31; 325)	*1.08*	*1.08*	*1.08*	*1.05*	*1.05*	*1.04*	*1.04*	*1.04*	*1.04*
Lincoln (1)	0.19	0.18	0.17	0.17	0.18	0.17	0.17	0.17	0.17
(45; 116)	*0.50*	*0.44*	*0.42*	*0.41*	*0.42*	*0.42*	*0.42*	*0.42*	*0.42*
Lincoln (2)	1.11	0.96	0.88	0.83	0.81	0.79	0.79	0.79	0.79
(12; 101)	*1.64*	*1.31*	*1.09*	*0.99*	*0.96*	*0.98*	*0.98*	*0.98*	*0.98*
Longfellow	0.20	0.18	0.16	0.16	0.16	0.16	0.16	0.16	0.15
(165; 303)	*0.54*	*0.42*	*0.41*	*0.40*	*0.41*	*0.40*	*0.40*	*0.40*	*0.40*
Lucas	0.71	0.67	0.62	0.61	0.61	0.60	0.60	0.60	0.59
(43; 465)	*1.44*	*1.41*	*1.20*	*1.20*	*1.22*	*1.20*	*1.20*	*1.20*	*1.18*
Mann	0.26	0.23	0.23	0.22	0.22	0.22	0.22	0.22	0.22
(101; 274)	*0.73*	*0.57*	*0.55*	*0.55*	*0.56*	*0.56*	*0.57*	*0.57*	*0.57*
Penn	0.51	0.47	0.47	0.46	0.46	0.45	0.44	0.45	0.45
(39; 340)	*0.84*	*0.76*	*0.75*	*0.75*	*0.74*	*0.72*	*0.71*	*0.73*	*0.73*
Roosevelt (1)	0.34	0.30	0.28	0.29	0.29	0.28	0.27	0.27	0.27
(63; 214)	*0.67*	*0.55*	*0.56*	*0.55*	*0.55*	*0.53*	*0.53*	*0.53*	*0.53*
Roosevelt (2)	0.70	0.60	0.51	0.49	0.49	0.47	0.47	0.46	0.47
(34; 253)	*2.09*	*1.12*	*0.87*	*0.80*	*0.81*	*0.77*	*0.76*	*0.76*	*0.76*
Shimek	0.50	0.42	0.35	0.35	0.32	0.32	0.31	0.31	0.31
(33; 177)	*1.09*	*0.78*	*0.65*	*0.65*	*0.60*	*0.60*	*0.60*	*0.60*	*0.59*
Twain	0.34	0.32	0.31	0.30	0.29	0.28	0.28	0.29	0.29
(76; 364)	*0.77*	*0.74*	*0.74*	*0.72*	*0.66*	*0.62*	*0.62*	*0.64*	*0.64*
Wood	0.98	0.93	0.88	0.85	0.85	0.84	0.85	0.85	0.83
(26; 407)	*1.27*	*1.22*	*1.22*	*1.20*	*1.21*	*1.21*	*1.21*	*1.18*	*1.17*

Notes:
Numbers in *italics* are sample standard deviations.
The numbers within parentheses below each school name show the census blocks and the 1991-92 total enrollment for that school.

DISCUSSION AND CONCLUSIONS

These analyses suggest that the accuracy of the MSF projections improves as the number of students used to make estimates of progression parameters increases from zero to fifteen. This improvement is generally monotonic and is observed in the trends of the total projected enrollments (Table 1), average absolute error (Table 2), and average relative error (Table 3). The trends are also evident when observations are grouped with respect to grade (Table 5) and space (Table 6). The rate of improvement, however, decreases as M increases and is negligible for values of M greater than fifteen. In the analyses performed here, there is no indication that there are intervals in the values of M for which accuracy fluctuates systematically.

ACKNOWLEDGMENTS

We would like to thank David Smyth for his assistance in developing the database on which the analyses in this paper are based and Trudy Meyers for assistance in data entry.

REFERENCES

Caliper Corporation, 1990, *GisPlus: Geographic Information System*, Newton, MA.

Ericksen, E.P. 1974, A regression method for estimating population changes of local areas: *Journal of the American Statistical Association*, Vol. 69, pp. 867-875.

Isserman, A. 1977, The accuracy of population projections for subcounty areas: *Journal of the American Institute of Planners*, Vol. 43, pp. 247-259.

Isserman, A. 1984, Projection, forecast, and plan: On the future of population forecasting: *APA Journal*, Vol. 50, pp. 208-221.

Lee, E.S. and Goldsmith, H.F. 1982, ed. *Population Estimates: Methods for Small Area Analyses*, Sage Publications, Beverly Hills.

Lolonis, P. 1992, Procedures and Programs for Implementing the Modifiable Spatial Filter (MFA) Method for Making Small-Area Population Projections, (forthcoming, Discussion Paper, Department of Geography, University of Iowa, Iowa City, IA).

National Research Council. 1980, *Estimating Population and Income of Small Areas*, (Panel on Small-Area Estimates of Population and Income), National Academy Press, Washington D.C.

O'Hare, W. 1976, Report on a multiple regression method for making population estimates: *Demography*, Vol. 13, pp. 369-379.

Rushton, G., Armstrong, M.P., Lolonis, P. 1992, Small Area Student Projections Based on a Modifiable Spatial Filter: *Socio-Economic Planning Sciences* (forthcoming).

Tayman, J. 1992, An Evaluation of Subcounty Population Forecasts: *URISA Proceedings*, Vol. IV, pp. 172-179.

ON THE INCORPORATION OF UNCERTAINTY
INTO SPATIAL DATA SYSTEMS

K.E. Lowell
Centre de recherche en géomatique
Pavillon Casault, Université Laval
Ste-Foy, Québec, Canada G1K 7P4
(418) 656-7998
Internet: 3703KLOW@VM1.ULAVAL.CA

ABSTRACT

Existing digital spatial systems function under the implicit assumption that boundaries are real, identifiable, locatable, and widthless, and that new boundaries may appear or old ones disappear, but existing boundaries do not change position. While this may be true in socio-political geography, in those subjects involving natural phenomena, this model of space is inappropriate. In these subjects, there may be a considerable amount of uncertainty associated with each boundary and each map category. Thus a new model of space which shows transition zones for boundaries, and polygon attributes as indefinite, is required. The use of uncertainty in digital representations of "the real world" will impact two areas directly: map representations and spatial operators. To address these issues, uncertainty-based cartographic representations and spatial operators are discussed.

INTRODUCTION

In existing digital spatial technologies (DSTs), there are implicit limitations that are often not recognized. These DSTs, especially geographic information systems (GIS), are based on deterministic maps which use definite fixed boundaries. Thus, while different polygons may be added to others by dissolving boundary lines, rarely are inidividual spatial units subdivided. Furthermore, fixed attributes -- which are often categorical -- are attached to each polygon giving the result that polygons either are or are not considered to be a given map type; that is, there is no uncertainty associated with the attribute. Admittedly, if attributes are continuous -- e.g., population density -- one may attach a confidence level to the attribute. This is rarely done, however, and even if it is, the confidence level, or uncertainty, is generally not used in subsequent analysis.

Nonetheless, such an approach to mapping makes a fair amount of sense intuitively. Certainly on paper maps it is difficult to conceive of a different system. Moreover, such an approach lends itself well to digital computer systems as reflected in the wide-spread use of two spatial data structures -- raster and vector -- which are based on these concepts. However, the result of this conceptual spatial mapping model within DSTs is that these technologies are effectively limiting themselves to the manipulation of "electronic paper." That is, the computer is merely performing operations more quickly than a human cartographer can. For example, the computer can change map scales and projections quickly and precisely, and can rapidly overlay and summarize numerous maps; humans cannot. But the advanced computational power of the modern computer is not being used to perform new "humanly impossible" analysis.

There are a considerable number of even more basic problems with this fairly intuitive approach. Principal among these is that the use of spatially deterministic maps with fixed categorical attributes in geographic analysis will result in a deterministic response to a given question. Thus if one seeks a given combination of n desired factors -- e.g., Soil Type 1, Forest A, within 2 km of a road and 10 km of a major population center -- one will receive a deterministic answer. Usually, this will be in the form of a binary map that shows areas that do (or do not) have the desired combination of characteristics. If the analyst is a bit more creative, perhaps one might produce a map that shows areas having all n factors, $n-1$ factors, etc. But this will still produce a deterministic map product -- "yes/no a given combination does/does not exist at a given location." The problem with this approach is that, in reality, a decision maker is more likely to want to know "*how_likely* am I to find certain desired characteristics at a given location?"

This is particularly evident with naturally-occurring phenomena for which the conventional model of space functions poorly. With such phenomena, boundaries are not widthless, nor are attributes necessarily very definite. For example, the boundary between a swamp and a lake is extremely "fuzzy" and the difference between a forest of 80% density and one of 100% density may merely be in the eye of the beholder. With such features and conditions, it is evident that the identification of "suitable areas" using conventional deterministic overlay analysis may be highly flawed and erroneous (MacDougall 1975, Newcomer and Szajgin 1984, Bailey 1988). Moreover, in addition to incorrect location, one might identify a number of areas as having all of the desired characteristics but, in reality, some are likely to be "highly suitable" while others are "barely acceptable." That is, without consideration of the real-world-to-map uncertainty associated with each map, one cannot obtain even an ordinal subdivision of the "acceptability" of areas.

It is clear that a failure to include some provision for uncertainty in existing DSTs has the potential to give questionable, and relatively imprecise results. It is stressed here that by "uncertainty" it is not errors associated with the difference between a digital and a paper map that are being discussed. Examples of these are digitizing errors, or errors in satellite image registration. Instead, "uncertainty" is being used to mean those simplifications necessarily introduced when one wishes to represent "the real world" on a map and/or in the computer. It is the belief of this researcher that it is necessary to include uncertainty concerning "real world to map" simplifications if one is to move beyond merely manipulating electronic paper within DSTs. Such a step has the potential to impact two areas directly: computer-based map representations, and associated spatial operators.

ALTERNATIVE DIGITAL MAP REPRESENTATIONS

Traditionally, spatial analysis has been constrained to the use of deterministic tools principally because of two factors. First, pen-and-paper technology does not lend itself very well to complex representations of spatial surfaces. Second, humans are limited in their abilities to comprehend complex spatial processes. For example, every person who has taught map interpretation to beginning students can attest to the problems that some students have in creating and/or understanding topographic contour maps. However, computers are not subject to the same set of constraints as humans and are capable of storing and analyzing much more complex representations of "the real world." Therefore, the first question concerning how to use uncertainty in DSTs must be "How does one represent uncertainty within the computer?" Two conceptual ideas are offered for consideration: individual feature uncertainty and uncertainty surfaces.

Figure 1. Existing vector data-base design and
conceptualization of the inclusion of
uncertainty.

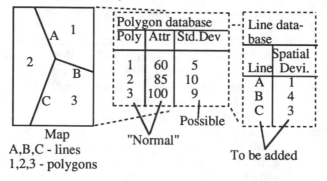

Map
A,B,C - lines
1,2,3 - polygons

Individual Feature Uncertainty

This concept is best-suited to the vector data structure. In such a structure, one has a series of lines from which polygons are "created" -- or, more accurately, identified -- by "building topology." In its current form, the lines are assumed to be error-free (Figure 1, left). Thus currently, the only features which can incorporate uncertainty are the polygon attributes. This can be done by assigning an attribute to each polygon on the polygon list and assigning a "certainty value" to each attribute. In the case of continuous attributes, this is likely to be a standard deviation or statistical confidence level (Fig. 1, center). For categorical attributes, one could give a value for the certainty of the type assigned to a particular polygon, and also the certainty of the other types on the map for that polygon. Thus, one might determine that an individual polygon is Type A by virtue of having a "70% chance" of being A, 25% of being B, and 5% of being C. Recall that here it is being assumed that a categorical attribute is considered to have only a

486

"high chance" of being the type assigned, not an absolute certainty. It also, therefore, has a ("low") chance of being some other type(s).

Such an approach is readily extendible to the lines in the data base as well. Intuitively, the spatial certainty -- i.e., location -- of the line between a forest clearcut and a mature forest is much higher than the line between a shallow lake and a swamp. Thus when one enters lines into a spatial data base by digitizing them, one may or may not be entering the correct location of the lines. Certainly one is entering the "most likely" position. But spatial uncertainty can be incoporporated into the data base -- at least conceptually -- by assigning a standard deviation or range for each line digitized. (Fig. 1, right).

Figure 2. Hypothetical example of raster data-base representation of uncertainty.

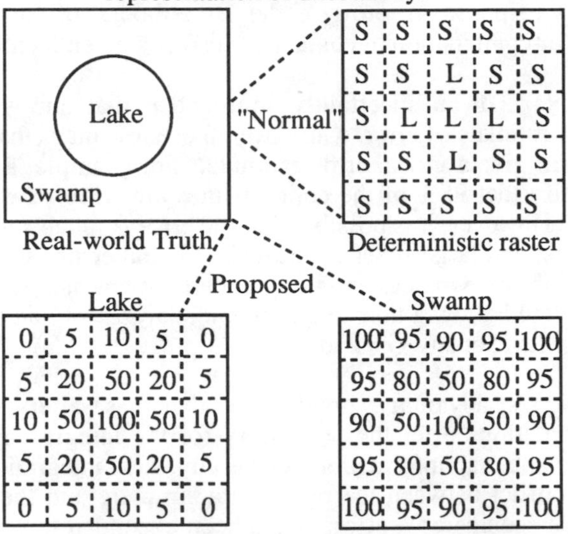

S	S	S	S	S
S	S	L	S	S
S	L	L	L	S
S	S	L	S	S
S	S	S	S	S

"Normal"

Real-world Truth

Deterministic raster

Lake

0	5	10	5	0
5	20	50	20	5
10	50	100	50	10
5	20	50	20	5
0	5	10	5	0

Proposed

Swamp

100	95	90	95	100
95	80	50	80	95
90	50	100	50	90
95	80	50	80	95
100	95	90	95	100

Proposed Representation Showing Estimated Certainty of Having a Given Type in A Cell

Uncertainty Surfaces

This concept of uncertainty representation in digital computers is well-suited to the raster data structure. In this, a "map" is no longer a single map, but is a series of surfaces. If there are k categories on the map, then there will be k surfaces. Thus each map category is represented by a single surface. The values on each surface may range from 0 (zero) to 100 and the value on a given surface represents the "certainty" of having that type at that given location. For example, if one has a perfectly circular lake surrounded by a swamp, presumably the center of the lake has the highest certainty of actually being "Lake" (Figure 2). Thus a value of 100 might be assigned to this central location. But as one moves towards the edge of the lake, the certainty of being "Lake" diminishes in favour of being "Swamp."

Certainties over all surfaces will sum to 100 since the total certainty for each cell must be equal.

Advantages of Alternative Representations

The biggest advantage of such representations is that one will have moved from the relative analytic poverty of a deterministic approach, to the richness of a "fuzzy" system which more realistically reflects real-world conditions. Indeed, some researchers have begun to discuss the implications of such spatial systems (Robinson 1988). While the mechanics of a fuzzy approach are much more difficult to comprehend and implement, as long as the computer is handling the details of manipulating such a system, this computational complexity should not be a problem for applications-oriented users. In the use of such a system, analysts would be able to make queries of the map-based system and, instead of getting a binary "yes/no" response, could obtain a response in terms of the "likelihood" of finding a given characteristic (or combination of characteristics) at a given location.

Another advantage of an uncertainty approach is that one would have a locally-reliable estimate of error. Currently, those few maps that provide an estimate of error give a global error estimate. For example, a topographic map may specify that 95% of the contour lines are within 2 metres of the true elevation. However, it is possible that the error is much smaller in relatively flat errors, and larger in steeper areas. Global error estimates do not reflect this possibility, whereas assigning uncertainty measures to individual map features would allow a user to query and analyze both an attribute and associated error level at any given location.

Perhaps the biggest advantage, however, is that these alternative "uncertainty" representations offer the possibility for the integration of satellite image and thematic map information. Generally, such integration is considered to be a conflict between the raster data structure and the vector data structure. Thus considerable efforts have been expended to convert one to the other. However, the use of uncertainty offers a possibility for circumventing this artificiality.

Consider that one of the standard operations that an image analyst performs on a satellite image is that of classification. That is, one attempts to convert the raw digital scores of b data bands (7 for the Thematic Mapper) and g gray-levels (256 for the Thematic Mapper) into a thematic map having less than 10 or so categories. This is done largely because one wishes to integrate such information into a GIS which contains thematic map-based data. However, intermediate between the raw image and the final classified map/image are a set of probabilistic surfaces identical in principle to the uncertainty surfaces described in the previous section. That is, each surface shows the probability -- or certainty -- of having a given cover type at a given location. Generally, these are simply used internally within the computer to determine which map category has the greatest likelihood of occurring at a given location; they are not usually actually mapped, though some researchers have done so (Skidmore and Turner 1988). But if real-world-to-map uncertainty becomes incorporated into spatial analysis, one can use the probabilistic image surfaces within the GIS in exactly the same fashion as the thematic map uncertainty surfaces. Thus one could truly integrate the two types of information rather than merely converting one data structure to another.

Difficulties With the Use of Uncertainty for Map Representations

The primary difficulty with the ideas discussed thus far is that one must obtain information about spatial (and attribute) uncertainty. In the case of satellite images, this is a relatively straightforward process as one can define (un)certainties based on statistical analysis. But if one wishes to use the proposed model(s) of uncertainty for thematic map data, one must quantify the concepts discussed relative to the (un)certainty or "fuzziness" of various boundaries, and convert these to "(un)certainty scores" representing the likelihood of observing a given type at a given location. Though no comprehensive technique exists for determining such scores, such work might start with the use of ordinal rather than interval/ratio measures. Already, some multi-factor spatial analysis is conducted using "weighting factors." Such analysis is usually formulated as an interval/ratio statement -- e.g., "soil type is *twice* as important as distance to water in the location of a factory." In reality, however, one has merely been able to identify that some factor is *more important* than another -- an ordinal concept. Only after determining the relative importance of factors does one assign an interval/ratio number to the ordinal concept based on some unquantifiable "feeling." A similar approach may prove to be a useful starting place for the work discussed.

Another problem is that, assuming that one has information about (un)certainty for each of the boundary lines on a map, one must still determine the frequency distribution of the error around each. While it may seem reasonable to think that the error will be normally distributed about a boundary line, the polygon type on either side of the line may affect this error distribution. In fact, it is likely that error conditions may depend on the characteristics of the two polygon types on either side of the boundary line. Certainly it is reasonable to believe that the boundary between two highly variable types is fuzzier than the boundary between two uniform types. The error distribution could be different as well.

Another problem related to computer technology itself is that the storage of the information described may become excessive. This is especially true for the certainty surfaces discussed relative to the raster data structure. In its current form, an 8-bit 400-by-400 raster having 10 categories will take 160 kb of storage. But if one must store one surface for each of the 10 categories, such a map will require 1.6 mb of hard-disk storage. Thus it may become necessary in such situations to make greater use of run-length encoding, particularly if many of the values on a given map-category surface are 0 (zero) as would seem to be likely. In vector uncertainty representations, it is more likely that RAM -- rather than disk storage -- could become a problem because of the large amount of real-time operations that such a system is anticipated to require.

The final problem identified herein is that, assuming that one can measure the (un)certainty of each feature and store it in a computer, one must determine how to conduct spatial analysis using this information. Most existing spatial analyses assume that map-based information is known without error. Indeed, a basic tenet of (aspatial) statistical analysis is that observations have been measured without error. Thus there is relatively little quantitative theory that exists for how one can utilize the uncertainty of measurements for analysis in addition to the measurements themselves. Thus there is a need for a new family of spatial operators based on uncertainty.

UNCERTAINTY-BASED SPATIAL OPERATORS

A critical factor in understanding uncertainty-based spatial operators is the formulation of an appropriate question. With deterministic maps, queries are always specified as, for example, "Show all of the areas having a given combination of characteristics" or "What type of geologic parent material underlays this lake?" These questions assume that each area mapped has definite, known characteristics and spatial boundaries. The use of uncertainty in DSTs will necessitate a reformulation of these questions which reflects this uncertainty: "Show the *likelihood* of having a certain combination of characteristics over the map area" or "What is the *most likely* parent material to underlay this lake?" This is not to suggest that overlay is the only spatial analysis that can and will be impacted by the incorporation of uncertainty into DSTs. However, it is a very common operation with such technologies and provides the simplest example for the explanation of the ideas presented herein.

It is admitted here that the exact nature of uncertainty-based spatial operators which must be developed is unknown. However, a simple example can help to provide a conceptual basis for the development of these. In this example, a raster-based data structure is used instead of a vector structure strictly for simplicity's sake. The same concepts apply equally to the vector data structure though the software implementation for vector would be considerably more difficult. The top of Figure 3 presents a (hypothetical) 3-by-3 raster soils map having three soil types (A, B, and C) and a raster forest map for the same area containing two forest types (1 and 2). Suppose that a user poses the query "Where are all the areas of forest type 1 which are underlain by soil B?" The upper right of Figure 3 shows the results of conventional deterministic overlay techniques and identifies a single cell as having the desired B1 combination.

But now consider an uncertainty-based approach. Beneath each deterministic map is an uncertainty-based representation of the same two maps similar to that found in Figure 2. (As acknowledged in the previous section,

490

Figure 3. Hypothetical example of uncertainty-based "overlay" analysis.
Query: Show the area(s) having Forest Type 1 underlain by Soil Type B.

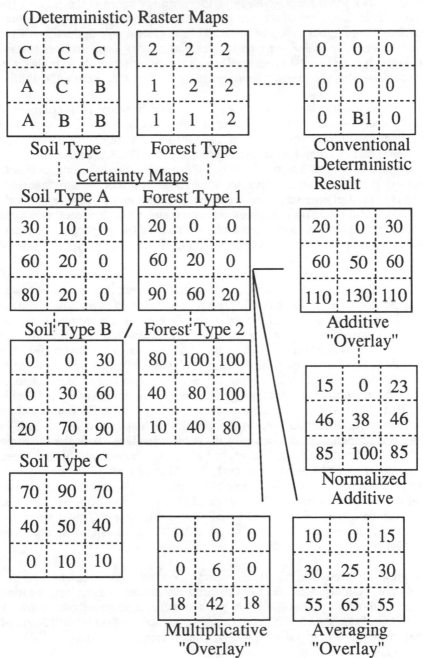

(Deterministic) Raster Maps

Soil Type

Forest Type

Conventional Deterministic Result

Certainty Maps

Soil Type A

Forest Type 1

Additive "Overlay"

Soil Type B / Forest Type 2

Normalized Additive

Soil Type C

Multiplicative "Overlay"

Averaging "Overlay"

the actual method for determining these scores has yet to be developed.) Note that the highest "certainty score" for each cell on the soils map is the highest for the type represented on the deterministic map. The same is true of the forest map. Thus, for example, the cell in the upper-right corner of the deterministic soils map is type C and the certainty score for type C is 70 for that cell. Note that for a given type, the certainty score is the highest for those cells in the centre of a given polygon (which on a raster map is represented as a contiguous group of cells). Thus for forest type 2, the upper right corner has a value of 100 with certainties diminishing as one moves toward the lower left and forest type 1. Though this is a fictional example, it is expected that this is a realistic representation of what will occur in "real life."

After the certainty scores have been determined, it remains to combine -- i.e., "overlay" -- them to provide the information desired. One way is to simply add the scores for the soil B and forest 1 certainty surfaces and then normalize them so that the maximum score is 100. This has been done in the lower right of Figure 3 with scores being normalized by dividing by 1.3 (a factor of the maximum additive score of 130). Another way to "overlay" them is to multiply the scores on the respective certainty surfaces and divide by 100 (in order to maintain scores between 0 and 100). Yet another way is to simply take the average score over all cells for the respective surfaces. (Each of these possibilities has been represented in the lower right section of Figure 3.)

Note that all three methods produce somewhat different results. In all cases, the most likely cell to have the desired B1 combination is the lower central cell. Therefore the results of the deterministic overlay are seemingly reproduced. However, recall that the question to be answered is something like "What is the *likelihood* of finding B1 at a given area?" Given this, the multiplicative method of combining maps would say that only four cells have *any* chance of actually having B1, and only one has a "high" chance. But both the additive and averaging methods of combining cells would indicate that all cells save one have a chance of having the B1 combination. While the actual method used to combine these maps must be studied, in many ways the multiplicative seems the most appealing. It is the only one of the three that has the desired property that, if the certainty of any given factor at any given location is 0 (zero), the cell combination will be 0 (zero). This seems to make sense -- if there is no chance of B appearing at a location, there would seem to be no chance of B1 at that location. All three of the methods discussed have the equally desirable property that, when the value for a location on all feature maps is 100, the composite certainty score will also be 100.

The previous discussion highlights what is a considerable problem with this approach. Namely, what do the (un)certainty scores at each cell location mean? Are they "probabilities" in the sense that a score of 60 for soil A means that if one goes to 100 locations having a value of 60 for soil A, 60 of those locations will actually be A? Or does it mean that at a given location the soil will have "60% of the characteristics of A?" (whatever this means). Or will the (un)certainty scores simply indicate that 60 for soil A means that an area has a higher chance of being A than any other type? If so, then does a value of 60 actually indicate exactly two times "more likely" than a value of 30? Or are the scores merely ordinal indications of something that is not precisely measureable? As this research moves forward, these are questions that must be answered.

CONCLUSIONS

To move beyond the mere manipulation of "electronic paper," real-world-to-map uncertainty must be incorporated into digital spatial technologies (DSTs). The inclusion of uncertainty will impact how maps are represented and manipulated within a computer, and how they are analyzed by humans. Though there are many unanswered questions about how to measure and manipulate uncertainty, this paper has advanced a set of concepts which may prove useful for future work on this subject.

ACKNOWLEDGEMENTS

The author wishes to acknowledge the Association of Forest Industries of Québec and the Natural Sciences and Engineering Research Council of Canada for funding this work. The author also gratefully acknowledges discussions with G. Edwards and C. Gold which helped improve the clarity of the ideas presented.

REFERENCES

Bailey, R.G. 1988. Problems with using overlay mapping for planning and their implications for geographci information systems. Environmental Management, Vol. 12, pp. 11-17.

MacDougall, E.B. 1975. The accuracy of map overlays. Landscape Planning, Vol. 2, pp. 23-30.

Newcomer, J.A., Szajgin, J. 1984. Accumulation of thematic map errors in digital overlay analysis. The American Cartographer, Vol. 11, pp. 58-62.

Robinson, V.B. 1988. Some implications of fuzzy set theory applied to geographic databases. Computer, Environment, and Urban Systems, Vol. 12, pp. 89-98.

Skidmore, A.K., turner, B.J. 1988. Forest mapping accuracies are improved using a supervised nonparametric classifier with SPOT data. Photogrammetric Engineering and Remote Sensing, Vol. 54, pp. 1415-1421.

DEVELOPMENT OF AN INTEGRATED KNOWLEDGEBASED SYSTEM FOR MANAGING SPATIOTEMPORAL ECOLOGICAL SIMULATIONS

D. Scott Mackay, Research Assistant
scott@tworivers.erin.utoronto.ca

Vincent B. Robinson, Director
vbr@geophagia.erin.utoronto.ca

Institute for Land Information Management
University of Toronto, Erindale Campus
Mississauga, Ontario CANADA L5L 1C6
(416) 828-5459

Lawrence E. Band, Associate Professor
lband@eos.geog.utoronto.ca
Department of Geography & Institute for Land Information Management
University of Toronto
Toronto, Ontario CANADA M5S 1A1
(416) 978-4975

ABSTRACT

The Knowledgebased Land Information Manager and Simulator (KBLIMS) is a system for managing spatiotemporal simulations of ecological processes organized around a watershed-based model of terrain. It demonstrates that an object-oriented spatial database for watersheds can be easily organized as a graph and exploited as such for building a query system. KBLIMS includes modules for the extraction of a watershed representation directly from grid digital elevation models, and an object-based information system allowing the selection, browsing, navigation among, and query of watershed objects using a graphical user interface. Procedural calls to the simulation system provide answers based on simulation results. Transitive closure queries on relations and constraints on the database graph provide a means for defining complex, or higher order, objects. KBLIMS provides a unique decision support system for integrated resource management in mountainous forested watersheds.

INTRODUCTION

This paper reports on work towards developing an integrated knowledgebased-ecological simulation system for managing spatiotemporal simulation experiments. Development of simulations of ecological processes is organized around a watershed-based model of terrain. Spatial organization of terrain in this model is an infinite hierarchy of watersheds, hillslopes and stream links, representing an infinite number of spatial resolutions. Spatial representation at each level of spatial resolution is organized into a network, consisting of a set of stream links, stream-to-hillslope adjacency links, and hillslope-to-hillslope adjacency links. This network, or graph structure, is the basic organizing principle for the object-oriented spatial database. A user interface exploits this framework to present a graphical object-oriented view of the database.

The interface allows one to browse objects, navigate from object to object, and specify queries on objects using a graphical user interface (GUI). The query system exploits procedural calls to the simulation system to provide answers based on simulation results. Transitive closure queries on specified relations and constraints on the database graph provides a means for defining complex, or higher order, objects. For example, consider the query - *what is the incremental volumetric growth for all upstream areas in 1988.* Processing this query involves finding the transitive closure of the *upstream* relation, identifying all objects in the transitive closure, organizing the database information for each respective object in terms of requirements for the simulation system, running the simulation system on each object, and providing a report on the incremental growth for each object. In addition, some form of aggregation may be required as part of query, such as finding the total incremental growth for the specified region. It is significant that this system manages such complex queries on ecological/spatial objects in a transparent manner throughout the simulation experiment.

MOTIVATION

A major goal of current ecological research is to develop the ability to compute ecological and hydrological flux processes over large land areas. However, this is hampered by the extreme variability of land cover, topography and soils. Thus the motivation for a knowledgebased ecological simulation system is driven by two needs:

- First, a large part of the effort in design of simulation experiments is centred on being able to relate high-level concepts such as forest stands on hillslopes, to low-level information derived from digital terrain models and satellite imagery.

- Second, it is important that the domain-specific knowledge be organized in terms of distinct and identifiable domain features, such as hillslopes, ridges, valleys, watersheds, etc.

KNOWLEDGEBASED INFORMATION MANAGEMENT

A brief overview of the principles upon which the system is based is provided. We refer the reader to Mackay *et al* (1991) for details on the motivation and database organization requirements for the system.

Knowledgebases and Databases

Central to designing a knowledgebased system for managing land information and simulation is the need to provide seamless integration of terrain knowledge, database facilities and simulators. These three components should be accessible through the query language. A database management system (DBMS) provides support for storage, retrieval and query on collections of data. Basic semantic and structural aspects of this data are defined by a database scheme. On the other hand, knowledgebase management systems (KBMS) are sophisticated programs developed with a greater emphasis on incorporating information into schemes called classes. KBMSs provide facilities for describing and reasoning about the classes as well as the data. Roughly speaking a KBMS is class-intensive whereas a DBMS is data-intensive. This distinction is blurred with development of object-oriented DBMSs which have a strong notion of both class hierarchy and tools for managing large quantities of data. The system described in this paper is designed in terms of class hierarchies (e.g. see Mackay et al, 1991).

Relational databases can be viewed as labelled, directed hypergraphs with nodes representing domain elements, and the labelled edges representing tuples. If the relations are binary, then the database is treated as a directed labelled graph (Maier, 1983; Yannakakis, 1990) which is analogous to semantic networks (Roussopoulos and Mylopoulos 1975). Formally, a graph is defined as:

> **Definition 1.** A graph G consists of n nodes and e edges. Each edge has a label $L(e)$ belonging to some domain having two operations, + (sum) and × (product), where for every path in G we can define its label as the product of the labels of its edges, and label $L(u,v)$ of a pair of nodes u, v is defined as the sum of the labels of all paths from u to v (Yannakakis, 1990).

Reachability between two nodes in a graph is defined as a disjunction of paths between the nodes, where each path is a conjunction of the labels on each of its edges. A traversal recursion on graph G defines the reachable subgraph, and possibly its computed node and edge labels from some specified starting node in G. A traversal recursion, or transitive closure on a graph database is a query that generates a graph consisting of derived entities (Rosenthal *et al*, 1986). Paths are traversed by recursively matching the terminus of one edge with the origin of another. Graphical systems have recently attracted attention for interfacing with databases (e.g., Ioannidis, 1992). We show that a database for watersheds can be easily organized as a graph and exploited as such for building a query system.

Arguments favouring the use of topography as an organizing principle for a ecological simulation system knowledgebase are: (1) watersheds have well-defined boundary conditions determined by the systems of drainage and ridge lines (Band, 1989a,b), and (2) topography is relatively stable compared with vegetation and soils. Assumptions are made regarding the nature of topography in order to define a functional model of the 3-dimensional surface.

> **Assumption 1.** *A minimal set of topographic object classes is adequate to represent the terrain "structure". Topographic features are assumed spatially mutually exclusive, and the set of all extracted topographic features fully exhaust the terrestrial surface.*

> **Assumption 2.** *The discrete surface of the terrain can be represented as a tree.*

> **Assumption 3.** *A continuous topographic surface is generalizable as a finite set of homogeneous patches or facets (or hillslopes) and links (or streams).*

Using Assumptions (1-3) a watershed is partitioned into a set of stream links and hillslopes using the technique based on Band (1989). Using this method a topographic surface, within watershed boundaries, becomes a *drainage area transform* (DAT) tree defined as:

Definition 2. For each watershed on a digital elevation model, a DAT tree exists such that each grid cell corresponds with a node N_i in the tree and has as its value the number of nodes $N_j...N_n$, $i \leq j \leq n$, corresponding to all upstream cells on which flowing water eventually travels down to a cell at node N_i.

Pruning the DAT forms an acyclic stream network. Each link in the stream network drains exactly 2 hillslope areas which are attached as leafs representing the space once occupied by the pruned areas of the DAT. The result is a database of the topographic surface, organized as areal objects attached to links in the tree. Generalization of topography as a less complex surface with a less extensive stream network and fewer hillslopes is easily accomplished by changing the extent to which the DAT tree is pruned. The greater the pruning, the coarser the partitioning, and the less dense is the final database graph.

Stream networks consist of streams and tributary streams, each of which consists of further tributary streams, *ad infinitum*. Topological structures such as stream networks lend themselves to representation as hierarchical structures. However, the depth of this hierarchy cannot be pre-determined. Complex objects are needed to allow arbitrarily complex structures to be represented within databases. These recursive structures do not easily translate into flat tables, nor are they easily queried using traditional query languages which do not support recursion. To exploit the graphical nature of the stream network a graphical view of the database is developed which incorporates knowledgebase tools to express queries in terms of these graphical structures. This graph structure is illustrated in Figure 1.

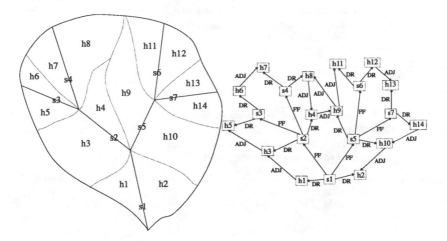

Figure 1. A simple watershed database for use with Queries 1-3, consisting of 7 stream links, $s_1...s_7$ and 14 hillslopes, $h_1...h_{14}$. Three base relations are represented: (1) FF (flows_from), (2) DR (drains), and (3) ADJ (adjacent).

The topographic database is used to drive a set of ecological simulation models which operate on hillslope objects (Band *et al*, 1991;1992). It is used for two purposes: (1) organization of the spatial domain for object retrieval, and (2) classification of terrain features for identifying sets of common features for running simulations. A central concern is provision of ability for browsing, navigating, and retrieving objects for use in the simulations. Browsing and navigational features are provided through point-and-click routines on the stream network.

Selection of objects for running simulations can be made using one of two approaches: (1) direct selection by pointing, or (2) selecting a single object, then querying about its relationship to other objects. We have focused on queries of topological relationships, specifically on path-based queries requiring transitive closure on one or more spatial relations. Example transitive closure queries include simple reachability such as finding all paths that are contained within the area drained by a selected object. More complex queries such as aggregating along paths in order to classify complex features have also been addressed. The latter type of query is a non-traditional use of traversal queries on graph databases, but does represent work that should be performed by a database that has knowledge about the physical organization of the domain values. This is one very good reason for having the database do transitive closure queries (Dayal and Smith, 1986). The following programs illustrate the kinds of queries on the topographic database that are used to select objects for running the simulation:

Query 1. Consider a topographic database with base relations flows_from(S,T) and drains(T,H) representing that stream link S flows from stream link T and stream link T drains hillslope H. The following program, written in Datalog (Ullman, 1988), defines the query upstream(S,H) that asks for hillslopes that are upstream from stream link S:

```
upstream(S,H) ← flows_from(S,T), drains(T,H).
upstream(S,H) ← flows_from(S,T), upstream(T,H).
```

Query 2. Consider a topographic database with base relations flows_from(L,T) and source(L) respectively denoting that stream L flows from stream T, and stream L is a source (having no tributaries). The following monotone program uses aggregation to define stream magnitudes:

```
magnitude(L, 1) ← source(L).
magnitude(L, SUM(M))[L] ← flows_from(L, T),
                          magnitude(T, M).
```

This program terminates because there are no cycles in the stream network database. A variation on this model is to aggregate the total area draining into a stream. This gives a truer measure of the size of a stream as it reflects the total area of runoff that can contribute water volume to a particular stream, and is important in setting up simulations useful in managing water resources.

Query 3. Consider a topographic database with base relations flows_from(*L, T*), drains(*L, H*), and area(*H, A*), representing stream *L* flowing from stream *T*, stream *L* draining hillslope *H*, and hillslope *H* having area *A*. The following program gives the total area drained by a particular stream:

```
tc_flows_from(L, T) ← flows_from(L, T).
tc_flows_from(L, T) ← tc_flows_from(L, U),
                      flows_from(U, T).
drain_area(L, SUM(A))[L] ← tc_flows_from(L, T),
                           drains(L, H), area(H, A).
```

Query 1 can be used to retrieve hillslopes and then calculate the total area in Query 3. These query models allow objects to be selected based on similar properties. For example, once a classification program is performed on the database, all objects of a given class can be selected for running a simulation by simply selecting a particular instance of that class. In this way the graphical query system exploits the user's ability to identify visual clues that aid in formulating queries.

Knowledgebased Simulation

Simulation generally refers to numerical experimentation and numerical modelling. Numerical simulations are abstractions of reality that incorporate heuristic knowledge about specific domains. They represent incomplete knowledge by simulating natural systems to an acceptable level of error, using an acceptable simplifying assumptions. As such, numerical models represent significant sources of knowledge.

Knowledgebased simulation has its roots in the 1950's with the development of discrete-event and continuous simulation. Simulation developed for the purpose of predicting the future state of a system given a model of the system and its current state (Round, 1989). Advantages of knowledgebased techniques include improvements in generating the simulation code, incorporating knowledge of the system into the model, preventing and detecting errors in the model formulation, and manipulating the numerical simulations (Loehle, 1987). Integration of knowledgebase and simulation systems typically follows one of four models: (1) sequential integrated systems, (2) parallel integrated systems, (3) knowledgebased front-end systems, and (4) rule-driven systems. Sequential integrated systems use a one-way mapping between a knowledgebase and a numerical simulation model, typically from knowledgebase to model (see Abelson *et al*, 1989 for a review).

In parallel integrated systems, knowledgebase and numerical simulation components pass information back and forth, maintaining both assertional and numerical components of the system throughout the simulation run.

Sequential and parallel integrated models take an existing numerical model and add a knowledgebase component. In contrast, knowledgebased front-end systems are an attempt at exploiting a knowledgebase to generate numerical simulations(e.g. Robertson *et al*, 1989).

Rule-based systems are generally used when knowledge in the problem domain is heuristic, vague, or incomplete, and where numerical simulations cannot be generated. Systems built exclusively using a rule-based paradigm are generally limited to solving simple, contrived problems.

Our approach closely follows the parallel integration model. It is our belief that the simulation system should be a natural extension of the knowledgebase and that queries on the system should not explicitly distinguish a database component and a simulation component; these should be transparent in the query. This was accomplished by having three types of predicates in the knowledgebase: (1) extensional (or base relations), (2) intensional (rules or query programs), and (3) simulation predicates. The first two are described implicitly in the above discussion, and are considered in more detail in Mackay *et al* (1991). Simulation predicates are generated from class descriptions which have links to appropriate simulators. For example, the following sequence shows how volume incremental growth is determined for a simulation object:

Class SimulationObject with *annual-growth*: VolumeIncrementalGrowth
Class VolumeIncrementalGrowth with *yearend*: Growth
Class Growth with *derived_from*: PSN in Yearday
Class Yearday with *simulation_system*: PHsimulator

The result is a sequence of generated queries (Mackay *et al*, 1991) resulting in the predicate, *yearend(growth, SimulationObjectId, Value)*, being added to the knowledgebase for each queried simulation object.

IMPLEMENTATION

The above ideas are implemented in the Knowledgebased Land Information Manager and Simulator (KBLIMS) (See Figure 2) which provides systematic organization and query of watershed information, including the computation of watershed-level forest productivity and hydrologic processes in mountainous environments. In KBLIMS, emphasis is placed on terrain information capabilities for forested mountainous regions. KBLIMS is specifically designed around the construct of *nested watersheds*, thereby facilitating the handling of integrated forestry/watershed processes and concerns.

KBLIMS includes modules for the extraction of a watershed representation directly from grid digital elevation models, and an object-based information system allowing the selection and query of the watershed objects. We have incorporated a set of watershed simulation modules that provide the ability to predict the distribution of forest productivity as well as hydrologic response of the surface over specified years.

KBLIMS consists of five main software components as illustrated in Figure 2. TPS (after Band, 1989), TESS (Lammers and Band, 1990), and TESS2PRO are considered low-level image and feature processing tools written in C. They have been integrated into GRASS (version 4.0). TOI is an object-based system providing a graphical interface and query system written in Prolog. PHS is a set of productivity and hydrology simulators that are accessible through the query system of TOI.

TOI provides the ability to both query the database for static information (e.g., watershed structure, land cover information) and dynamic information using PHS (e.g., expected forest productivity, snowmelt, runoff production). The parameter files for PHS can be manipulated using the TOI GUI such that the effects of various forest management activities may be queried by altering object specific information. Thinning or clear cutting can be simulated by altering stand LAI, standing biomass and the first-order effects of climate change can be explored through manipulation of the meteorological files.

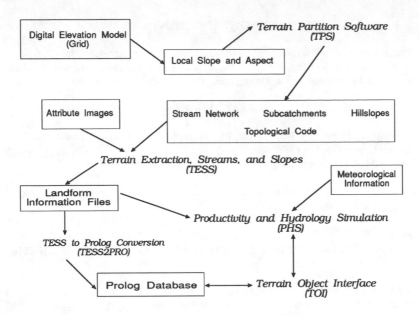

Figure 2 Connections between the main components of KBLIMS showing the main inputs and outputs of each module.

CONCLUSIONS

Robinson et al (1989) discussed areas that knowledgebased systems could be applied to GIS so that problems in resource management might more readily and effectively be addressed. Among these areas were automated terrain and feature extraction, incorporation of simulation and modeling abilities, management of object-oriented databases, and intelligent query facilities which would support a graphical user interface. KBLIMS has successfully addressed many of those areas. Furthermore, it is significant that KBLIMS manages quite complex queries on ecological/spatial objects in a transparent manner throughout a simulation experiment.

It is shown that a database for watersheds can be easily organized as a graph and exploited as such for building a query system. Transitive closure queries on specified relations and constraints on the database graph provides a means for defining complex, or higher order, objects

KBLIMS is based on our belief that the simulation system should be a natural extension of the knowledgebase and that queries on the system should not explicitly distinguish a database component and a simulation component; these should be transparent in the query. This was accomplished by having three types of predicates in the knowledgebase: (1) extensional (or base relations), (2) intensional (rules or query programs), and (3) simulation predicates.

The models in KBLIMS provide the user an ability to explore the spatial pattern of forest productivity, watershed potential response to climate change. As such it could be a useful tool to explore various forest management strategies and their impacts on both forest productivity and hydrologic response. In this respect, KBLIMS provides a unique decision support system to support integrated resource management in forested watersheds.

ACKNOWLEDGEMENTS

Partial support from the Natural Sciences and Engineering Research Council, Canada Centre for Remote Sensing, National Aeronautics and Space Administration, and the Institute for Land Information Management is gratefully acknowledged.

REFERENCES

Abelson, H., Eisenberg, M., Halfant, M., Katzenelson, J., Sacks, E., Sussman, G.J., Wisdom, J., and Yip, K. 1989, Intelligence in scientific computing: Communications of the ACM, Vol, 32, pp. 546-562.

Band, L.E. 1989a, A terrain-based watershed information system: Hydrological Processes, Vol. 3, pp. 151-162.

Band, L.E. 1989b, Automating topographic and ecounit extraction from mountainous forested watersheds: AI Applications in Natural Resource Management, Vol. 3, No. 4, pp. 1-11.

Band, L.E., Patterson, J.P., Nemani, R., and Running, S.W. 1992, Forest ecosystem processes at the watershed scale: Incorporating hillslope hydrology: Agricultural and Forest Meteorology, in press.

Band, L.E., Peterson, D.L., Running, S.W., Coughlan, J.C., Lammers, R.B., Dungan, J., and Nemani, R. 1991, Forest ecosystem processes at the watershed scale: basis for distributed simulation: Ecological Modelling, Vol. 56, pp. 171-196.

Dayal, U., and Smith, J.M. 1986, PROBE: a knowledge-oriented database management system: On Knowledge Base Management Systems: Integrating Artificial Intelligence and Database Technologies, Springer-Verlag, New York, pp. 227-257.

Ioannidis, Y.E. 1992, ACM SIGMOD Record Special Issue: Advanced User Interfaces for Database Systems.

Lammers, R.B., and Band, L.E. 1990, Automating object representation of drainage basins: Computers and Geosciences, Vol. 16, pp. 787-810.

Loehle, C. 1987, Applying artificial intelligence techniques to ecological modeling: Ecological Modelling, Vol. 38, pp. 191-212.

Mackay, D.S., Band, L.E., and Robinson, V.B. 1991, An object-oriented system for the organization and representation of terrain knowledge for forested ecosystems: GIS/LIS'91 Proceedings, Vol. 2, pp. 617-626.

Maier, D. 1983, The Theory of Relational Databases, Computer Science Press, Rockville, MD.

Robertson, D., Bundy, A., Uschold, M., and Muetzelfeldt, R. 1989, The ECO program construction system: ways of increasing its representational power and their effects on the user interface: International Journal of Man-Machine Studies, Vol. 31, pp. 1-26.

Robinson, V.B., and Frank, A.U., and Karimi, H.A. 1989, Expert systems for geographic information systems in resource management: Fundamentals of Geographic Information Systems: A Compendium, pp. 155-166.

Rosenthal, A., Heiler, S., Dayal, U., and Manola, F. 1986, Traversal recursion: a practical approach to supporting recursive applications: Proceedings of ACM SIGMOD International Conference on Management of Data, pp. 166-176.

Round, A. 1989, Knowledge-based simulation: The Handbook of Artificial Intelligence, Vol. 4, Addison-Wesley, Reading, MA., pp. 415-518.

Roussopoulos, N., and Mylopoulos, J. 1975, Using semantic networks for database management: Artificial Intelligence and Databases, Morgan Kaufmann, San Mateo, CA., pp. 112-137.

Ullman, J.D. 1988, Principles of Database and Knowledge-Base Systems, Vol. 1, Computer Science Press, Potomac, MD.

Yannakakis, M. 1990, Graph-theoretic models in database theory: Proceedings of the 9th SIGACT-SIGMOD-SIGART Symposium on Principle of Database Systems, pp. 230-242.

POLARIS - A LIVING BREATHING PROJECT - PROPERTY TITLE DATA
CAPTURE AND CONVERSION AND DIGITAL PROPERTY MAP DEVELOPMENT

A. M. (ANDY) MACKENZIE, OLS., P. ENG.
TERANET LAND INFORMATION SERVICES INC.
141 ADELAIDE ST. WEST, SUITE 700
TORONTO, ONTARIO, CANADA M5H 3L5

ABSTRACT

POLARIS, Province of Ontario LAnd Registration Information
System, provides an accurate property map base and automates
the 4 million plus properties in the Province of Ontario.
Teranet is a hybrid company, exhibiting hybrid vigour, was
spawned by the Province and the private sector and is charged
with completing the automation of the land registry records
and of developing an internationally competitive GIS industry.

The resulting automation product is of significant interest to
organizations that use information having a geographic
component. The accurate map product caters to the needs of a
wide spectrum of users.

Methods utilized to collect and convert title data, write
property mapping instructions, collect horizontal coordinate
values and other cadastral survey data and to prepare the
digital property map-s (including the use of coordinate
geometry) are described.

An overview of the directions taken to improve productivity,
enhance the utility of the Registration System and develop
markets for the information products is provided.

DESCRIPTION OF THE PROVINCE OF ONTARIO

The Province of Ontario is one of 10 provinces and two
territories in Canada. It has an area of just over 400,000
square miles. The southernmost part lies south of the north
border of the state of California and the northernmost part
lies north of the south end of the Alaska panhandle. About 20%
of the Province is covered by water. The Province has 250,000
freshwater lakes. The population is just over 9 millions with
over 80% of those souls living south of the 49th parallel.

The Province has just over 4 million properties. The title
records for those properties are maintained in about 60
Provincially administered Registry Offices.

DESCRIPTION OF THE LAND REGISTRATION SYSTEM

The Registration System in Ontario was established in 1795.
The Act establishing the system was one of the earliest pieces
of Canadian colonial legislation.

In the 1860's changes were made to the Act to record title
records on a geographic basis rather than in an alphabetical
index which had become very cumbersome to use because of
increasing population and common names. Amendments to the
legislation formed the basis of the modern land registration
system, both under the Registry Act and the Land Titles Act by
giving registered documents priority.documents.

A series of changes in the legislation affecting search rules, certification of titles and new standard, statutory forms for title documents were introduced. The new forms are part of the changes introduced to make way for the automation of the entire system.

A Comparison of the Land Titles and Registry Systems

The major features of the two systems compared as follows:

(1) In the Land Titles System, the title records are organized by ownership, so that there is one page or series of pages for each owner's title. In the Registry system, the records are organized by geographic location. Books are organized according to lot and plan or concession. As a result, on the record for a concession lot, since documents are registered in the order in which they are received, documents relating to dozens or even hundreds of titles may be mixed up.

(2) In Land Titles, ownership and the effect of charges and discharges is guaranteed. The Government says who the owner is, that the mortgage has been properly discharged. Searchers can rely on that information. In the Registry system, documents are simply recorded and it is up to lawyers and other searchers to look at the historical record of documents to determine which still apply to the property. The documents are then examined to determine their effect on the property.

(3) Both systems now have basically the same rules regarding priority. Registered documents rank in the order in which they are registered, and are effective over unregistered documents. In the Land Titles system an unregistered document cannot be effective at all, because the theory of the system is that the entire title is shown on the parcel register.

(4) Because of the nature of the two systems, the search and registration processes are different. In the Registry System, the lawyer for the purchaser must research the title history to satisfy himself/herself that the vendor has title. On registration staff conduct a quick review of registration requirements. With no Government guarantee of title, registration can be performed quickly. In the Land Titles system, since the Government has guaranteed the vendor's title, the search process is quick and straightforward. Upon registration, since the Government must now guarantee the purchaser's title, a more complete registration review is required.

(5) The introduction of common forms of documents under both systems has brought the systems closer together. Many affidavits have been removed.

The purpose of the Registry Act is to give public notice of interests claimed in land and to establish priorities between claimants. The purpose of a title search under the Registry Act is to determine if the vendor has the title promised to be conveyed in the Agreement of Purchase and Sale and what interests the title is subject to.

In addition to the above determinations, there are many other potential interests relating to occupation, use, liens, and even actual title that are recorded or maintained in other locations, often other offices of the Provincial or Municipal Government.

It is also necessary to determine if certain sections of the Planning Act have been violated. This determination is especially important since if a violation has occurred, the transfer of title becomes a nullity.

DESCRIPTION OF TERANET

This strategic alliance of government and the private sector is a unique arrangement in Ontario. It is probably unique in the world for land information systems management. It is an equal partnership in the form of a private sector corporation under the laws of Ontario. The partners (shareholders) equally contribute equity capital and equally share in the profits. The corporation is staffed with experts from both sectors achieving a strong blend of technical expertise, business awareness and marketing experience. The board of directors reflects equal partnership in numbers and representatives skills and experience. The Company is responsible to the Government of Ontario for the implementation of POLARIS and its operation, and to the shareholders for good business performance.

PRIMARY MANDATE

The primary mission of Teranet is to implement and operate (under a limited term renewable license) "an automated land registration system and an automated mapping system (collectively know as the POLARIS System), in the Province of Ontario and developing a land related information system (LRIS) industry in Ontario to market LRIS products and expertise world-wide".

ADDITIONAL VENTURES

Teranet is examining markets for the automated products developed under its primary mandate along with such other information as is of interest to those markets within the Province of Ontario . The accurate digital map product, the linkages between the property identifier numbers and other data such as the title records, the assessment roll number and the non confidential portions of the property assessment data, non confidential municipal government data (zoning, official plan, existing land use), other non confidential Provincial data, digital topographic data, data held by utilities, demographic data, marketing related data and many more data bases not yet defined will all form a pool of data of interest to a variety of organizations. Teranet is therefore a data utility.

DESCRIPTION OF POLARIS

POLARIS - **P**rovince of **O**ntario **LA**nd **R**egistration **I**nformation **S**ystem - is a computerized system that stores information on all real properties in Ontario. The information is contained in a Title Index - a computerized version of the Abstract Index and Parcel Register, together with automated digital property maps - organized on a property ownership basis.

Conversion of Registry Act properties to Land Titles Act properties is a major improvement to the land registration system in the Province of Ontario. With the title to virtually every property guaranteed by the Province, the very extensive reduction in the volume of data maintained in the active files, the precisely defined and highly accurate property index maps combined with remote access capabilities, the new POLARIS will prove to be the most modern and progressive to be found in any jurisdiction world wide.

UNIQUE PROPERTY IDENTIFIER NUMBER (PIN)

BLOCK NUMBER - first five digits

Five digits allows for 100,000 blocks in the Province.

A block is an area containing at least one property and is usually bounded by limits such as registry division boundaries, roads, railway right of way limits, major utility corridor limits, water boundaries or property limits. There can be no less than one property per block and no property can be located in more than one block.

PROPERTY NUMBER - last four digits

Within each block, properties are identified by a unique (within the block) four digit property number.

The combination of the five digit block number and the four digit property number constitutes the Property Identifier Number or **PIN**. The **PIN** is unique for each property in the Province. Theoretically, provision has been made for a maximum of 1000 million properties. From a practical standpoint, up to 2,000 properties per POLARIS block are created when the records are converted. The smallest block to date contains 29 properties.

For a severance, the parent property number is retired and two daughter property numbers are created. For a consolidation, the two parent property numbers are retired and a new daughter property number is created. A record of the chronology of the numbers is kept so that the situation as it existed at any point in time may be determined.

TITLE RECORDS

ABSTRACT FROM DOCUMENTS

The POLARIS title data bases are abstracts of selected fields of information found on the various types of documents. These fields include:

Document Registration Number;
Date and Time of Day of Registration of the Document Type;
"Parties to";
Book Number Where Property Previously Abstracted (Registry Act);
Parcel Number (under the Land Titles Act);

Capacity by which title is held;
Consideration or Monies Paid;
Property Street Number and Name (if one exists);
Municipal Jurisdiction in which the property is Property
Assessment Roll Number;
"Thumb Nail" description of property, and;
Special Remarks.

DIGITAL PROPERTY INDEX MAPS

BASIC MAP BUILDING PHILOSOPHY

1. The maps are built as accurately as possible using all
registered deposited and other readily available control and
survey information;

2. The maps are NOT treated as a simple mathematical entity.
Cadastral data are entered based on surveys which represent
the opinion of a trained and licensed land surveyor. Opinions
can be incorrect and are subject to change over time.

Adjustments are not made blindly.The data are analysed in
accordance with survey principles and adjustments are made in
the most 'survey appropriate' manner. This is often quite
different from the results of a 'blind' mathematical adjustment. The addition of new survey data during map maintenance
may cause changes in how the adjustment should have been made.
Appropriate action is taken at that time;

3. The maps are based on a "last registered description"
search of all properties;

4. Although it is possible to construct good maps using a few
ties between control and cadastral corners, experience has
demonstrated that the overall cost of map production is
significantly reduced by well designed control traverses
(especially in areas where recent survey plans are lacking);

5. Data from surveys are entered using coordinate geometry
techniques so that the mathematical accuracy of the survey is
preserved;

6. Digitising is only used when no survey information is
available. Only topographic maps are digitised. Survey plans
are NEVER digitised, and;

7. The maps are compiled, as far as possible, to reflect the
situation on the ground. That is, apparent gaps and overlaps
caused by errors in descriptions are NOT illustrated.

THE MAP BUILDING PROCESS

1. On the basis of a search in the Registry Office, the last
registered description is identified for each property. This
process includes the extraction of ALL survey data from the
files, including sketches attached to documents.

The search staff analyse the data to determine the relationships between adjoining properties and resolve, on the basis
of the available information, any apparent gaps or overlaps

due to errors in description. The result of this analysis is a set of "mapping instructions" for the staff who will build the digital maps.

2. Additional survey information is acquired from sources such as the Ministry of Transportation (Ontario), the Ministry of Natural Resources, local municipal authorities and local surveyors. The Ministry of Natural Resources also provides control information from its COSINE database (a database of coordinate values of some 48,000 points).

3. A basic ground relationship is established.

This is by far the most critical phase of the entire process. If done improperly, the effects will ripple through the rest of the map building process, the on-going maintenance process and attempts to use the database. It is vital that a maximum effort be applied to utilising all available control and survey data and to making this basic ground relationship as accurate as possible. Failure to do so will increase the one-time cost of building the database and the on-going costs of map maintenance.

The ground relationship is generally established for a large area, extending over an entire town or township or even lager area. The size of the area depends on the availability of control to which cadastral surveys can be tied. If no such control is available in the area to be mapped, the area must be extended outwards until such control is found. The process must start with fixed points. There can be no uncertainty about this basic ground relationship.

A top-down approach is followed in building the ground relationship. For primary adjustments, centre lines of roads and outer limits of large plans such as subdivisions are used. Once these data have been adjusted and fixed, related data such as the outer limits of the roads is entered. Survey data which are tied to these plans are then added.

In rural areas the process begins with a determination of what control exists and what corridor plans can be used to link to the control and to other corridor plans. The information from these corridor plans is then entered through coordinate geometry programs. Adjustment software is used to adjust the plans to the control and to each other. Any plans tied to these initial plans are then entered.

The scarcity of control, and survey information means that this process provides coordinates for very few township lot corners. The remaining corners are defined by digitising from Ontario Base Map topographic sheets. At the same time, any other available survey information is used to accurately define the width across a lot, the angle between a road allowance and a lot line, etc. This provides small pockets of high RELATIVE accuracy with low ABSOLUTE accuracy. All available survey information is incorporated into the digitising process. These data are entered by coordinate geometry techniques, NOT by digitising.

Part of the digitising process involves capturing those fences which define property lines since, especially in the case of aliquot part descriptions, the fences are probably the best definition of the property lines available.

In urban areas, the task is made simpler by the existence of more control and survey data. The process starts by defining an overall network by using corridor plans, large subdivision and reference plans or groups of subdivision and reference plans to link to the control. It is at this stage that a decision to establish additional ties between cadastral corners and control may be made in order to reduce mapping costs. Adjustment programs are used to make any necessary shifts to coordinate values. Each area created by this network is similarly processed by fitting the limits of subdivisions and other plans within them. This can be continued until the limits of all subdivision plans (and any other plans of survey which define lots or blocks of land) have been fitted into the picture.

4. The data entry is completed.

In rural areas, this involves adding (by coordinate geometry techniques) any survey information not yet entered and defining the property lines by reference to plans, metes and bounds description, fences, the township fabric and aliquot part descriptions. For practical considerations, aliquot part descriptions are input on the basis of linear dimensions rather than areas.

In urban areas, the internal fabric of subdivision and other plans is entered together with any other survey information which is available. (This is all done by coordinate geometry). Property lines are then defined, almost entirely by reference to the subdivision and other plan fabric.

In both cases, the sequence of entering the data is determined by an analysis of all of the available data to order it from most reliable to least reliable. THE MOST RELIABLE DATA ARE ENTERED FIRST.

It is important that any conflicts between adjoining and (apparently) overlapping plans are resolved according to good survey practice. Failure to do so will result in increased map building costs, reduced usefulness of the data and increased maintenance costs.

In both cases, data related to roads, railways and easements are then added, mostly by duplicating already existing data.
5. Quality control checks are performed.

6. The text is then entered.

This simply consists of adding the required property iden- tifiers, plan descriptions, road and railway names, easement descriptions, etc. The text is entered by a different person from the one who completed the data entry. This provides a check on the work of the data entry person.

7. Additional quality control checks are performed.

8. The map is then prepared for plotting.

Plots are sent to the Registry Office and the map enters into its maintenance mode.

The ultimate aim is to produce a continuous and seamless map across the Province of Ontario. Property boundaries will eventually be shown in relation to the UTM NAD '83 datum. Currently mapping is produced on either the 1927 datum for MTM or the 1974 adjustment for UTM. As soon as algorithms for conversion to UTM NAD '83 have been proven and become available, all existing map related data will be converted to that datum. At that time all new mapping will also be done on that datum.

METHODS USED TO COLLECT TITLE DATA

The record books in the Registry office are examined by the searcher. For each property, all documents affecting title are identified, each registration number is recorded on a property work sheet, the property identified number is created, other pertinent data are added to the property work sheet, certain checks are made to ensure that specific acts have not been violated, boundary related information is recorded on a sketch map obtained from the Assessment Branch of the Ministry of Revenue (known as the mapping instructions) and a determination is made as to whether each Registry Act property can be converted to a Land Titles Act property.

The work requires attendance at a special intensive training course which lasts for three weeks. New staff must be experienced in searching title and interpreting property descriptions before being recruited to work at POLARIS searching.

CONVERSION OF TITLE DATA TO DIGITAL FORMAT

Once the title data have been recorded on the property work sheet, data entry staff take those sheets along with the appropriate abstract index books (in the case of Registry Act properties) or the Parcel Register (in the case of Land Titles Act properties) or the Condominium Parcel Register and set up for the conversion to digital format.

A series of digital files containing the following information are created: (BBBBB refers to the five digit block number)

    1) Property Files:

    BBBBB.C1 - Document Number/PIN Correlation

    BBBBB.B1 - Property Description (thumbnail)
    BBBBB.B2 - Assessment Numbers
    BBBBB.B3 - Book Numbers
    BBBBB.B4 - Municipal Jurisdiction
    BBBBB.B5 - Street Address
    BBBBB.B6 - Land Titles PIN/Parcel - Section Correlation

    BBBBB.01 - Document Pool

The next step involves a process called data assembly. It is broken into four discreet steps named "BUILD", "SPRAY", "POLISH" and "CHECKBLOCK". These steps are carried out on "AT" type microcomputers. "BUILD" builds a file of all PINs with property descriptions. It combines the B1 & B6 files. "SPRAY" applies document numbers to all properties. Abstracted document information (.01 files) is distributed to individual property records under the control provided through the C1 file. "POLISH" adds the remaining property information from the B2, B3, B4 AND B5 files. "CHECKBLOCK" checks for format errors in the data and produces the upload file (.UPL).

The final step in this process involves the "uploading" of the assembled data obtained through the process just outlined. The final destination of the data is to the mainframe computer. This computer currently handles the title data under the control of "CICS" or "Consumer Information Control System" - an IBM product.

NEW DIRECTIONS

Three studies are currently being undertaken by Teranet:

1. PRODUCTIVITY IMPROVEMENT

The methods and procedures developed in the prototype office and as listed above are being examined to determine where improvements in the process can be identified. Modified procedures will be developed and incorporated into the automation process. It is interesting to note that saving of one minute per property represents a saving of over 40 person years when spread among the 4 million plus properties in the Province of Ontario.

2. ENHANCING THE UTILITY OF THE REGISTRATION SYSTEM

Access to the records stored in the registration system has been limited to examination of information "trapped" on paper. The liberation of this information through the POLARIS automation process will be further improved through the development of document imaging, linking the digital map product to the title records and through access to the resulting digital products through remotely located terminals.

3. MARKET DEVELOPMENT FOR LAND RELATED INFORMATION FOUND IN GOVERNMENT FILES

It is expected that remote access will produce significant sources of revenue - both to Teranet and to the originators of data bases not developed by Teranet (through development of a royalty payment system). Development of this marketing expertise is expected to produce a revenue stream for data originators as well as relieve those originators of the responsibility for responding to non-standard requests for data - a process that requires additional time, staff and infrastructure resources to satisfy.

An Evaluation of the 9-Intersection for Region-Line Relations

David M. Mark
National Center for Geographic Information and Analysis
Department of Geography, SUNY Buffalo, Buffalo, NY 14261
geodmm@ubvms.cc.buffalo.edu

and

Max J. Egenhofer
National Center for Geographic Information and Analysis
Department of Surveying Engineering and Department of Computer Science
University of Maine, Orono, ME 04469
max@mecan1.maine.edu

ABSTRACT

In this paper, we describe a research project which is using human subjects testing to evaluate the 9-intersection model, a topological model of spatial relations between a line and a region. Preliminary results appear to confirm the validity of the distinctions made in the 9-intersection, and also indicate that many 'natural' concepts of spatial relations can be formed by aggregating the 19 line-to-region relations. However, there were a few cases of subjects making finer topological distinctions, and also cases of subjects classifying spatial relations by geometric as well as topological criteria. Implications and further experiments are discussed.

INTRODUCTION

The study of spatial relations has received increasing attention over the last few years. Initially considered a purely academic exercise, investigations into spatial relations have demonstrated enormous practical relevance in the design and use of geographic information systems. Most obvious is the use of spatial relations in a query language where users describe spatial conditions that must be met by the objects recorded in a database in order to retrieved and presented as the query result. These spatial conditions are commonly expressed as Boolean combinations (and, or, not) of *spatial relations* (Frank and Mark, 1991) or *spatial prepositions* (Herskovits, 1986). Some examples are *inside*, *north*, and *far* (Freeman, 1975; Peuquet, 1986).

Spatial relations in geographic information systems have several different perspectives. First, it is important to understand the concepts that people employ when they think about space and about the spatial relations among the objects. This is an area of study heavily influenced by cognitive and psychological aspects. Second, there is the use of natural language expressions to describe spatial relations in different natural languages. This is the level with which humans are most familiar as they use predicates explicitly in their everyday life. Finally, there is the concern about formal descriptions of spatial relations such that they can be implemented into a query processor of a geographic information system.

It is important to note that all three levels of investigations are crucial to the success of finding useful sets of spatial relations. Unfortunately, in most efforts to investigate spatial relations, only one component has been addressed. Designers of spatial query languages are only concerned with terminology, as they provide most often only a list of spatial predicates without addressing the semantics or their formal descriptions. On the other hand, formalists usually have been only concerned with the mathematics, without assessing how humans actually use their concepts develop. Finally, studies by linguists generally have not led to formalisms rigid enough to be implemented in a computer.

There is a strong need for research that exploits the interplay between these different aspects. At a methodology level, this paper shows how such an interplay can be accomplished. We start with a formalization of some spatial relations and examine the relations found in this model with human subject tests. At a more concrete level, our investigations focus on *topological relations*, particularly the ones that result from a recent categorization region-line relations (Egenhofer and Herring, 1992). It is an extension of an earlier model for region-region relationships (Egenhofer and Franzosa, 1991). This model is very popular as it has been extended to describe more complex spatial relations (Herring, 1991; Pigot, 1991) and used for a number of applications in query languages (Svensson and Zhexue, 1991; de Hoop and van Oosterom, 1992). The extended model identified 19 different topological relations between a region—a connected 2-dimensional object in R^2 with connected boundaries—and a simple line—a 1-dimensional object with exactly 2 end points that has no self-intersections.

In this paper, we explore the categorization for spatial relations between a line and a region using human subjects. For each of the 19 cases of line-region relations, we produced two identical drawings of a region representing a park; for each of these two we added geometrically-different lines to represent roads. The 38 drawings, 2 for each of the topologically-distinct line-to-region relations, were shown to several native speakers of English, and the subjects were asked to group them such that they would use the same terminology to describe the spatial relation. After completing the grouping task, each subject was asked to describe each set of drawings with a brief phrase or sentence, and to identify the prototypical member for each group.

PREVIOUS RESEARCH

Previous Published Definitions of Spatial Relations Between Lines and Regions

Most categorizations of spatial relations distinguish between topological relations, directions, and distances (Pullar and Egenhofer, 1988; Worboys and Deen, 1991), for which spatial query languages contain a large number of spatial predicates (Frank, 1982; Roussopoulos et. al, Herring et al.; Egenhofer, 1991; Raper and Bundock, 1991); however most of them lack any formal definition of the terminology.

While a great deal of attention has been paid in the GIS literature to spatial relations between regions (Freeman, 1975; Claire and Guptill, 1982; Peuquet, 1986; Egenhofer and Franzosa, 1991; Hernández, 1992), and point-region relations (the classic "point-in-polygon" problem, for example), there has been relatively little published work on relations between lines *and* regions.

Cox, Aldred, and Rhind (1980) just three pairs of Boolean relations, which they called equality, sharing, and exclusivity, together with their notations. Cox et al. did not give definitions of these relations, other than to note that their arguments could be points, lines, or areas, that the result is Boolean, and that equality and sharing are symmetric whereas exclusivity is not. They give point-in-polygon(area) as a special case of 'sharing', which implies that 'sharing' is true if the objects have one or more points in common. Gütting (1988) listed three Boolean spatial relations between a line and a region: inside, outside, and intersect. Roussopoulos *et al.* (1988) list three pairs of Boolean spatial relation operators between a line and a region: within, not-within; cross, not-cross. However, as with Cox, Aldred, and Rhind's paper, the published paper does not contain any detail of the exact topological definition of these predicates. Menon and Smith (1989) included metric spatial relations between points and lines (distance, direction), but no Boolean predicates. Bennis *et al.* (1991) distinguished spatial relation between line objects and region objects in two directions, that is, the spatial relations are order-dependent. For example, a region can be left-of a line, but a line cannot be left-of (or right-of) a region. A region can be right of or left of a line (Boolean). The other

Boolean spatial relations presented by Bennis *et al.* were overlap and inclusion, and to these they added distance and direction.

The '9-Intersection' Definition of Spatial Relations

Recently, Egenhofer and Herring (1992) extended a previous formal categorization of topological spatial relations between spatial objects (Egenhofer and Franzosa, 1991) to account for relations between objects other than regions, such as two lines in 2-D or two regions with holds. The relationships of concern in this paper are the ones between a region and a line.

A spatial region is a connected, homogeneously 2-dimensional 2-cell. The definition of a line is based on 1-cells, i.e., the connections between two geometrically independent nodes. A line is a sequence of 1...n connected 1-cells such that they neither cross themselves nor form cycles. Nodes at which exactly one 1-cell ends will be referred to as the *boundary* of the line. Nodes that are an endpoint of more than one 1-cell are *interior nodes*. The *interior* of a line is the union of all interior nodes and all connections between the nodes. The *closure* of a line is the union of its interior and boundary. Finally, the exterior is the difference between the embedding space and the closure of the lines (Figure 1).

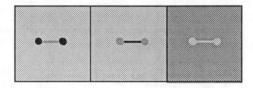

Figure 1: Boundary, interior, and exterior of a line in R^2.

We will call a sequence of 1-cells a *simple line* if it has exactly two boundary nodes. A *complex line* is a line with more than two boundary nodes. (Lines that would have less than two boundary nodes would include cycles, which are excluded by definition).

The binary topological relationship R between a region, A, and a line, B, is based upon the comparison of A's boundary, interior, and exterior with B's boundary, interior, and exterior. These six object parts can be combined such that they form nine fundamental descriptions of a topological relationship between two lines, called the *9-intersection* (Egenhofer and Herring, 1992). With each of these nine intersections being empty or non-empty, the model has 512 possible relations between the objects. Egenhofer and Herring go on, however, to show that most of these combinations of 9-intersections are impossible for connected objects in the Cartesian plane: the categorization shows 8 distinct topological relations between two simple regions (regions without holes); 33 relations between two simple (unbranched) lines; and 19 relations between a simple unbranched line and a region. More detailed distinctions would be possible if further criteria were employed to evaluate the non-empty intersections such as the dimension of the intersections or the number of separate components.

THE EXPERIMENT

In this research, we are exploring the categorization of spatial relations between a simple line and a simple region using human subjects. We produced 40 identical drawings of a region said to be a 'park', and for each of these drawings we added a 'road'. The position of the road relative to the park was different in each case, and the roads were positioned so as to provide two (or, in two cases, three) geometrically distinct examples of each of the 19 topologically-distinct cases of line-region relations. Some examples are shown in Figures 2, 3, and 4. Two examples are shown in Figure 2.

Figure 2: Two examples of the stimuli; these two cases were isolated as categories by all subjects.

In the experiment, the 40 drawings were shown to five native speakers of English and to one native speaker of Chinese[1], and the subjects were asked to group the drawings so that a single natural language phrase in their native language could describe each drawing in the group. To the maximum extent possible, the instructions were given and responses recorded in the native language of the subject. In English, the instructions were:

> "Here are 40 different sketches of a road and a State Park. Please arrange the sketches into several groups, such that you would use the same verbal description for the spatial relationship between the road and the Park in each case."

When the subject completed the task, the experimenter recorded the groups, and elicited a descriptive phrase for the spatial relation for each group. Lastly, the subject was asked to select the 'best example' from each group, as a prototype. Background information on each subject was limited to age, gender, language spoken, and prior experience with other languages, cartography, map-reading, and geographic information systems.

Results

Validity of the 9-Intersection. If the distinctions made in the 9-intersection model were irrelevant to the way people think about spatial relations, then we could expect that the two drawings for each of the 19 relations would be assigned to categories independently. If a subject divided the examples into, say, 10 categories, then the chance that both members of a pair would end up in a common class would be 0.1 (1 out of 10). If this probability is multiplied by the number of relations (19), then we would anticipate about 1.9 pairs to be grouped together. More generally, the expected number of such cases would be 19/N, where N is the number of categories assigned by the subject.

In the experimental results, topologically-identical cases under the 9-intersection model were grouped together far more often that chance would suggest. For five subjects, a randomness hypothesis would predict about 10 co-classified 9-intersection relations in total. The 81 observed cases across these five subjects indicate that topologically-identical cases were put together 8 times more often than randomness would predict. This is not at all surprising, but is a strong confirmation that for most subjects and situations, topological similarity or identity is a more powerful basis for categorizing spatial relations that is geometry. (One exception is documented below.)

---

1   At the time that articles for these conference proceedings were due, we had data for only six subjects. We expect to have data from larger samples available for presentation at the conference, and to publish those in later papers on this topic.

Table 1:

Expected (Under Randomness) and Observed Numbers of
9-Intersection Cases Grouped Together by the Subjects

Subject	Number of classes	Cases grouped together expected	observed
C1	10	1.9	13
E1	7	2.7	17
E2	10	1.9	15
E3	13	1.4	18
E4	9	2.1	18
E5	19	1.0	19
Total(mean)	(11.3)	11.0	100

The 9-intersection 'twins' were grouped together by the subjects about 9 times as often as a randomness hypothesis predicts. Thus, while a rigorous statistical test was not performed, it seems obvious that the distinctions made by the 9-intersection model are cognitively salient.

Prototype effects. Recall that each subject was asked to identify a prototype ('best example') for each category they identified. With 6 subjects averaging 11.3 classes, there were a total of 68 category prototypes. One testable hypothesis is that all 19 of the 9-intersection cases were equally likely to be selected as category prototypes. This leads to an expectation of 69/19, or 3.57 prototype selections per case. Using a Poisson distribution, we can predict the number of cases that would be prototypes exactly 0, 1, 2, 3, ... 6, 7+ times, and compare these to the observed frequencies:

Table 2:

Expected (Under Randomness) And Observed Frequencies of
Selection of Each 9-Intersection Case as a Class Prototype,
Computed Using the Poisson Distribution
(Chi-square=4.89 with 8 d.f.)

n	O	E	E-O
0	0	0.53	0.53
1	2	1.91	-0.09
2	4	3.41	-0.59
3	4	4.06	0.06
4	2	3.62	1.62
5	4	2.59	-1.41
6	3	1.53	-1.47
7+	0	1.35	1.35

The table shows a strong excess of cases selected 5 or 6 times, and a sharp deficiency of 4-time selections. However, with 8 degrees of freedom, the value of Chi-square is not statistically significant. However, the prototypes seem not to be distributed at random among the examples, but rather more often are 'simple' cases in which the body of the line is entirely inside or entirely outside the region.

517

The two cases shown in Figure 2 were grouped with their topologically-identical 'twins' by each of the 6 subjects, and furthermore were identified as isolated classes with no other members. Every other 9-intersection case was combined with at least one other case by one or more subjects. Five other examples were selected 4 or more times as category prototypes (Figure 3). All six subjects identified the right-hand case (Figure 3C) as a category prototype, but 4 of the 6 subjects grouped it with other cases. The other four examples shown in Figure 3 were identified as prototypes by 5 of the 6 subjects, but were combined with other patterns by all but one subject.

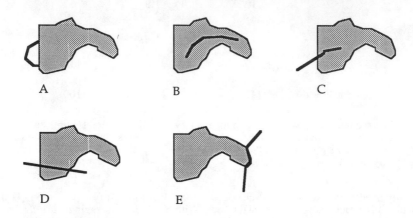

Figure 3: Five other cases that were frequently selected as prototypes for spatial relations.

A Rare Example Of Discrimination By Geometry

Whereas almost all subjects appeared to emphasize topological factors in their responses, one of the English-language subjects apparently used a geometric criterion to classify some of the stimuli. Figure 4 shows this exception. It seems likely that the particular geometry of the four cases on the upper right-hand side of figure 4 that caused them to be grouped together by this subject.

SUMMARY

As the differences between the two members of each topologically-similar pair of road-park examples were geometric—different direction of the line, different shape, different length—the outcome suggests that people often ignore such quantitative differences and are primarily concerned with qualitative differences. The results of the experiments also suggest that many of the qualitative differences humans make about spatial relations are captured by the 9-intersection.

Future Work. Clearly, larger sample sizes are needed. Once more subjects have been studied, additional statistical tests will be performed. We believe that the cross-linguistic dimension of the problem also is worth pursuing, and will test additional native speakers of English and Chinese, also adding Spanish and German speakers.

Also worth investigating is the possible influence of the hypothetical phenomena in the drawings. Would the results be significantly different if the line-region relation were described as a storm track and an island? Or a road and a gas cloud? And does scale (scope) matter, that is, would the categorization be different if the line and region were things on a table-top, or at continental scales.

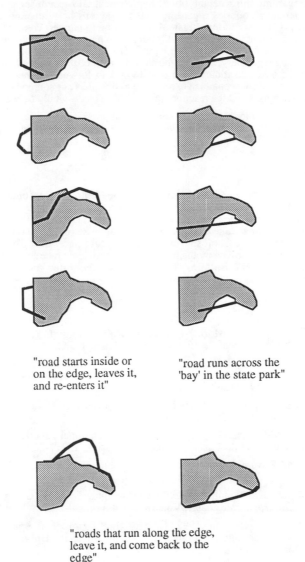

"road starts inside or
on the edge, leaves it,
and re-enters it"

"road runs across the
'bay' in the state park"

"roads that run along the edge,
leave it, and come back to the
edge"

Figure 4: Three groups of stimuli according to one of the English-
language subjects. The four stimuli in the upper left part of the
diagram were described by the sentence "starts inside or on the edge,
leaves it, and re-enters it", whereas the four in the upper right were
grouped under "road runs across the 'bay' in the state park". The
lower two examples were grouped together and described by the
phrase "roads that run along the edge, leave it, and come back to the
edge". Each of the left-right pairs of park drawings above represents
two realizations of the same 9-intersection relation.

It also would be interesting and potentially valuable to perform human subjects experiments regarding the acceptability of hypothetical GIS responses to hypothetical quasi-natural-language queries regarding spatial relations between line features and region features. One way to investigate this would be to convert some of the phrases obtained in the experiment described in this paper into natural-language queries, such as "Find all roads that enter Thoreau State Park." The subjects would be presented with such a natural-language query, and then be shown a map of the park and one road. The subject would be asked to indicate, on a scale of 1 to 5, whether the map is a fully acceptable (5) or fully unacceptable (1) response to the query, or somewhere in between (2-4); both the responses and the reaction times of the subjects would be recorded.

Lastly, we feel that the model described herein would provide a good basis for analyzing line-region queries provided in GIS software or in descriptions of designs for GIS query languages.

Acknowledgments

This paper is a part of Research Initiative 10, "Spatio-Temporal Reasoning in GIS", of the U. S. National Center for Geographic Information and Analysis (NCGIA), supported by a grant from the National Science Foundation (SES-88-10917); support by NSF is gratefully acknowledged. Feibing Zhang administered the test to the Chinese subjects, and Hsueh-cheng Chou suggested some useful references. The experiments described herein were approved by the human subjects review procedures of the Faculty of Social Sciences, SUNY Buffalo. Thanks are especially due to the subjects who grouped the examples for us.

References

Bennis, K., B. David, I. Morize-Quilio, J. Thévenin, and Y. Viémon, 1991. GéoGraph: A Topological Storage Model for Extensible GIS. Proceedings, Auto Carto 10, pp. 349-367.

Berlin, B., and P. Kay, 1969, Basic Color Terms: Their Universality and Evolution. Berkeley: University of California Press.

Claire, R., and S. Guptill, 1982, Spatial Operators for Selected Data Structures. *Auto Carto V*, Crystal City, VI, pp. 189-200.

Cox, N. J., B. K. Aldred, and D. W. Rhind, 1980. A relational data base system and a proposal for a geographical data type. Geo-Processing, 1, 217-229.

De Hoop, S., and P. van Oosterom, 1992. Storage and Manipulation of Topology in Postgres. *EGIS '92*, Munich, Germany.

Egenhofer, M., 1991. Extending SQL for Cartographic Display. *Cartography and Geographic Information Systems*, Vol. 18, No. 4, pp. 230-245.

Egenhofer, M., and J. Herring, 1992, Categorizing Topological Spatial Relations Between Point, Line, and Area Objects. Technical Report, University of Maine, Submitted for Publication.

Egenhofer, M., and R. Franzosa, 1991. Point-Set Topological Spatial Relations. *International Journal of Geographical Information Systems*, Vol. 5, No. 2, pp. 161-174.

Frank, A., 1982, Mapquery—Database Query Language for Retrieval of Geometric Data and its Graphical Representation, ACM Computer Graphics, Vol. 16, pp. 199-207.

Frank, A. and D. M. Mark, 1991. Language Issues for GIS, in: D. Maguire, M. Goodchild, and D. Rhind, editors, *Geographical Information Systems: Principles and Applications*, Longman, London, Vol. 1, pp. 147-163.

Freeman, J. 1975. The Modeling of Spatial Relations, *Computer Graphics and Image Processing*, Vol. 4, pp. 156-171.

Gütting, R. H., 1988. Geo-relational algebra: a model and query language for geometric database systems. In Noltemeier, H., editor, Computational Geometry and Its Applications. New York: Springer-Verlag.

Hernández, D., 1991. Relative Representation of Spatial Knowledge: The 2-D Case, in: D. Mark and A. Frank, editors, *Cognitive and Linguistic Aspects of Geographic Space*, Kluwer Academic Publishers, Dordrecht, pp. 373-386.

Herring, J., 1991. The Mathematical Modeling of Spatial and Non-Spatial Information in Geographic Information Systems, in: D. Mark and A. Frank, editors, *Cognitive and Linguistic Aspects of Geographic Space*, Kluwer Academic Publishers, Dordrecht, pp. 313-350.

Herring, J., R. Larsen, and J. Shivakumar, 1988. Extensions to the SQL language to support spatial analysis in a topological database. In: GIS/LIS '88, San Antonio, TX. pp. 741-750.

Herskovits, A.,1986. *Language and Spatial Cognition*, Cambridge University Press, Cambridge, MA.

Menon, S., and T. R. Smith, 1989. A declarative spatial query processor for geographic information systems. Photogrammetric Engineering and Remote Sensing, 55(11), 1593-1600.

Peuquet, D., 1986. The Use of Spatial Relationships to Aid Spatial Database Retrieval, in: D. Marble, editor, *Second International Symposium on Spatial Data Handling*, Seattle, WA, pp. 459-471.

Pigot, S., 1991. Topological Models for 3D Spatial Information Systems, in: D. Mark and D. White, editors, *Auto Carto 10*, Baltimore, MD, pp. 368-392.

Pullar, D., and M. Egenhofer, 1988. Towards Formal Definitions of Topological Relations Among Spatial Objects, *Third International Symposium on Spatial Data Handling*, Sydney, Australia, pp. 225-243.

Raper, J., and M. Bundock, 1991. UGIX: A Layer-based Model for a GIS User Interface, in: D. Mark and A. Frank, editors, *Cognitive and Linguistic Aspects of Geographic Space*, Kluwer Academic Publishers, Dordrecht, pp. 449-476

Roussopoulos, N., C. Faloutsos, and T. Sellis, 1988, An Efficient Pictorial Database System for PSQL, IEEE Transactions on Software Engineering, 14, 630-638.

Svensson, P. and H. Zhexue, 1991. A Query Language for Spatial Data Analysis, in: O. Günther and H.-J. Schek, editors, *Advances in Spatial Databases—Second Symposium, SSD '91, Lecture Notes in Computer Science,* Vol. 525, Springer-Verlag, New York, NY, pp. 119-140.

Worboys, M., and S. Deen, 1991. Semantic Heterogeneity in Geographic Databases, *SIGMOD RECORD*, Vol. 20, No. 4, pp. 30-34.

DIGITIZATION and DATABASE ACCURACY

Ewan Masters
Conrad Blucher Institute for Surveying and Science [1]
Corpus Christi State University
6300 Ocean Dr
Corpus Christi
TX 78412

ABSTRACT

As organizations endeavor to come to grips with new digital mapping products, aspects of accuracy and database quality are recurring themes. This paper investigates the many processes that drive database development and how these fit in with the typical GIS/LIS vision of community accessible data. A few problems with defining spatial database accuracy are also examined.

INTRODUCTION

One of the common visions for Geographical and Land Information Systems (GIS/LIS) sees many disparate digital map data sets being linked together to provide new types of information to the community and especially being made available for decision makers (McLaughlin, 1991). We have all seen the typical layered view of the world as shown in figure 1, but how often do we examine the spatial assumptions that are made before such a grand vision can be brought to fruition. In the real world, layers of information will reside with data trustees who are geographically, functionally and technologically disparate (eg. O'Callaghan & Garner,1991).

This vision of GIS assumes that disparately sourced spatial data can actually be combined, which in turn assumes that the original function, scale and coordinate system of each data source can be reliably juxtaposed. Also, it assumes that the way objects are defined for each data source is understood by both data providers and users, and finally that the data exchange standards and communications links are available to permit such linking of databases.

These somewhat technical problems are also compounded by the business forces that drive the development of many of these databases in individual organizations. This paper examines these issues and uses the Integrated Facilities Information System (IFIS) database of the Sydney WaterBoard to illustrate the driving forces of database development and to show how these interplay with the development of the broader GIS/LIS vision.

---

[1]On leave from The School of Surveying, University of New South Wales, PO Box 1 Kensington 2033, Australia.

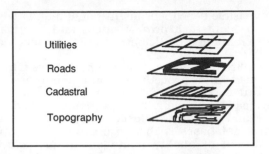

Figure 1:   The GIS Vision: Integrated Databases

DRIVING FORCES OF SPATIAL DATABASES

Limited funding and calls for accountability of government and semi-government institutions means that organizations responsible for managing the land and public infrastructure are increasingly required to improve existing business processes. Many of these organizations are also encumbered with ever-expanding legal requirements to deal with environmental concerns such as the quality of air, water and natural habitats. In many places, organizations are also being required to achieve some type of cost recovery for the GIS/LIS products they are developing. Products must be marketed, which in turn implies that the organizations must become more efficient and customer oriented (Rhind,1992). It is not unreasonable to expect that with the change of focus from "public service" to "market driven" operations, there will be an increasing requirement to provide land information products with standard descriptions and quality definitions.

The development of databases is often undertaken to improve the efficiency and cost-effectiveness of existing administrative functions. Marketing is a new and unknown function for many organizations involved in GIS/LIS database development. New products and functions are often unquantifiable and therefore not easily considered in the economic justification of any GIS business case. This paper surmises that the development of GIS databases is based mainly on satisfying existing business functions and these functions may or may not be in accordance with the general layered vision of GIS as described earlier. In this case, quality documentation will be essential for data to be widely used in the community.

Data Requirements

Traditional systems analysis generally aims to make existing systems within an organization more efficient and for good reason therefore tends to analyze an organization in isolation. The result being that the specified functional and data requirements will suit the primary business requirements of the individual organization, but may not necessarily fit the broad based picture of GIS/LIS. Even when there is a corporate appreciation of the wider benefits of LIS/GIS, there may not be any visible benefits or institutional mandate for individual organizations to adapt their own business and information system specifications to fulfill and participate in such broader visions.

As part of a quality management program, data requirements must also be specified and procedures implemented for assessing whether products conform to those requirements or not. Unfortunately, cadastral mapping is often compiled from varied, uncoordinated and disparate sources. Specifying the quality of digital cadastral databases is therefore extremely complex, making the implementation of quality management procedures difficult. The organizations creating these databases usually understand the idiosyncrasies and complexities of their own data very well and may therefore see little need to provide good quality documentation. These types of problems should be little different for other types of digital mapping databases.

Generally, no matter what data requirements are specified, there will usually be little choice with regard to the accuracy of acquired data. Everyone would like the most accurate data possible. However, for pragmatic reasons, completion of an operational database is usually of the utmost importance, rather than quality management (Ryan & Masters, 1991a). Usually, the "best available" mapping data is digitized. It would seem that the accuracy of databases is therefore driven by the available source data and the digitizing processes, rather than by the user requirements and specifications. Fortunately, the source data is most often used for existing operations and will therefore satisfy most of the data requirements of the organization, as long as existing map accuracy is maintained.

The Resulting Databases

The result of all these competing business forces and processes is that GIS databases may be developed around existing business processes and functions without necessarily looking to the broader GIS/LIS market. They will be created to satisfy a majority of users within a specific organization, but probably not all users. They will have minimal spatial quality assurance performed and they will have accuracies essentially defined by the source data and digitizing processes.

Typically, after a database is well developed, other organizations discover the product and become interested in acquiring the data. All organizations then need to be able to specify their own data requirements in terms of some standard data definition and accuracy definition.

Any decision maker hopefully wants to know the reliability of the information that they base their decisions upon. Accordingly, there is a growing concern in the GIS/LIS community about the effect of errors on the results of all types of analyses. Various approaches can be undertaken. For example, the error sources can be identified and quantified, the lineage of processes used to form information tracked, errors propagated through the functional algorithms and finally the errors presented in some understandable form (see Veregin, 1989; Lanter, 1990; Hunter & Beard,1992; Leung et al,1992). This paper examines the fundamental assumptions that are used to define position and how these affect any spatial analysis and documentation of quality.

With a lack of definitive GIS accuracy standards and appropriate data on accuracy, there is a noticeable tendency to ascribe conventional map accuracy standards to any type of digital mapping. For example, a commonly used standard for conventional mapping is that "*90% of well defined points shall fall within 0.5mm of their true position*" (Merchant, 1987). This standard really only applies to mapping produced using photogrammetric techniques under specific controlled conditions. The statement is a little misleading, because "true position" sounds very "absolute", but actually means the position defined on a specific geodetic datum, which itself may not be very accurate. In the context of digital map data taken from many sources, seamless digital maps and the availability of ever-expanding numbers of spatial analysis techniques, conventional definitions of positional accuracy may be of limited use.

The view of GIS/LIS described in Figure 1 assumes that we live in an ideal world, where coordinate systems are unique, can be absolutely defined and never change. Any concept of accuracy must be considered in the context of coordinate systems defined by geodetic datums and map projections and also by the originally intended function of existing mapping, its scale and its geographical coverage. Many GIS systems unfortunately assume that the coordinate systems that maps are based on are all the same and all have the same accuracy. We may think that these are fairly basic problems. However, such issues continue to present difficulties for GIS projects (eg. Manning & Harvey,1992).

Adoption of improved geodetic networks like NAD83 or WGS84 will hopefully overcome these problems for future data. However, much of the data that has been put into databases already is on old geodetic datums and various map projections. Some of the lineage information may not be documented or may only reside in the minds of operational staff. Under these conditions the problem of mixing and matching coordinate systems will probably never go away. Any conventional map accuracy statement may not adequately describe how well these data can be used for many types of spatial analysis.

We also need to consider that positioning techniques are advancing as quickly as GIS technology. It is often more effective to actually resurvey rather than using existing maps. New data, on the latest coordinate systems, must therefore be integrated with older datasets. Specifying the accuracy of data in these contexts is an extremely complex task and will be handled to varying degrees of sophistication and reliability by various software packages.

Relative and Absolute Accuracy

The definition of absolute accuracy is usually something like *the difference between the actual real world location of a point on the surface of the Earth as against its mathematically assigned geographic coordinate* (eg. Noonan,1992). We can probably assume that most GIS technology would treat coordinates according to this definition. Relative accuracy is essentially *the accuracy of coordinate differences between points or of distances between points*. Relative accuracy can be spatially dependent, but is more often assumed constant and expressed as a part per million or per thousand of the distance between points. Unfortunately, it is difficult to account for relative accuracy in any algorithm. Most surveying/geodesy programs account for relative accuracy by incorporating rigorous statistical data into any computations. However, these processes require a high overhead in storage and computational complexity, which may be impossible to implement in GIS software.

As a slight digression, we should realize that no practical coordinate system is really absolute as has just been defined or as most GIS systems treat coordinates. We all like to think the world actually has lines of latitude and longitude marked out, against which we can determine our coordinates in some absolute sense. For example, I tend to think of the Greenwich meridian as the line through London against which we can determine our longitude. Whereas, real world coordinate systems are dependent on the mathematical idiosyncrasies of the geodetic measuring systems that are used to create and maintain them. That is, they are mathematical abstractions rather than physical realities. Fortunately, the mathematical models behind the modern space-based reference systems produce highly stable coordinate systems, that can be physically reproduced to much higher accuracies than required for most GIS/LIS systems. The Global Positioning System coordinate system WGS84 is an example of such a coordinate system and is really the only practical coordinate system that could be considered absolute in any sense.

The formal statistical measurements of accuracy for geodetic networks only have practical meaning in a relative sense. Geodetic networks generally have very good relative accuracy, though this statement must be balanced against when, where and how the coordinates were derived. On the other hand, the absolute accuracy of geodetic coordinates is difficult to define. Positioning equipment enables us to relatively locate ourselves with respect to some other object, even though the position is usually expressed as absolute coordinates. It is somewhat unfortunate that this is a difficult concept to portray and communicate. We tend therefore to simplify it by saying that the coordinates of a specific point on the ground are accurate to 2 meters or 5 cm or whatever, whereas what we actually mean is that we know our position with respect to another point to 2m or whatever.

For GIS technology, there is therefore a fundamental problem with the mixed use of the concepts of relative and absolute accuracy. GIS algorithms and the accuracy statements we commonly use infer absolute position and accuracy, whereas the coordinate systems are actually relative. Similarly, a conventional map accuracy statement gives the accuracy of points on a map with respect to the specified geodetic framework, which says little about absolute accuracy. We therefore need to examine how relative or absolute accuracy will affect the results of any spatial analysis, whether any derived problems can be minimized and finally how this information should be communicated.

Specifying Positional Accuracy

Data accuracy should be specified to satisfy user requirements and the functional capabilities of GIS systems, which requires accuracies to be provided for all aspects of lineage of the database coordinates and may be required for all cartographic objects. The quality statement for the Spatial Data Transfer Standard envisages storing the nature of the geodetic datum and map projection (ACSM,1988). However, the accuracy of an overlay operation will also depend on the absolute accuracy of these datums themselves. In practice, satisfactory estimates of these accuracy components can usually be made.

Masters (1991) analyzed the relative or absolute accuracy requirements of a basic set of spatial operations. These analyses indicated that most functions require relative accuracy rather than absolute accuracy. Generally we can say that operations within a layer depend on relative accuracy, while operations between layers require absolute accuracy. Even then, the accuracy of maps required for overlay operations can be reduced to relative accuracy, as long as identifiable common points occur on separate layers. Separate maps can then be registered to each other, which only requires relative accuracy rather than absolute accuracy. We should note that map projections need to be treated very carefully as they actually do require absolute accuracy.

As most operations depend on relative accuracy, it seems that an accuracy statement expressing the positional accuracy of spatial objects against a geodetic network will be adequate. In cases where many disparate datasets are to be combined, accuracy statements should include quantitative information about the geodetic networks. As this information becomes more complex, it is probably also useful to have some generalization or simplification of accuracy. Graphical presentation of accuracy would also help to overcome the complexity of accuracy statements.

AN EXAMPLE DATABASE

The Integrated Facilities Information System (IFIS) for the Sydney Water Board contains a Digital Cadastral Database for the Sydney region along with utility information on waste water, water, proposed subdivision and stormwater. The cadastral database contains over 1.2 million parcels of land while the WaterBoard is responsible for thousands of kilometers of water and sewer infrastructure. The core business of the WaterBoard could be classified as customer services, asset management and environmental management. As with most land and infrastructure oriented organizations there is a requirement for all business operations to have databases describing the position and nature of land features and infrastructure. The difficult steps are to determine the accuracy requirements for this information and to eventually assess the operational database accuracy.

Data Process	Asset Management.	Customer Services	Environment Management
capture	1m	1m	1m
maintenance	<0.5m	1m	>1m
upgrade	<0.5m	<0.5m	1m
analysis	<0.5m	1m	10m

Table 1: **Positional Accuracy for Business and Data Processes** (after Ryan & Masters 1991a)

The processes required to manage the IFIS database can be classified as data capture, data maintenance, quality improvement or upgrading, and analysis. Table 1 describes the data accuracy requirements for these same business processes in the context of the data management processes of the Waterboard. These values indicate how the data requirements are perceived to vary through an organization and even within the development of a specific database. Even though the Board participates in the development of the New South Wales statewide Land Information System, the requirements are not designed for that system, and any usefulness to the broader GIS/LIS system are probably fortuitous rather than being designed that way.

The positional accuracy of the IFIS database has been assessed by survey in small regions (Ryan & Masters, 1991b, Smith & McLeod, 1991). These assessments have generally indicated that the data has a positional accuracy better than one meter, meaning with respect to the base geodetic coordinates and therefore satisfies the stated accuracy requirements in table 1. However, these accuracy assessments cover only a small fraction of the complete database. Whether the database as a whole complies with any specific requirements is therefore largely unknown. Any such accuracy statement also oversimplifies the fact that the database has been derived from many varied sources and digitizing processes. We should also remember that these accuracy statements do not indicate how well this data will match data from other organizations.

The designers of IFIS, not knowing what the real accuracy of their data was, had the foresight to include a lineage attribute for graphic data objects in the database. Each graphic element is attributed with a pointer to session information describing the source material and digitizing technique. An example of the type of codes is given in Table 2. However, the problem still remains of finding a reliable method of transforming these session codes into quantifiable positional accuracy indicators.

PLAN or INPUT METHOD	COGO ENTRY	Larger than 1:1000	1:1000 to 1:4000	Less than 1:4000
ISG SURVEYS	B			
SURVEY PLANS				
controlled	B			
transformed control	C			
PLAN TYPES		D	F	
Compilations		D	F	
R.E.Maps		D		
Watermain Record Maps		D	D	D
SPAD Data		E	F	
SRS Sheet		E	E	E
Parish Maps				G
Other Maps		G	G	G

Table 2:   **IFIS Accuracy Codes**
(after Ryan & Masters 1991a)

The IFIS database has essentially undergone all the processes and takes on all the characteristics described earlier in this paper. The database was developed to improve the efficiency of the operational processes and to satisfy the business functions of the WaterBoard. The usefulness of this data for the State Land Information System or any other organization is an additional benefit rather than being a core driving force behind the database. Quality assurance processes for the database are more concerned with attribute matching than positional accuracy (see Ryan & Masters, 1991b). Because the database has been specifically designed for the WaterBoard's business, data may not be complete or to the required specifications of other organizations. The lineage of the data is extremely complex and has undergone a long evolution, making the positional accuracy of the data difficult to quantify and track in the database. The data is extremely useful to other organizations and has been marketed widely to local governments and other organizations. However, no standard documentation of data quality is provided in such cases.

CONCLUSION

The broad vision of GIS/LIS requires many disparate data sets to be accessed and integrated to produce new information. Many databases are now operational. However, many organizations are just beginning the process of computerization. The evolution of existing databases provides a useful basis for analyzing the types of problems that will occur when we try to integrate disparate data sets and also provide quality documentation. In any case, data trustees should develop a broad view of corporate information, that goes beyond the primary business requirements of their own organization, so that any database developments will not preclude the broader use of spatial information. Methods need to be developed for communicating what data is stored in these databases and also also how accurate it is. Positional accuracy is but one component of data quality and should be documented to enable users to both select appropriate data for their own requirements and also to determine how accurate any information derived from analysis of that data is. The documentation of positional accuracy should be both understandable to general data users and be specific enough to predict the accuracy of any decision making process. It is interesting to now see the exploding but healthy interest in digital exchange standards as organizations grapple with the problem of integrating data from disparate sources. However, a lot of work will be required to bring the currently broad conceptual models of quality documentation to specific levels of definition and detail that can actually be encoded with digital map databases.

ACKNOWLEDGEMENTS

The views expressed in this paper are those of the author and do not claim to represent any of the organizations mentioned within this paper. Many thanks are expressed to Terry Ryan and staff at the Sydney Water Board and Bill Hirst at the New South Wales Roads and Traffic authority for the many helpful discussions held on quality and accuracy issues.

REFERENCES

ACSM, 1988. "The proposed Standard for Digital Cartographic Data". The American Cartographer, Vol 15, No.1,7-140.

Hunter G., & Beard K., 1992. "Understanding Error in Spatial Databases"., The Australian Surveyor, Vol.27, No.2 June.

Lanter D.P. 1990. "The Problem of Lineage in GIS", Tech. Paper 90-6, U.S. National Center for Geographic Information and Analysis, University of California, Santa Barbara.

Leung Y., Goodchild M.F. & Lin C. 1992. "Visualization of Fuzzy Scenes and Probability Fields", proc. of 5th Int. Spatial Data Handling Symposium, Charleston, South Carolina, 3-7 August.

Manning J. & Harvey B.,1992. "A National Geodetic Fiducial Network", The Australian Surveyor, Vol. 37 No. 2, June.

McLaughlin J., 1991. "National Spatial Data Infrastructures: The Next LIM Challenge", in Land Information Management, Monograph 14, School of Surveying, University of New South Wales, 159-166.

Masters E.G., 1991. "Defining Spatial Accuracy" proc. Symposium on Spatial Database Accuracy, University of Melbourne, Victoria, Australia,19-20 June, 215-224.

Merchant D., 1987. "Spatial Accuracy Specification for Large Scale Topographic Maps", Photogrammetric Engineering and Remote Sensing, 53(7), 958-961.

Noonan P., 1992. "Staged Land Base Precision Development Strategy", URISA92, Vol1, 67-78.

O'Callaghan J. & Garner B., 1991. "Land Informations Systems in Australia", in Geographical Information Systems: Principles and Applications, ed. D.J.Maguire, M.F. Goodchild & D.W. Rhind, Longman, 640pp.

Rhind D., 1992. "The Information Infrastructure and GIS", proc. 5th International Symposium on Spatial Data Handling, Charleston, South Carolina, USA, August 3-7,

Ryan T., & Masters E.G., 1991a. "Spatial Accuracy as a Quality Indicator", proc. Symposium on Spatial Database Accuracy, University of Melbourne, Victoria, Australia,19-20 June, 175-185.

Ryan T., & Masters E.G., 1991b. "Quality Issues in Land Information Management", Land Information Management, Mono. 14., School of Surveying, University of NSW, Australia.

Smith F. & McLeod R.,1991. "Using GPS to assess the Accuracy of IFIS", proc. Symposium on Spatial Database Accuracy, University of Melbourne, Victoria, Australia,19-20 June.

Veregin H. 1989. "A Taxonomy of Error in Spatial Databases", Tech. Paper 89-12, U.S. National Center for Geographic Information and Analysis, University of California, Santa Barbara.

WASTEWATER FLOW ESTIMATING MODEL

James W. McKibben
Joseph M. Gautsch
CH2M HILL
2510 Red Hill, Santa Ana, CA 92705

ABSTRACT

Wastewater flow estimates can be prepared for any location in the Los Angeles sewer system for any year between 1990 and 2090. Estimates are being prepared for the Los Angeles sewer system total wastewater flow and its three major components: 1) Residential Flow, 2) Industrial and Commercial Flow, and 3) Groundwater Infiltration. The primary model inputs are: 1) Residential Population, 2) Residential Flow Rates, 3) Employment Population, 4) Industrial and Commercial Flow Rates, 5) Major Industrial Flows, and 6) Groundwater Infiltration Rates. Maps of the sewer system are used to define service areas or subbasins that are used by the wastewater flow model to estimate wastewater flows. The maps of the service areas flow estimates aid the engineers in developing an understanding of the wastewater system and the location of the wastewater flows. The output maps are designed to answer the following questions:

- What is the projected growth and where is the growth expected to occur?

- What is the makeup of the current and future wastewater flows?

- What areas yield the highest wastewater flows?

- What areas are projected to have the highest increase in wastewater flows?

Developing a model to estimate sewage flows for the City of Los Angeles requires the use of new GIS technology combined with proven sanitary engineering principles.

INTRODUCTION

Developing sewage flow estimates for the City of Los Angeles (City) is a major challenge and the foundation of sound planning for wastewater collection
and treatment. The Los Angeles wastewater collection and treatment system serves a large area with a diverse mixture of land uses and dynamic growth patterns. The amount of sewage that must be collected, transported, and treated is also influenced by such factors as the area's drought and weather conditions, changing lifestyles, and evolving industrial base. Developing a model that will predict sewage flow and account for all these influences requires new technology, combined with proven sanitary engineering principles.

Role of Geographic Information System . The combination of Geographic Information System (GIS) technology and a dynamic sewer system flow routing computer model provides simple use of geography-based data, superior output graphics, and sophisticated dynamic sewer flow routing. GIS technology uses a wide range of geographically based data to estimate current and future wastewater flows. The GIS outputs high-quality, adaptable maps that develop an understanding of the current and future wastewater systems. This information is not possible with other, more conventional presentation techniques. The computer flow routing models allow the engineer to estimate sewer flow, taking into account routing effects. In addition, the computer flow routing model can estimate other hydraulic characteristics that are needed to adequately plan sewer system improvements and operation strategies. The combination of GIS technology and sewer system flow routing computer models extends the capabilities of both programs and provides a unique opportunity to plan wastewater collection system modifications to carry Los Angeles into the next century.

Credits The wastewater flow estimating model described in this paper is being developed by a joint venture of James M. Montgomery and CH2M HILL for the Bureau of Engineering for the City of Los Angeles. The Collection System Task Force, which is part of the Wastewater Program Management Division, is staffed by a combination of City employees and members of the joint venture firms. The project manager and GIS specialist are members of CH2M HILL.

KEY INFLUENCES ON THE LOS ANGELES SEWER SYSTEM

Because the wastewater collection system model and one of its components, the wastewater flow estimating model, have been designed for the Los Angeles sewer system, several key features or characteristics of the Los Angeles sewer system have had a major influence on the development of these models. Some of the more significant features are summarized below:

Size Although the Collection System Task Force has reviewed both sewer systems, this discussion of system features focuses on HSA because HSA is the dominant system. The wastewater systems shown in Figure 1 (Project Area) provide service to nearly 4 million people. This population is expected to double during the next century. The wastewater collection system service areas can be divided into two major systems. The area around Long Beach and the Los Angeles Harbor is served by the Terminal Island Treatment Plant. The Terminal Island Service Area (TISA) is a relatively small area of only 20 square miles. The larger of the two major wastewater service areas is the Hyperion Service Area (HSA), which covers the majority of the City of Los Angeles. Although the Collection System Task Force had reviewed both sewer systems, this discussion of system features focuses on HSA because HSA is the dominant system. HSA has a tributary area of nearly 540 square miles. Wastewater treatment is provided at five different sewage treatment plants.

The sewer system has between 6,000 and 7,000 miles of sewers, ranging from 8 to 150 inches in diameter, and over 150,000 maintenance holes. The system can be compared to a tree, with the sewers in one of three different classifications. The first classification contains the major outfall sewers or interceptors, which are the largest sewers and form the trunk of the tree. In the HSA, there are 100 miles of major outfall sewers that deliver the sewage to the sewage treatment plants. The second classification of sewers contains the primary sewers and forms the major branches of the collection system tree. The primary sewers are 18 inches and larger in diameter and transport the sewage to the major outfall sewers or interceptors. There are an estimated 600 miles of primary sewers. The major outfall sewers and the primary sewers represent about 10 percent of the total length of the HSA sewer system. Figure 1 presents both the major outfall sewers and the primary sewer system. The third classification of sewers contains the secondary sewers (8 to 18 inches in diameter, which are the small branches of the collection system tree. The majority of the sewers are in this group, and they collect the sewage from the individual houses and businesses.

Complexity One of the most noticeable features of the sewers in the HSA is the large number of flow splits and the construction of many parallel sewers. Several sets of parallel sewers can be seen in Figure 1. Flow splits are defined as a maintenance hole with more than one outlet sewer. Currently, over 2,200 flow splits have been identified and are used to transfer sewage to parallel relief sewers and to sewers in other subbasins. The flow splits occur throughout the system and can be found in the major outfall or interceptors, primary sewers, and secondary sewers. Flow splits make the definition of

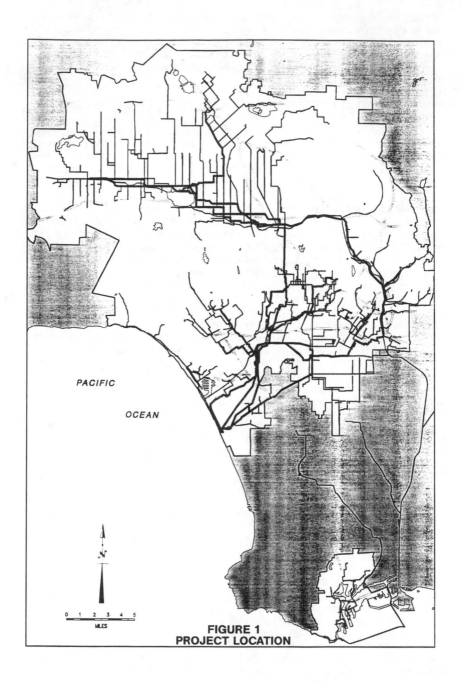

PACIFIC

OCEAN

0 1 2 3 4 5
MILES

FIGURE 1
PROJECT LOCATION

the tributary sewer system difficult and modeling of the sewer system much more complex.

Length of Planning Horizon Sewers can be expected to last over 100 years. The first sewers in Los Angeles were constructed near the turn of the century. A planning period or "horizon" of 100 years is necessary to allow an understanding of the long-term impacts of constructing sewers. Developing long-range population projections is difficult, and the models should be able to define the impacts of several different long-term population projections.

Immediate Relief Needed Overloaded sewers currently exist, so relief sewers must be developed now. Planning for the relief sewers cannot wait for the development of a sophisticated model to estimate flows and evaluate sewer system capacity and is being completed now. The model to estimate wastewater flows and evaluate sewer system capacity must be developed in phases, and the results of the initial efforts will be and are being used in the ongoing relief sewer planning process.

Growth Patterns The population in the Los Angeles area is continuing to grow and is the subject of a major controversy. The models must be able to analyze different growth distributions and determine their effect on the wastewater collection and treatment system. The 1990 census data are now available and must be incorporated into the growth projections. The impacts of new growth projections on the future flow estimates must be defined.

Contract Agencies The HSA includes most of the City of Los Angeles, as well as the Cities of Santa Monica, Beverly Hills, Culver City, Burbank, and Glendale, and several other areas, contract agencies, and cities outside the City of Los Angeles. The model must be able to estimate flows in these areas, even though information is often lacking. The wastewater flow estimates should be at the same level of accuracy as the rest of the system.

Information Base The City of Los Angeles has a massive amount of data concerning the sewer system. Currently, there are between 5,500 and 6,500 sewer wye maps at a scale of 1"=50', 130 S-maps at a scale of 1'=400', 7 I/I maps at a scale of 1"=2,000', and thousands of construction as-built drawings. Much of this information is available only on microfiche, and none of the maps are available in any computer-aided drafting (CAD) format. In addition, the City has developed a computer data base of the sewer system inventory data for all 150,000 reaches of the sewer system.

WASTEWATER COLLECTION SYSTEM MODEL COMPONENTS

The wastewater flow estimating model is one component of a model designed to evaluate the sewer system capacity. The wastewater collection system model is composed of several parts or components. These are summarized below.

Population Projection The population projection module, which has already been completed, automates the population and employment projection procedure developed by the advance planning report (APR). Population and employment projections are two of the primary input for estimating wastewater flow in the sewers. The APR population estimates adjusted upward the 2010 population and employment projections developed by the Southern California Association of Governments. In addition, long-term population and employment projections for 2050 and 2090 were developed in the APR. Several long-term growth scenarios were developed, and the wastewater model is designed to define the impacts of the different scenarios.

Wastewater Flow Estimating This module, which is also complete and is the focus of this paper, estimates wastewater flows for the HSA and TISA for any year between 1987 and 2090. Wastewater flows are estimated for small subbasins that will be used by the flow accumulation module to estimate flow in the sewer system. In addition, flow can be estimated for any area in the HSA or TISA, which will allow an estimate of the flows generated by the various contracting agencies.

Wastewater Flow Accumulation The wastewater flow accumulation module, which is complete, takes the wastewater flow estimated for the various subbasins and accumulates the flow through the sewer system. The flow accumulation model adds the average daily dry weather flows from the small subbasins and uses the City's peaking factor curve to estimate peak dry weather flows. The flows from the wastewater flow accumulation module are passed to the collection system capacity analysis module.

The dynamic flow routing model for the major outfalls or interceptors, LASAM, is complete. LASAM was developed and calibrated with sewer system flow data. Flows developed by the flow accumulation module are checked with this sophisticated flow routing model.

Collection System Capacity Analysis The collection system capacity analysis, which is under development, will use the sewer flow estimates developed by the wastewater flow accumulation module and check the capacity of the existing sewers. Physical data, such as pipe diameter, slope, length, materials, and age, will be obtained for an existing sewer system inventory data base. This data base has data on almost every one of the 150,000 reaches of sewers in the City of Los Angeles. The sewer inventory data do not contain information about sewers in the contracting agencies outside the City of Los Angeles. The schedule of when sewers will be overloaded will be estimated by developing flows for several different years and mapping when sewers become overloaded.

Link to Dynamic Flow Routing Model LASAM is a dynamic sewer system flow routing model and is complete for the major outfalls or interceptors. The flow estimates and the existing sewer system inventory data base will be linked to LASAM. The wastewater flow accumulation model will develop the needed input files for LASAM and run LASAM. The results will be mapped for presentation. The program to link LASAM to the wastewater flow estimating model will be expanded to include the estimating of wet weather flows. The use of the GIS-based wet weather wastewater flow estimating model will allow the simulation of observed storms to further calibrate the models.

Model Calibration Procedures
The model calibration module, yet to be developed, will allow the use of an extensive set of in-sewer monitoring data the City is currently collecting. The sewer monitoring data are being collected continuously at 33 points in the sewer system. The City plans to expand the permanent sewer monitoring network to over 100 points in the near future. The numerous flow splits make use of these data difficult. A procedure and program will be developed to allow the model to be frequently updated and refined with the data for the in-sewer monitoring.

OBJECTIVES FOR WASTEWATER FLOW ESTIMATING MODEL

The wastewater flow estimating model is designed to accomplish the following objectives:

1) Develop a wastewater flow estimating model to be used for all wastewater collection improvements projects. The wastewater flow estimating model must be able to estimate:

- Flows for the entire HSA and TISA and any subbasins or areas in these service areas even if they are not in the City of Los Angeles.

- Flows for a wide range of subbasins or areas to allow planning for collection system improvement projects, ranging from master plans for large areas to detailed studies and designs for local relief sewers. Wastewater flow estimates are needed for the major outfall sewers, primary sewers, and the smaller secondary sewers.

- Flows that are consistent with the existing sewer system model (LASAM) and use the APR flow estimating procedures.

2) Implement simple and flexible wastewater flow estimating procedures to allow the following:

   - Changes in population and employment patterns
   - Must be able to simply incorporate 1990 census data and new employment data
   - Must be able to incorporate adjustments in population and employment data in subbasins or local areas to reflect actual local development and areas with growth patterns that are different than those in the APR projections

   - Changes in wastewater system flow data as a result of water conservation and continued model calibration. The model calibration is expected to result in the development and refinement of area-specific values such as the following factors:

   - Residential per capita flow
   - Employment per capita flow
   - Area-specific groundwater infiltration
   - Area-specific wet weather flow

3) Develop a wastewater flow estimating model that is compatible with the new City GIS being developed by the Automated Wye Mapping Project and does not duplicate any work that will be completed by this project. The Automated Wye Mapping Project will develop a detailed, parcel level base map and a description of the sewer system with all house connection wyes located. The flow estimating procedures, flow accumulation procedures, sewer capacity analysis, and linking to LASAM will be transferred to the larger and more accurate mapping of the sewer system.

WASTEWATER FLOW ESTIMATING MODEL

The wastewater flow estimating model uses ARC/Info by ESRI to estimate the residential, industrial and commercial, groundwater infiltration, and municipal point source average dry weather wastewater flows. The model combines the four wastewater flows and develops total dry wastewater flow. Wastewater flow models are always prepared for a specific area and for a specific year. The wastewater flow model is designed to estimate the total dry weather wastewater flow for any area in the HSA and TISA for any year between 1990 and 2090.

Wastewater flow is often estimated by multiplying the tributary population by a flow rate usually measured in gallons per capita per day (gpcd). More sophisticated wastewater flow estimating models account for some of the major wastewater components. The wastewater flow model developed by the APR uses four major wastewater components: 1) residential wastewater flow, 2) industrial and commercial wastewater flow, 3) groundwater infiltration flow, and 4) municipal point sources. These four components were selected partly because of availability of data, such as the population and employment

537

data for the entire HSA and TISA. The use of these four wastewater flow components accounts for areas with very high employment such as the downtown area.

Model Diagram The wastewater flow estimating model is diagramed in Figure 2. The shaded part of the diagram represents the part of the model that is in the current GIS. The open boxes are the information sources used to construct the model and the supporting data. The model was constructed entirely with existing maps, information, reports, and data.

Wastewater Flow Estimates Total average daily dry weather wastewater flow is the sum of the four wastewater flow components as shown in the diagram in Figure 2 and in Equation 1 in Table 1, Equations to Estimate Dry Wastewater Flow. As noted earlier and in Table 1, total dry weather wastewater flow is estimated for a specific area and for a designated year.

The equations and procedures to estimate wastewater flow have been modified to use GIS to its fullest advantage. All the equations presented in Table 1 could be easily combined into one simple equation; however, the equations for each major wastewater component are kept separate to allow the development of maps and data to document the origin of wastewater and to assist in developing a better understanding of the wastewater system. Developing maps for each major wastewater component uses the power of GIS to better visualize results of the model. Additionally, the procedure developed by the project team to calibrate the wastewater flow model with GIS requires accumulating the individual wastewater components.

The four major flow components used in this wastewater flow estimating model are defined below.

Residential Flow. Residential flow is the wastewater that originates in the households of the residents. Residential flow is estimated by taking the residential population of an area and multiplying it by a individual flow rate factor as shown in Equation 2 in Table 1. The residential flow rate of 90 gpcd was developed by the APR in calibrating the dynamic flow routing model (LASAM) for the major outfalls or interceptors in the Los Angeles sewer system. This calibration was based on monitoring flows in the major outfalls and at the wastewater treatment facilities. The residential flow rate was assumed to be uniform across the HSA.

Industrial and Commercial Flow. Industrial and commercial flow is the wastewater that originates from the different industrial and commercial facilities such as shopping areas, office complexes, hospitals, schools, and small and large manufacturing facilities. The estimated industrial and commercial flow is composed of two terms as shown in Equation 3 in Table 1. The first term is the existing major industrial wastewater sources, which are defined as those wastewater sources greater than 40,000 gallons per day (gpd). These wastewater sources were identified from the records of the Bureau of Sanitation Enforcement Division. An average of 6 months of data is used. The second term in Equation 3 is the employment population multiplied by the estimated industrial and commercial employment rate. The industrial and commercial employment flow rate is 30 gpd per employee and was developed by the APR project team as part of the LASAM calibration. The equation for industrial and commercial wastewater flow does not project any increases in the major industrial wastewater sources. The industrial and commercial employment flow rate was set high enough to account for the future development of major industrial and commercial wastewater sources. This projection procedure appears reasonable because the Los Angeles area is short of water, and the development of new major industrial water users will be limited.

Groundwater Infiltration. Groundwater infiltration is the flow that leaks into sewers throughout the year and is normally found in dry weather flow. This term is not the

538

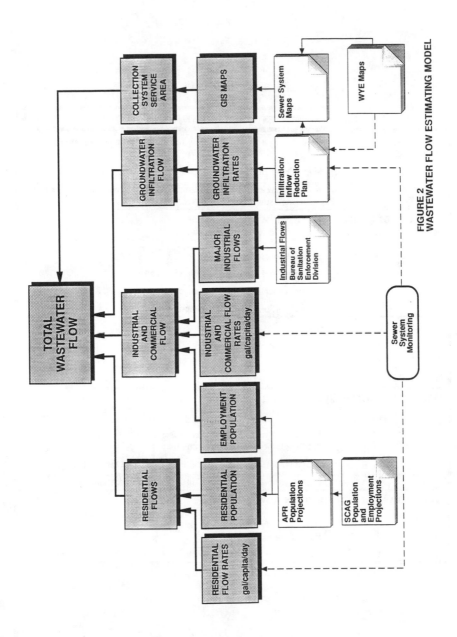

FIGURE 2
WASTEWATER FLOW ESTIMATING MODEL

Table 1
BASE MAP LAYERS

EQUATIONS TO ESTIMATE WASTEWATER FLOW

The following flow equations will be used to develop flow for a given year in a subarea or sewer catchment.

y	=	Year of flow estimate
N	=	Name of subarea or catchment for which sewage flows will be estimated.

AVERAGE DRY WEATHER FLOW (DWF)

DWF	=	$RF + ICF + GWI + MPS$	Equation #1
RF	=	Residential Flow in mgd	
ICF	=	Industrial and Commercial Flow in mgd	
GWI	=	Groundwater Infiltration in mgd	
MPS	=	Municipal Point Sources in mgd	

RESIDENTIAL FLOW (RF)

RF	=	$P * F_1 / 10^6$	Equation #2
P	=	Population of Year Y in area N measured in people	
F_1	=	Flow Factor #1 for Year Y in area N in gal/capita/day	

INDUSTRIAL/COMMERCIAL FLOW (ICF)

ICF	=	$[M + E * F_2] / 10^6$	Equation #3
M	=	Major Industrial/Commercial Waste Sources found in Area N in gpd(Single point sources larger than 40,000)	
E	=	Employment Population for Year Y in area N in people	
F_2	=	Flow Factor #2 for Year Y in Area N in gal/capita/day	

GROUNDWATER INFILTRATION (GWI)

GWI	=	$A_N * F_3$	Equation #4
A_N	=	Basin area in acres	
F_3	=	Groundwater Infiltration Factor in Basin for Year Y in mgd/acre	

inflow that enters the sewer during and following storms and causes elevated wet weather flows in the sewers. Groundwater infiltration is not a large flow contributor in most areas. However, in some areas such as the one along the coast, groundwater infiltration can be one of the largest wastewater components. The Infiltration and Inflow Reduction Plan, an existing source, provided groundwater infiltration amounts for the HSA. The groundwater amounts and subbasins used by the wastewater flow estimating model are taken from this plan. Groundwater infiltration for any area is the area multiplied by the appropriate groundwater infiltration rate as shown in Equation 4 in Table 1. Groundwater infiltration generally increases with age and the length of sewer. The APR assumed that the groundwater rates would increase at a rate of 0.5 percent per year to account for the increased age of the sewer system and the construction of new sewers.

Municipal Point Sources. Municipal point source flow is the wastewater from municipal sources that is out of the study area and that is not included in the wastewater flow estimating model. This term has been added to the wastewater flow estimating equations to account for sewers entering a study area. For the HSA, there are no municipal point sources. However, several of the study areas that have been defined inside the HSA have sewers with upstream flows. This term will be used only for study areas that are not large enough to account for all the tributary flows and will be used to account for wastewater flow entering the study area from outside areas. For example, several municipalities that receive wastewater flow originate from upstream parts of Los Angeles. The municipalities add wastewater flow, which is discharged to the Los Angeles sewers. The municipal point source term will be used to account for the upstream flow that originates outside the city limits. The equations for estimating the municipal point sources are the same as the equations for estimating flows in the HSA.

Model Inputs The wastewater flow estimating model is designed to estimate wastewater flow for any area in the HSA and TISA for any year between 1990 and 2090. As noted earlier, wastewater flows are estimated for a designated date and a specific area. Two sets of inputs are needed: 1) input data sets for a designated year and 2) sewer system service area definition. The data of the flow estimate are established from the input data set.

Input Data Sets. The model inputs for a designated year are listed below:

- Residential Population
- Residential Flow Rates
- Employment Population
- Industrial and Commercial Flow Rates
- Major Industrial Flows
- Groundwater Infiltration Rates
- Municipal Point Source Flows (if required)

Each model input, with the exception of Municipal Point Source Flow, is represented by a shaded box on the wastewater flow model diagram in Figure 2. To reflect changes in land use, growth patterns, economic activity, drought and water conservation, and a changing industrial base, a model must be developed to change one of these model inputs.

Service Area Definition. Determining the sewer system service area is one of the most important parts of the wastewater flow estimating process. Los Angeles's complex sewer system and numerous sources of information make this a challenge. Sewer system service areas for a point in the sewer system are based on the extent of the existing tributary sewer system. Two methods are used to define the tributary wastewater estimating subbasins, depending on the level of detail needed and the level of sewer system mapping used. The most detailed and preferred method requires mapping the entire sewer network in GIS. The size of the HSA sewer system combined with the need to obtain quick results limits this approach. The second method requires mapping the primary sewer network and using the infiltration and inflow basins as the tributary sewer system service areas. Each of these methods is summarized below.

Detailed Sewer Mapping. This method, which is reserved for analysis of specific relief sewers with small service areas, requires entering the complete sewer system into the GIS. First, all public sewers are mapped. After the sewers have been entered into the GIS, they are checked for continuity. All sewer vectors always point downstream, so GIS sewer mapping utilizes the vector notation used by ARC/Info to establish flow direction and determining sewer system connections. Also during detailed sewer mapping, all flow splits (maintenance holes with two or more outlets) are defined and become accumulation centers or points for defining tributary subbasins. Additional accumulation centers are added by the engineer for locations that are important to that project. ARC/Info is used to define the sewer system subbasins by allocating sewers to centers and setting line symbols based on the accumulation center item (ACI) output by write allocate. The subarea boundaries are drafted onto a plot of sewers and digitized. Wastewater flows are estimated for each area. If additional flow splits are uncovered, new sewers are added, or the engineer changes the location of the accumulation centers, this process is repeated.

Primary Sewer Mapping. The determination of the sewer system service area for all the primary sewer systems uses the infiltration and inflow (I/I) basin maps that were prepared as part of the Infiltration and Inflow Reduction Plan. The I/I basins are being split to have only one outlet sewer per basin. Additional subbasins have been added to complete the coverage of the HSA and to accommodate the flow splits on the primary sewer network. As the detail mapping is completed, the tributary basin, defined as a result of the detailed sewer system mapping, replaces the I/I basins. Once the entire HSA has been mapped, the I/I basins will not be used to estimate flow.

GIS DATA LAYERS

The GIS data layers required to estimate the wastewater flows can be placed into five major groups, each for specific uses. Most of these layers must be established before wastewater flows can be estimated. The quality of the final analysis is determined by the quality of these foundation layers. The GIS data layers used by the wastewater flow estimating model are summarized below.

Base Map The base map is probably the most important data layer in the GIS wastewater flow estimating model. The base map establishes the overall accuracy of the project mapping. The base map used for the Los Angeles wastewater flow estimating model was the USGS DLG maps. These maps were selected because they were readily available at a price the project could afford. Moreover, the maps covered the entire HSA, which included areas outside the City of Los Angeles, and they were accurate enough to meet the needs of the project. A summary of the information presented in the base map layers is presented on Table 2.

The base map serves two very important functions. First, the base map becomes the primary means of entry for information into the GIS. The maps must have enough detail to allow the sewer system to be drafted, entered into the GIS, and checked.

The second function of the base maps is to provide a base or foundation for the output graphics. The base map should provide the viewer with a geographic reference to assist in understanding the presentation. Too much detail clouds the presentation and inhibits communication. As a result, the base maps should be able to print several different levels of detail. Maps of the entire HSA that were used in formal presentation show only major

Table 2
BASE MAP LAYERS

NAME	TYPE	DATA	FUNCTION
FWYS	Lines	Location of freeways	Provide base map features used on all maps
MAJROAD	Lines	Location of major roads	Provide base map features used on medium-scale maps
ROADS	Lines	Location of all roads	Provide base map features used on large-scale maps and on manuscripts for inputting and checking data
RAILS	Lines	Location of railroads	Provide base map features used only on limited presentation maps and check plots
WATBODY	Polygons	Location of water bodies and major streams (buffered)	Provide base map features used on all maps
STREAMS	Lines	Location of streams	Provide base map features on a limited number of maps

543

map features. In our project, the major map feature included the freeways, major streams, and water bodies. Additional base map detail is used for smaller projects. The layers on the base map must be able to respond to these different needs.

Reference Layers Reference layers are maps that are useful in getting and checking the data or in presenting the results of the project. They are not used in the computation of the wastewater flows. The reference layers used are presented in Table 3. Three map grids are used and are important for inputting and checking data. The political boundaries are used mainly for presentations. However, flows have been developed for some contracting agencies and municipalities. Sewer system planning area boundaries are used to limit the analysis and graphic presentations.

Descriptive Layers Descriptive layers are used to describe the wastewater collection and treatment system. A summary of the descriptive layers is presented in Table 4. The descriptive layers are not used directly by the wastewater flow estimating model, with one exception. The I/I basins are used to estimate wastewater flow in areas where detail mapping of the entire sewer system is not complete. The descriptive layers are used to define the subbasins that are used to estimate wastewater flows. These layers are used by the wastewater accumulation module to add wastewater flows and link the GIS wastewater model to other data sources used to estimate sewer capacity and define overloaded sewers.

Derived Layers Derived layers are from the descriptive layers and are used in the wastewater flow estimating computations. Most of the derived layers, which are listed in Table 5, are built on the descriptive layers. The primary function of these layers is to define the subbasins that will be used to compute wastewater flows. The second important function is to define how the sewer network is connected to assist in the accumulation of flows.

Thematic Layers The thematic layers, which are listed in Table 6, contain the data used by the model to estimate wastewater flows. These layers contain the all the input data listed above and are the bottom shaded row of boxes on the wastewater flow diagram in Figure 2.

OUTPUT

The output for the wastewater flow estimating model is a series of computer data files and maps used to present the wastewater flows.

Computed Output The computation of wastewater flows follows the diagram presented in Figure 2 and uses the equations presented in Table 1. As stated previously, the wastewater flows are estimated for specific areas and for a specific year. The model will compute the wastewater flow for 6 years in the population and employment and groundwater infiltration data bases. Currently, 6 years are in the data base: 1990, 1995, 2000, 2010, 2050, and 2090. Other years can be used if the population and groundwater infiltration data bases are changed. The subbasin or areas for which wastewater flow estimates will be prepared are contained in the Derived GIS data layer SUBAREA. The following is a short summary of the procedure and GIS data layers used to compute flows.

Data Distribution Assumptions. The computation process relies on two very important assumptions.

- Population and employment data are evenly distributed to the census tracts with areas such as lakes, large parks, and cemeteries with no or zero population deleted. It should be noted that census block data could be used in this model. The use of block data would greatly increase the data base size but was not used owing to current computer limitations.

544

Table 3
REFERENCE LAYERS

NAME	TYPE	DATA	FUNCTION
INIGRID	Polygons	Sheet layout of Infiltration and Inflow Maps	Assist in data input and quality control checking
S_MAP	Polygons	Sheet layout of S-Maps	Assist in data input and quality control checking
S_QUAD	Polygons	Sheet layout of S-Maps quadrants	Assist in data input and quality control checking and the presentation of some sewer system inventory data
MUNI_BND	Polygons	Location of municipal boundaries	Define areas of responsibilities and to estimate wastewater flows
MSTR_PLN	Polygons	Location of sewer system master planning areas	Define study limits, assisting in monitoring progress, and assigning priorities
COM_PLN	Polygons	Location of Community Planning Areas and Contracting Agencies	Define long-term growth factors of population projections, areas of responsibilities, and estimating flows
NORS-SA	Polygons	Location of NORS II Concept Report study area	Define study area limits to extract data and print maps
HOLLY-SA	Polygons	Location of Hollywood Concept Report study area	Define study area limits to extract data and print maps
WIL-SA	Polygons	Location of Hollywood Concept Report study area	Define study area limits to extract data and print maps
EAGLE-SA	Polygons	Location of Eagle Rock Concept Report study area	Define study area limits to extract data

545

Table 4
DESCRIPTIVE LAYERS

NAME	TYPE	DATA	FUNCTION
PRIMARY	Lines	Location of major outfalls and primary sewers in entire HSA and TISA (18" in diameter and larger)	Describe location of major sewers to assist in defining service areas and accumulating wastewater flow; also used as a major map feature on most maps
SWRNET	Lines	Location of all sewers in limited areas	Describe location of all sewers to define service areas and accumulate wastewater flows
MAINT_HOLE	Points	Location of key maintenance holes (MH) on major outfalls and primary sewers	Provide link to sewer system inventory data and other data sources and maps
ELEV_MH	Points	Location of high point MH with two or more outlets but no in-flow	Assist in accumulating flows
TREAT_PLT	Points	Location of wastewater treatment plants	Define essential sinks for accumulating flow and major flow exit points
PUMP_SWR	Lines	Location of pumping station sewers	Define service area upstream of pumping stations to estimate wastewater flows
PUMPS	Points	Location of pumping stations	Define flow accumulation centers for pumping stations
GAGE_STN	Points	Location of sewer system wastewater flow gauging stations	Define existing sewer flows for model calibration
PRIM_BSN	Polygons	Location of sewer service areas defined by I/I Reduction Plan that have been modified to have one outlet and cover the entire HSA	Define sewersheds that will be used to develop flow estimates for the primary sewer system

546

Table 5
DERIVED LAYERS

NAME	TYPE	DATA	FUNCTION
HSA	Polygons	Location of Hyperion Service Area (HSA) defined from sewer system mapping and political boundaries	Define limits of study area and GIS data, and used as a major map feature on most maps
TISA	Polygons	Location of Terminal Island Service Area (TISA)	Define limits of study area and GIS data, and used as a major map feature on most maps
FLOW_SPLT	Points	Location of flow splits or MHs with two or more outlets	Aid in defining local service area and assist in accumulating flows
A_CENTERS	Points	Composite of all flow splits, pumping plants, and strategically located maintenance holes	Locate flow accumulation points in the sewer system
SUBAREA	Polygons	Flow subareas defined from the sewer system in SWRNET	Sewer system subareas used to estimate wastewater flows and to allocate flow to the sewer network
CONECT	Lines	Simplified diagram of sewer system that connects the flow accumulation points	Aid in understanding sewer network and define the flow accumulation logic used by the flow accumulation module

Table 6
THEMATIC LAYERS

NAME	TYPE	DATA	FUNCTION
RES-RATES	Polygons	Residential per capita flow rates	Residential per capita flow rates are a major wastewater flow model input variable.
CENSUS-80	Polygons	1980 census tracts with the APR population and employment projections (Currently: 1990, 1995, 2000, 2010, 2050, and 2090)	Population and employment projections are major wastewater flow model input variables.
EMP_RATES	Polygons	Employment per capita flow rates	Employment per capita flow rates are a major wastewater flow model input variable.
MAJ_INDS	Points	Major industrial wastewater sources (industrial point sources greater than 40,000 gpd)	Industrial flows are a major wastewater flow model input variable.
GWBASIN	Polygons	Subbasins with groundwater infiltration data	Groundwater infiltration flows are a major wastewater flow model input variable.
INFLOW	Points	Location and wastewater flow data of municipal sources	Point wastewater sources of flow entering a study area that are routed through the study area (primarily used for study areas that are part of the HSA or TISA).
CENSUS-90	Polygons	1990 census tracts	1990 census population is used to check APR population projections.

- All wastewater flow in a hydrologic subbasin is evenly distributed in the hydrologic subbasin. The hydrologic subbasins can be subdivided if the engineer determines that this assumption is causing problems.

Residential Flow. The residential flow is computed by first determining the equivalent population for each subarea with the layers CENSUS-80 and its data base overlayed with SUBAREA. The residential flow is simply the residential population multiplied by the flow rate from RES-RATES. Both the residential population and residential flow for each subbasin in SUBAREA are maintained for later use.

Industrial and Commercial Flow. The industrial and commercial flow is computed by first determining the equivalent employment population for each subarea with the layers CENSUS-80 and its data base overlayed with SUBAREA. The industrial and commercial flow is simply the employment population multiplied by the flow rate from EMP-RATES plus the major industrial waste sources from MAJ-INDS that are found in the subbasins. The data used for each basin in SUBAREA include the industrial and commercial flow, the major industrial wastewater sources, and the total employment population.

Groundwater Infiltration Flow. The groundwater infiltration flow is estimated by taking the years for the population data base and estimating the groundwater rate for each basin in GWBASIN. The equivalent groundwater rate is computed for each subbasin in SUBAREA. The groundwater flow is the area of the subbasins in SUBAREA multiplied by the computed groundwater infiltration rate for that subbasin. The only information used is the groundwater flow for each subbasin.

Municipal Point Flow. The municipal point flows are a sum of all the municipal point source flow from INFLOW found in each subbasin in SUBAREA. The total municipal flow must be compiled for each year in the population data base. The total municipal flow found in each subbasin in SUBAREA is used.

Total Flow. The total flow for each subbasin in SUBAREA is simply the sum of the residential flows, industrial and commercial flows, the groundwater flows, and the municipal point flows found in each subbasin in SUBAREA.

Computer Data Files.

The wastewater flow model yields two computer files that are used by other applications. The first is a listing of the population and flow data for each subbasin in SUBAREA. The information includes the following:

- Subbasin ID
- Population
- Residential flow
- Employment flow
- Major industrial flow
- Groundwater infiltration flow
- Municipal point source flow
- Total wastewater flow

The second file presents the flow accumulation logic derived from derived GIS data layer CONECT. This file is a simple listing of each accumulation center and all of its immediate upstream accumulation centers. This file is used by a data base program to accumulate flow in the sewer system.

VISUALIZATION

One of the most important features of GIS and results of the wastewater flow estimating model is the development of maps to aid the engineers and planners in developing an understanding of the wastewater system. The output maps are designed to answer the following questions:

- What is the projected growth and where is the growth expected to occur?
- What is the makeup of the current and future wastewater flows?
- What areas yield the highest wastewater flows?
- What areas are projected to have the highest increase in wastewater flows?

The presentation maps developed from the GIS data layers to present wastewater flow estimating model results used a common base map composed of the following:

- FWYS   Freeways
- WATBOD   Water bodies
- STREAMS   Major streams
- PRIMARY   Primary sewers and major outfall sewers
- HSA   Hyperion Service Area boundary

The output maps can be grouped in three sets of maps.

Description of Sewer System The existing sewer system can be presented by combining several of the descriptive and derived GIS data layers. Maps can be presented to show the following: 1) major outfalls and the primary sewers with any of the various boundary layers and 2) the primary and secondary sewers in limited areas where the secondary sewers have been mapped. These maps, combined with the flow splits, provide one of the most complete maps of the Los Angeles sewer system currently available.

Population and Employment Analysis The primary input to the wastewater flow estimating model is the population data. The amount and location of growth is currently the subject of a major controversy. The presentation of several maps showing the population densities in people per acre has greatly helped in understanding why sewage flows have increased dramatically in some areas. The maps presenting the population and employment data are:

1) Total population and employment densities by year. Maps were prepared for each of the 6 years of data in the population data base. These maps readily identify the areas with high population densities.

2) Changes in population and employment densities between 1990 and one of the future years. Population and employment increases can be easily identified by plotting the changes on these maps.

3) Comparison between 1990 population projection and 1990 census data. The 1990 census data provide a check on the population projection used by the APR. A comparison of the 1990 APR population projections and the 1990 census identified areas where the 1990 census exceeded the APR population projections. In fact, the 1990 census already exceeds the 2010 population projections in several areas. As a result, new population projections will be developed and used with a wastewater future flow estimating model calibration.

Flow Analysis The amount of wastewater present in a sewer combined with the sewer's capacity determines the need for a relief sewer and the size of the relief sewer. The amount of wastewater in an area is the driving force in extending the sewer system. Understanding where wastewater flows are being generated provides the engineer with valuable information in planning sewer system improvements. The presentation of wastewater flows focused on three areas.

Total Wastewater Flow Components. Total wastewater flow components for a specified year are presented in a series of four maps. One of the most common questions about wastewater flow in an area is "Why is the wastewater flow so high?" The maps presenting the three major wastewater components answer this question. Four flow maps are presented. Municipal point source flows are not presented in most cases, since these sources are to account for flow entering the study area. In the case of the total HSA, there are no municipal point sources.

1) Residential flows are mapped by presenting the residential flow values developed by the model for each subbasin divided by the area of the subbasin. Presenting the residential flow density allows identification of the areas with the highest residential flows.

2) Industrial and commercial flows are mapped in a manner similar to the residential flows. In addition, all the major industrial wastewater sources are mapped as points.

3) Groundwater infiltration is mapped in a manner similar to the other wastewater components.

4) Total wastewater flow is also mapped for each subbasin, and a new classification scheme is established. The major difference in the total wastewater flows map is the use of a different color scheme to present results.

Total Wastewater Flow by Year. Total wastewater flow by year is plotted for several different years and displayed side by side. The viewer can see how the amount of wastewater increases. However, the maps for the different years are similar. The total wastewater flow plot is the same as the total plot described above.

Changes in Total Wastewater Flow. Changes in total wastewater flow between 1990 and the future years are presented side by side. As with the total population density plots, the changes are best viewed with maps that present the changes from 1990 to 1995, 1990 to 2000, 1990 to 2010, 1990 to 2050, and 1990 to 2090.

CONCLUSION

Several conclusions can be stated as a result of the development and use of the wastewater flow estimating model for the City of Los Angeles.

1) Wastewater flows can be estimated for any area in the HSA and TISA. The amount of flow from any area varies considerably. The use of GIS has allowed the development of flow with the geographic distributed data. The model inputs can be easily changed.

2) Wastewater flows can be estimated for a wide range of areas varying from a small area to the total HSA using the same flow estimating procedure. Different relief sewer projects can be developed and integrated into the whole system. Changing service areas by the addition of a relief sewer can now be tested to define the impact on the whole sewer system.

3) The results of the wastewater flow model provide the City and sewer system planners and engineers with a unique view of the wastewater flows in the HSA and TISA.

   • The area with high flows can be easily identified.
   • The area with the greatest increase in projected future wastewater flow can be easily identified.
   • The impacts of the 1990 census can be defined. As a result of this analysis, a new set of population and employment projections is being prepared along with a new calibration of the wastewater flow estimating model and its related sewer system flow routing model.

4) The maps developed as part of the wastewater flow estimating model have proved to be a valuable communications tool.

5) Additional uses for the map continue to develop. The sewer system maps are being used to track the development of relief sewer projects to correct sewer capacity problems. Maps are being developed to plot the sewer system age and present an assessment of the sewer system condition.

BIBLIOGRAPHY

1992. Advanced Planning Report, Summary Report. Wastewater Program Management Division. Department of Public Works, Bureau of Engineering, City of Los Angeles.

Bellaman, W. B., D. A. Cannon, C. Joyce. August 1991. TM-9K Drainage Basin Master Plan and Concept Report Development. Clean Water Program Advanced Planning Report. City of Los Angeles Wastewater Program Management.

Cannon, D. A., D. Akagi, J. Esmay. January 13, 1990. TM-6A Hydraulic Analysis of the Major Interceptor, Outfall, and Relief Sewer System in the Hyperion Service Area. Clean Water Program Advanced Planning Report. City of Los Angeles Wastewater Program Management.

Cannon, D. A., M. S. Dimzon. January 11, 1989. TM-3L Condition Assessment of Collection System. Clean Water Program Advanced Planning Report. City of Los Angeles Wastewater Program Management.

Cannon, D. A. February 1989. TM-6E Future Population and Flow Workshop. Clean Water Program Advanced Planning Report. City of Los Angeles Wastewater Program Management.

Cannon, D. A. May 10, 1989. TM-6JA Flow Capacity Analysis Workshop. Clean Water Program Advanced Planning Report. City of Los Angeles Wastewater Program Management.

Cannon, D. A. February 16, 1989. TM-9A Systemwide Wastewater Collection Goal and Strategies. Clean Water Program Advanced Planning Report. City of Los Angeles Wastewater Program Management.

CH2M HILL. January 1992. Final Report Infiltration/Inflow Reduction Plan. Prepared for the City Los Angeles Infiltration/Inflow Reduction Program.

March 1992. Concept Report, North Outfall Relief Sewer II. Clean Water Program Advanced Planning Report. City of Los Angeles Wastewater Program Management.

Dimzon, M. S. July 12, 1988 (Revised July 25, 1989). TM-5E Future Wastewater Flow and Distribution. Clean Water Program Advanced Planning Report. City of Los Angeles Wastewater Program Management.

Esmay, J., S. Reiner. January 26, 1990. TM-9D Wastewater Flows in the Hyperion Service Area. Clean Water Program Advanced Planning Report. City of Los Angeles Wastewater Program Management.

Esmay, J., D. Akagi. January 26, 1990. TM-6B Infiltration/Inflow Allowance for Design of the Sanitary Sewer System. Clean Water Program Advanced Planning Report. City of Los Angeles Wastewater Program Management.

Holmes, K. T., R. T. October 10, 1989. Stillmunkes. TM-3K Existing Wastewater Collection System. Clean Water Program Advanced Planning Report. City of Los Angeles Wastewater Program Management.

Marske, D., D. A. Cannon. January 25, 1990 (Revised March 25, 1992). TM-17B Infiltration/Inflow Reduction Program. Clean Water Program Advanced Planning Report. City of Los Angeles Wastewater Program Management.

Wilson, H. O., A. Ratliff. March 5, 1992. *TM-9G* Description of the Ultimate Wastewater Collection Plan for the Hyperion System. Clean Water Program Advanced Planning Report. City of Los Angeles Wastewater Program Management.

1969. Joint Committee of the Water Pollution Control Federation and the American Society of Civil Engineers. WPCF Manual of Practice No. 9 (ASCE Manuals and Reports on Engineering Practice No. 37) Design and Construction of Sanitary and Storm Sewers.

LEGALITIES, BENEFITS, IMPACTS AND PERSONAL EFFORT IN CREATING A GIS TO REAPPORTION THE STATE OF WYOMING IN FOUR MONTHS

Richard C. Memmel, GIS Coordinator
State of Wyoming - Computer Technology Division
2001 Capitol Avenue   Cheyenne, Wyoming 82002

ABSTRACT

On October 15, 1991 the State of Wyoming lost a lawsuit in which the plaintiffs claimed that the current way the State of Wyoming was apportioned for its State Representatives and State Senators was unconstitutional. The U.S. District Court gave the State until February 21, 1992 at 5:00 p.m. to submit a new plan that would meet its guidelines or else the U.S. District court would create a plan for the State. As GIS Coordinator for the Computer Technology Division for the State of Wyoming I was contacted by the Legislative Service Office on October 16, 1991 and asked if I could "instantly" put together a GIS that could be used to reapportion the State and be functional by mid December, 1991. In the next few days I made numerous phone calls, put my GIS "life" on the line and said that I could do it. I will discuss how I negotiated for the personnel and equipment needed for the team I put together, the amount of hours (70 hour weeks for four months) and personal sacrifices required to accomplish the job and the impacts on my team and the State as a result of the reapportionment. The pros and cons as well as the benefits and impacts of changing the way our legislature is elected will be discussed based on the new law which creates single member house districts that are fully nested in the single member senate districts. The three phase project (passing a reapportionment plan, U.S. District court hearing and defining reapportionment lines for county clerks) has been completed. The environment in which this project took place was very visible (top news story in Wyoming in 1991), very political and very intense.

Background: On October 15, 1991 the State of Wyoming lost a lawsuit (GORIN ET AL V. KARPAN ET AL) in which the plaintiffs claimed that the current way the State of Wyoming was apportioned for its State Representatives and State Senators was unconstitutional. The U.S. District Court stated:
1.   The deviation of the number of people in the legislative districts in both the Wyoming House (83 percent) and the Wyoming Senate (58 percent) is so large that it is facially invalid. The deviation so greatly exceeds those previously found unacceptable by the United States Supreme Court that the population inequality resulting simply exceeds tolerable equal protection clause limits.
2.   The existing plan is unconstitutional and the Wyoming Legislature can disregard Wyoming Constitution Article 3, Section 3 when developing another plan (this

554

provision required that a Senator and Representative be elected from each county).
3.   A maximum percentage of deviation was not defined but did state that substantial population equality should be the overriding objective and if the deviation is over ten percent there is a heavy burden on the state to articulate and justify its non-population related considerations for that deviation.

The U.S. District Court gave the State of Wyoming until February 21, 1992 at 5:00 p.m. to submit a new plan that met the court's guidelines otherwise the U. S. District court would create a plan for the State using single member districts.

As GIS Coordinator for the Computer Technology Division for the State of Wyoming I was contacted by Wyoming's Legislative Service Office (LSO) on October 16, 1992 and asked if I could "instantly" put together a GIS that could be used to reapportion the State and be functional by mid December, 1991 when committees would begin to consider different reapportionment scenarios.   I was being asked to do for Wyoming in few weeks what it had taken other states years to accomplish.

In the next few days I made numerous phone calls concerning:
1.   if I could put all my other work on hold for 6 months
2.   if I could get consultants immediately to help me get started
3.   if I could get the necessary hardware and software within a week or two
4.   if I could get the internal personnel to help me for the length of the project.
The answers were yes and I put my GIS "life" on the line and said that I could do it.

Criteria: The Joint Corporations, Elections and Political Subdivisions Interim Committee of the Wyoming Legislature served as the Reapportionment Oversight Committee and at its October 31, 1991 meeting set the following criteria:

1.   Election districts should be contiguous, compact and reflect a community of interest.
2.   The population of election districts should be substantially equal with the range of deviation among districts not exceeding 10 percent.
3.   To the greatest extent possible, in establishing election districts:
a. county boundaries should be followed
b. The majority of the population of each county should be in one district.
c. Precinct boundaries should be followed.
4.   The plan should avoid diluting the voting power of minorities in violation of the Voting Rights Act.
5.   The size of each house should remain approximately the same with the House not less than two nor more than three times larger than the Senate.
6.   Consideration should be given to "nesting" by

which senatorial districts are established and 2 representatives are elected for each senator in the district.

7. Significant geographical features, e.g. mountain ranges, should be considered in establishing districts.

8. Both single and multi-member district plans should be considered.

9. Consideration of the residence of current legislators should be avoided.

Data: After discussions with Richard H. Miller, Director Legislative Service Office and Steve Furtney, Administrator Economic Analysis Division, it was decided to use the Tiger/Line and Public Law 94-171 data from the 1990 census by the U.S. Census Bureau based on the fact that the Wyoming constitution requires the use of census data.

Personnel and Equipment: I started working for the State of Wyoming, Computer Technology Division, as a Programmer/Analyst in 1975 and had worked almost exclusively since 1983 as a Senior Programmer/Analyst helping the Wyoming Department of Transportation implement a large Computer Aided Design and Drafting (CADD) System. The Wyoming Transportation Department chose Intergraph Corporation as the hardware and software vendor for their CADD system.

I knew I would need a team of personnel that were already trained and familiar with the Intergraph System as well as someone who had done redistricting before. I also needed the proper equipment and software to do the job. Rick Miller, Director of LSO, assisted me by arranging for equipment and personnel. I contacted Intergraph Corporation and it was decided that Intergraph would loan the State of Wyoming the necessary hardware and software. They also agreed to provide two consultants at a fixed rate per week of work. One consultant, Steve Riddle, was very familiar with the GIS products and tools and the other, Andrew Weatherington, had done redistricting and was familiar with the Tiger and Public Law data and had translated Census files to the Intergraph system. I also needed internal full time personnel for graphics and data base work. I had been working with Bryce Freeman, Principal Appraiser, from the Wyoming Department of Revenue and Tax - Ad Valorem Division on their GIS tax district project. I had also worked with David Clabaugh, Senior Engineer, from the Wyoming Transportation Department who was very knowledgeable with the Informix Data Base that we were going to use for the project. Bryce and David were temporarily assigned to me by their agencies for the duration of the project to do the graphics work and data base work respectively.

Thus the core team of personnel that did the majority of the work on this project consisted of Rick Miller who interfaced between the Joint Corporations, Elections and Political Subdivisions Interim Committee and the technical people consisting of Bryce Freeman, David Clabaugh, Steve Riddle and myself.

The software that Steve Riddle, Andrew Weatherington and I decided on was: nucleus software including unix operating system; Microstation 32, graphics; MGE (Modular GIS Environment) for the GIS; MGA (Modular GIS Analyst) for the analysis and color map creation; Projection Manager for the Lambert Conformal Conic projection we used; MGT-US (Modular GIS Translators) for the translation of the Tiger and Public Law data from Census Bureau CD roms; Informix Online for the data base; Informix SQL for structured query language; DBA (Data Base Access) for the creation of reports and easy access to the data base.

The hardware loaned to us by Intergraph and Wyoming State Agencies (as noted) included an Intergraph 6040 single screen workstation with 48 MB memory and 670 MB hard disk drive; a Co Comp HP 1 GB hard disk drive, Computer Technology Division; an Intergraph 2020 dual screen workstation (16 MB memory), Wyoming Highway Patrol; an Intergraph 6040 single screen workstation (16 MB memory), Revenue and Tax; an Intergraph 220 single screen workstation (16 MB memory), Wyoming Transportation Department; a 36" x 48" digitizer; an Intergraph 2217 laser plotter/printer, Revenue and Tax; an Intergraph Exabyte 8mm tape drive, Revenue and Tax; a Shinko CHC65 color thermal plotter and an IBM PS2/60 pc and Epson printer, Computer Technology Division. Large format laser and electrostatic plotters were also accessed at the Wyoming Transportation Department site as well as color electrostatic plotters at the Intergraph Headquarters in Huntsville, Alabama.

Creating the System: Amazingly within 2 weeks the personnel and equipment were gathered and a contract was in place with Intergraph. I also managed to attend the First Annual National States Geographic Information Council meeting and the GIS/LIS Conference that last week of October in Atlanta, Ga. While in Atlanta, I had the opportunity to visit with Linda Meggers, Director of Reapportionment Services with the University of Georgia. Linda has been through 3 reapportionments in the last 21 years for the State of Georgia and she provided a lot of insight into the process, the problems and gave a detailed demonstration of the Georgia system.

On November 3, Steve Riddle arrived from Huntsville and we started work on the system. First, we installed all the software on the workstation and repartitioned one of the disk drives for the raw partition needed by Informix Online. Next shell scripts were written to help automate the translation of the Tiger and Public Law Census data. These would allow us to run procedures during the night and get results the next morning. It is not bad to translate the data for one county but our problems were magnified 23 times by the number of counties in Wyoming. We needed to and were able to bring up all 23 counties of data at the same time on the workstation.

With additional help from Andrew Weatherington we customized the schema for the Public Law data to include additional fields of information. These fields were used to maintain the multiple scenarios for the legislators that were developing reapportionment plans for the upcoming legislative special session that was to convene on February 10, 1992.

Next we updated our schema with the public law data for the 23 counties. Then the public law data records were linked with the centroids that had been created for the 52,301 census blocks formed by the 253,000 Tiger data records. The data was "scrubbed" to make the data base and map graphics correspond exactly.

Additional menus, shell scripts and user commands were customized and written for the redistricting process. These commands would allow us to assign the census blocks to whichever legislative district the plan (scenario) called for. Steve Riddle was very instrumental in this process and Steve worked with us the equivalent of 6 or 7 weeks during the five calendar weeks from the first week of November through the first week of December. This was followed by phone support all the way though the end of the project. Andrew worked three or four days at the beginning to get us started with the Census data.

This process took about three weeks calendar time but probably was about the equivalent of seven work weeks. Our typical day started anywhere from 7:00 to 10:00 a.m. and went to 11:00 p.m. to 2:00 a.m. in the morning. Weekends and family life did not exist anymore. We did three things: work, eat and sleep. I thought I had pulled my last all nighter when I graduated from grad school which was quite a while ago. Well I was wrong. During the four months that ended with the legislature passing the bill and the governor signing it, I worked through the night 7 or 8 times including one stint where I worked 37 1/2 hours straight to meet a critical deadline. I averaged about 70 hours per week for four months straight. We were now ready to start entering scenarios for the legislators.

Census Blocks, Voter Districts and Precincts: Census blocks and Voter Districts were the two levels of Census information that we used for reapportionment. Wyoming is made up of 52,301 census blocks of which 34,665 contain no population. These blocks are grouped into Voter Districts (VTD's) which the Census Bureau called SAC3 (special area code 3) in the data base. We had two major problems with the VTD's. The first was trying to get all the people involved with the process to understand that VTD's and precincts, although similar are not exactly the same. Since the Census Bureau based boundaries on physical features and not the public land survey which is used for precincts these often did not match. Where it became critical is when a legislator would assume the VTD was the same as a city precinct that he was familiar with and he would find out after a plan was created that a group of people were not in the legislative district that he thought they would be in.

The second problem was that one county, Sweetwater, did not participate in the Census program back in the late 80's when VTD's were being defined. They did however create precincts after the program was over. The legislators and county

558

officials became familiar with these precincts and wanted
the equivalent VTD's incorporated into the Census data for
the reapportionment process. We digitized the precinct
lines and then creating VTD's by selecting the census block
lines that were the closest to the precinct lines. We then
had to update the Public Law data base to reflect these
changes. The original census data had 468 VTD's for Wyoming
and after the changes for Sweetwater County we had 476 VTD's
to work with.

Legislative Districts: A legislative district is a grouping
of census blocks. These blocks can be picked individually
or by taking entire VTD's and groups of VTD's. The current
Wyoming legislature has 64 house members and 30 senate
members. The various scenarios that were developed had the
number of house members at 20, 40, 60, 62 or 64 and the
number in the senate at 10, 20, 30 or 32. The population of
Wyoming according to the 1990 census is 453,588. The ideal
number of population in a given legislative district would
then be calculated by dividing the number of proposed
house/senate members into 453,588. For example if the house
were to have 60 members then 453,588 / 60 = 7560. So
ideally each house legislative district would contain 7560
people.

The deviation from that ideal is then calculated based on
the actual number of people allocated to that district.
These deviations are then summarized and the total of the
deviation which is farthest below the 7560 and the deviation
farthest above 7560 are added together. This total should
not exceed the 10% deviation guideline set forth by the U.S.
District Court. In order to get scenarios to balance within
the 10% deviation it was often necessary for a VTD to be
split and send part of the population to one legislative
district and the balance of the population to another
legislative district. In some instances a VTD was split
three ways. Most splits were made to balance populations
but in some cases they were made to maintain contiguity of
the legislative district.

Scenarios: By the December 3, 1991, we were able to put
together a "legislative redistricting kit" which was made up
of the following:
   1.   A black and white "basic" state map showing
   boundaries and numbers of the 476 VTD's in Wyoming as
   well as county boundaries and the names and location of
   the major cities in Wyoming.
   2.   Two black and white maps that showed the detailed
   location of the VTD's for the 21 larger municipalities
   in Wyoming at a scale that was readable.
   3.   A printout reflecting the total population for each
   VTD as specified by the Census Bureau in order by
   county. This included the VTD's that we created for
   Sweetwater county to approximate their existing
   precincts.
   4.   The reapportionment criteria as outlined above in
   the "criteria section".

Using these materials it was possible for the legislators to
develop a plan/scenario. Some legislators spent from dozens

of hours to hundreds of hours to put together one or more scenarios.

We started to enter the first of the proposed scenarios the last week of November. We had reports generated and back to the some legislators by the end of the first week of December. During the time period from the beginning of November to February 21, 1992 when the final reapportionment bill was passed we entered approximately 30 to 40 different scenarios and at least as many major amendments. Some scenarios used single-member districts and some used multi-member districts from which 2 to 5 members would be elected at large. There were scenarios that had a mixed combination of some single-member districts and some multi-member districts. The number of members in the house and senate varied as mentioned above. Some scenarios nested the house and the senate so that two contiguous house districts would form a senate district or a senate district would also be used to elect 2 house district members at large. Other plans had completely different house and senate plans. This type of plan would have caused additional problems for the county clerks. The number of split districts varied in the scenarios from some plans having no splits to some that had over 50 split VTD's. The bottom line for each of these scenarios was the "report". This was a data base report that we created which listed the VTD's contained in each legislative district for both the house and the senate. The legislative district number was followed by the county name, the VTD number, the VTD name, the population and an "S" or a blank to indicate whether the VTD was split or not.
At the bottom of the report was the total population in the district and the percent deviation from the ideal size of the district which was based on the number of representatives or senators for that particular scenario. There was also a summary page which listed the percent deviations for all legislative districts sorted from the largest to the smallest number of people in the districts with a total deviation at the bottom of the page.

Legislative Session: The one week special legislative session that was called by Governor Mike Sullivan to deal with reapportionment started February 10, 1992. This would give the legislature one week to pass a new reapportionment law and get the bill to the Governor by February 17-18. The Governor would have three days to study and sign the law before it was due to the U.S. District Court by February 21, 1992 at 5:00 p.m. MST. Things did not work out that way.

The House of Representatives passed a reapportionment bill and sent it to the Senate by Wednesday evening. Each day many scenarios were presented along with many amendments. The team worked from about 8:00 in the morning until 1:00 - 2:00 in the morning each day of the session. Long after the houses adjourned for the day the team would continue to work to prepare amendments for the next day. Sometimes legislators would be with us helping to adjust plans to meet the desires of their constituents. Everyone would anxiously await the report we would produce to see if the "tweaks" we had made to the plan had kept the scenario's total deviation under 10% and that the districts were contiguous.

The Senate proposed amendments to the bill from the House and introduced new scenarios. The result was a bill which the House again passed on Friday night, February 14, 1992. This bill was received by the Governor on Saturday morning February 15. On Monday morning, February 17, 1992, the Governor vetoed the bill for various reasons, some of which were:

    1. The bill contained single-member senate districts and both single-member and multi-member house districts and there was no justification given for what reasons the multi-member districts had been created.

    2. The senate and house districts were not nested.

    3. There were inequities in the plan due in part to the use of multi-member districts and also incumbent protectionism and political partisanship.

After a weekend with my family, a short lived celebration and an attempt to recover from a flu bug, I was called in at 2:00 a.m. Tuesday morning to resurrect a scenario that we had "archived" and make amendments to it for the session that was to begin at 9:00 a.m. that morning. After working through the night with Bryce Freeman and Rick Miller, we had a new scenario ready for debate in the Senate chambers on time that morning. We again worked all week with both the Senate and the House preparing new scenarios and amendments and by about mid-morning on Friday February 21, 1992, another bill was sent to the Governor. This reapportionment bill was a nested plan that had 60 single-member house districts with pairs of contiguous house districts forming the 30 single-member senate districts. This plan had 22 two way split districts and one three way split district.

At 3:45 p.m. I stood in the Governor's press conference room and watched him sign Senate bill #0049 that reapportioned the Wyoming Legislature. After the Governor answered questions the 3 inch thick bill was officially delivered to the U.S. District Court at 4:19 p.m. This gave us 41 minutes to spare before the court imposed deadline was to expire. I went home that weekend with a great feeling of accomplishment and relief.

U.S. District Court Hearing: The U.S. District Court had scheduled a hearing on March 6, 1992 to review the reapportionment plan. Bryce and I prepared a large color map which graphically portrayed the 60 legislative house districts. We also prepared a color set of 15 municipal maps and one state map showing the new legislative house districts and their numbers. Rick Miller testified for the defense and neither the plaintiffs nor defendants had any major objections to the plan. There were some plaintiff interveners who raised problems with how certain counties were split with the reapportionment bill. After the hearing, the three judges discussed it for about a month and although they pointed out that there were problems with political maneuvering they held that it complied with constitutional guidelines.

County Clerks: The final step in this process was to communicate to the 23 county clerks where exactly the legislative district boundaries had been drawn by the reapportionment bill. At this point, I was essentially finished providing service to the Legislative Service Office. I was now helping the Secretary of State's Office. For the next six weeks, I worked closely with Toni Hinton, the Elections Director for the Secretary of State. Toni and I scheduled appointments for all the county clerks to travel to Cheyenne so that using the computer system as well as their county maps we could map the boundaries of each legislative district with them.

The biggest problem we had with this process is that many of the census block boundaries used two track dirt roads which do not have names. We discovered that about 70% of the rural Tiger/Line lines for Wyoming do not have information in the name fields of the record type 1. These fields were FEDIRP, FENAME, FETYPE and FEDIRS. By plotting census block maps on color transparencies at the proper scale we were able to overlay county maps and show the county clerks where their district boundaries were.

The responsibilities of the clerks were to go back to their county and tell each person which district they were in and also inform candidates in which district they were eligible to run for office. We were able to do some counties in a few hours and the larger counties took two full days. Many times we had to fax additional maps and finalize details with phone conversations. One problem that the county clerks had was that the VTD's did not match their precinct lines. Once we explained to them why it was this way, they were receptive to working with the system.

Impact on State: As I write this we are approaching our first primary election (August 18, 1992) with the new reapportionment plan in place. We have more candidates running for office than any time in the history of the State. The State has the second highest primary voter registration rate in the last 15 years. There is confusion about the new system with the legislators, the county clerks and the voters. Many of these problems are caused by the differences between the Census Bureau VTD's and the precincts, the county clerks having combined and renamed some of the precincts, the polling places have changed for many of the voters and the resulting number of different ballots in each of the counties that the county clerks will have to handle.

In Laramie County, where the state capital, Cheyenne, is located, there are 128 different ballots that the voters will use in the primary election. There are many slivers caused by the differences between VTD's and precincts and in some cases these result in only one or two people voting for a particular ballot in a polling location. The county clerks must protect the privacy of these ballots so methods are being devised to "hide" the results of these ballots with other totals so that these individual's votes will not be disclosed.

Many of the legislative candidates this year will conduct their campaigns differently than in the past. Those who previously ran in a multi-member district where for example voters would choose 9 of 18 candidates will now have to address a much smaller number of voters but will instead have to go one on one against their opponent. This will call for different campaign strategies.

County clerks will not have final totals for legislative candidates since many of the new legislative districts span multiple counties. The results from each county will go to the Secretary of State's office and they will then be totaled to see who are the winning candidates in the new legislative districts. At least 10 of the 30 members of the Senate will be new and 26 of the 60 members of the House of Representatives will be new after this election. We are certain to see changes in the next legislative session because of the new blood.

REFERENCES

1. Richard H. Miller January 31, 1992, Memo To: Interested Persons; From: Richard H. Miller, Director Legislative Service Office; Subject: Reapportionment Information Packet

2. Janet Hanley 1992, Reapportionment - Providing the Tools Non-Political and Behind the Scenes, IS TECH NEWS

3. Governor Mike Sullivan, February 17, 1992, letter to the Honorable W.A. "Rory" Cross, Speaker of the House, State Capitol

AN INTEGRATED GEOGRAPHIC INFORMATION SYSTEM, GLOBAL POSITIONING SYSTEM, AND SPATIO-STATISTICAL APPROACH FOR ANALYZING ECOLOGICAL PATTERNS AT LANDSCAPE SCALES

William K. Michener, William H. Jefferson, and David A. Karinshak
Belle W. Baruch Institute for Marine Biology and Coastal Research
University of South Carolina
Columbia, SC 29208 USA
(803) 777-3926

Don Edwards
Department of Statistics *and*
Belle W. Baruch Institute for Marine Biology and Coastal Research
University of South Carolina
Columbia, SC 29208 USA
(803) 777-5073

ABSTRACT

Spatial heterogeneity has historically been either ignored or viewed as a hindrance to extrapolating results obtained from ecological studies conducted at centimeter and meter scales to broader ecosystem and landscape scales. Geographic Information System (GIS) technology has recently been adopted by both the research and resource management communities as a tool for assessing spatial variability in ecological patterns and processes. Although numerous important contributions have been made, many of the problems associated with extrapolating data collected at fine scales to broader scales still remain within the GIS environment. Realization of the full potential of GIS for ecological research will require that ecologists incorporate a variety of related technologies into their research and sampling plans.

This paper presents results from a study which employed Global Positioning System technology (GPS), GIS, innovative sampling strategies, and spatio-statistical algorithms (kriging) for assessment of oyster population recruitment in an urbanized estuary (Murrells Inlet, SC). GPS receivers were utilized to obtain position fixes for the intertidal oyster reef sampling sites. Kriging was used for estimation of spatial distributions from point-sampled data. All data were incorporated into a vector and cell based GIS, along with water quality, adjacent land use patterns, and several additional data layers. Oyster population recruitment patterns were found to be related to both spatial variability in water quality (recruitment was lowest in polluted waters) and adjacent land-use patterns (type and degree of urbanization). The integration of the various technologies as discussed in this paper represents a powerful research tool for future ecological studies.

INTRODUCTION

Point and non-point source pollution, increasing population and development pressure, sea level rise, and large natural disturbance events can affect areas ranging from the ecosystem to global scale. Ecologists and other environmental scientists are attempting to respond to these societal concerns by beginning to focus their research on long-term and broader scale questions (Callahan, 1984; IGBP, 1988; Franklin et al., 1990).

It is generally recognized that the data collected in many past environmental research and resource management projects may not be relevant to an entire ecosystem and likely do not address the regional ramifications of specific projects or resource management strategies (Kelly et al., 1987; Schaeffer et al., 1988). For example, Eastern oysters represent an important commercial and recreational resource along the eastern and Gulf coasts of the United States. Populations in many areas are declining and there is concern that pollution, overharvesting and disease may be affecting adult, juvenile, and larval oyster populations. Since the abundance and distribution of adult oysters may be closely related to the successful recruitment of larval oysters (attachment of the free-swimming larvae to a substratum and survival to reproductive maturity), it is important to understand the factors which regulate oyster recruitment patterns and processes. We would expect that these patterns would vary spatially in response to hydrography, pollution, natural and man-induced disturbances, and other factors. However, most studies designed to examine oyster recruitment patterns and processes are based upon data collected at only a single site or at most up to 12 sites (e.g. Hidu and Haskin, 1971). Furthermore, it has been demonstrated that observed recruitment patterns may bear little relationship to actual population distributions, depending upon the spatial and temporal observational scales employed in the study (Michener and Kenny, 1991).

The objectives of the present study were to examine oyster recruitment pattens throughout a high salinity, urbanized estuary and to relate those observed patterns to water quality and other factors which may affect oyster populations. Global Positioning System technology (GPS), GIS, innovative sampling strategies, and spatio-statistical algorithms (kriging) were utilized in the assessment of oyster population recruitment and the identification of factors which may affect observed patterns.

STUDY AREA

Murrells Inlet is a shallow, ebb tidal dominated coastal inlet located on the northern South Carolina coast ($33^\circ32'$N, $79^\circ2'$W) (Figure 1). There is no riverine input and the estuary is characterized by vertically homogenous high salinity water (mean salinity, 31.4 ppt; mean water temperature 20°C; Blood, unpubl. data). Murrells Inlet Estuary is surrounded on all sides by development, except the southeastern edge, which is adjacent to Huntington Beach State Park. Six marinas are located throughout the estuary. Periodic dredging has taken place to maintain navigability of the creek channels.

Murrells Inlet is heavily utilized by both commercial and recreational fishermen.

METHODS

Oyster Recruitment

Plastic tubes with lengthwise mottled grooves were used as artificial, vertical settling substrata during this study. The tubes were obtained from IEC Collaborative Marine Research and Development Limited (Broadley, 1989) in 2m sections and cut into 33 cm sections. The tubes were acclimated in aquaria with filtered, flowing seawater for at least one month prior to placement in the field. The tubes were hammered into the mid-intertidal regions of all oyster reefs, with 15 cm exposed above the reef surface. The tubes were placed approximately 50 m apart on the ocean side of every creek in the Murrells Inlet estuary.

The tubes were placed in the field during March 1990 and retrieved one year later, to encompass one complete spawning and growing season. Oyster recruitment and spat size were monitored by enumerating and measuring all oyster spat that occurred on each tube. Oyster spat densities were standardized to number per m^2. Lengths and widths of individual oysters and total spat densities were incorporated into a kriging (optimal mapping) program for analysis.

GPS

Settlement tubes were positioned using Global Positioning System (GPS) techniques at the time of collection (Michener, 1992). Tube positional data were collected from NAVSTAR satellites by a remote GPS receiver, along with a base GPS receiver located on a known benchmark. Differential corrections were applied to the data to remove any common errors and to correct for selective availability errors that may have been introduced by the Department of Defense. Average accuracies of 2.2 m were achieved. The tube positions were converted to seconds and projected in UTM coordinates in the ARC/INFO system.

Kriging

Most formal statistical procedures assume that observations on the phenomenon of interest are taken under identical conditions, and that each observation is taken independently of any other. Data collected in such a manner comprise a "random sample". Conventional statistical analyses such as regression, multiple regression, and factorial analysis of variance are built on this assumption of independence. Associated formal inferences (hypothesis tests, confidence intervals, prediction intervals) are grossly invalid when independence is strongly violated.

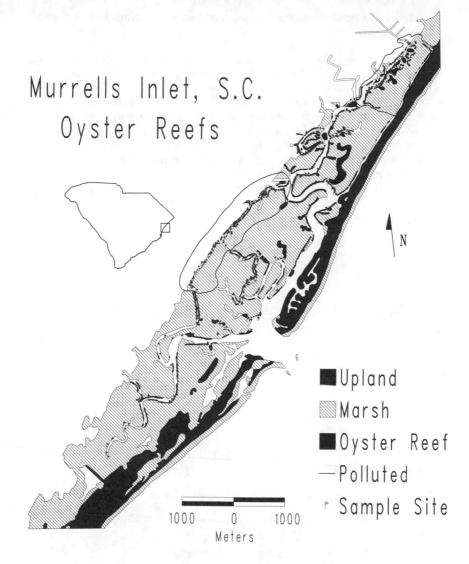

Figure 1. Map of Murrells Inlet Estuary showing polluted zones, all oyster reefs, and recruitment sampling sites.

In environmental data analysis, observations are rarely independent in space or in time. Valid data analysis requires a "regionalized variables" approach (Journel and Huijbregts 1978; Burrough 1987) and a statistical model which integrates three components: (1) deterministic patterns ("signals"), (2) spatially and/or temporally autocorrelated stochastic inputs, and (3) "high frequency" stochastic inputs ("noise"). The second of these three components is the primary focus of spatial statistics, which includes modeling of space-time dependencies in residuals (autocorrelation and semivariogram analyses) and optimal mapping (kriging).

We have used the simplest and by far most widely used form of kriging, ordinary kriging, in which the parameter to be mapped (for example, oyster recruitment per m²) is modeled as an isotropic stochastic process over the region of interest. For a fully detailed definition, the reader is referred to Cressie (1991). The central assumption is that the correlation between measurements made at any two locations depends solely on the distance between those locations; as the distance increases, the correlation decreases. The analysis proceeds in two steps: (1) modeling the rate at which correlation decreases with distance through the fitting of a semivariogram model to data at observation sites; and (2) kriging - the prediction of the parameter over a grid of sites at which it was not observed. Details of these steps will now be provided. All calculations were carried out using programs written in S-PLUS software (Rao, 1992).

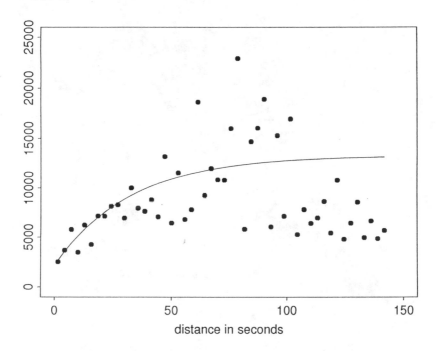

Figure 2. Empirical semivariogram values versus distance in seconds for the oyster recruitment measurements.

If Z1 and Z2 represent observations made at arbitrary sites separated by a distance r, the semivariogram is half the variance of (Z1-Z2) considered as a function of r. The empirical semivariogram uses all pairs of observation points, classifying these into distance classes and computing the sample variance of differences (Z1-Z2) over all pairs in each given distance class. As recommended by Journel and Huijbrechts (1978), only pairs at a distance of half the maximum possible distance were used, to assure a substantial number of pairs in each distance class. The points in Figure 2 represent the empirical semivariogram values versus distance in seconds for the oyster recruitment measurements.

The empirical semivariogram cannot be used directly for kriging. Instead, a smooth function of a certain parametric form is fit to the points of the empirical semivariogram. A number of popular forms exist, but we have been satisfied with the fit by weighted least squares (Cressie, 1985) exponential semivariogram:

$$\gamma(r) = Co + C_E(1-e^{-r/A_E}) \quad r \geq 0$$

This function increases conversely from a "nugget" Co at distance 0 to an asymptote C_E+Co (Figure 2). The nugget Co reflects measurement error variance plus microscale variability. The asymptote (Co+C_E) is the variance between repeated measurements well separated in space. The range A_E controls the rate at which the asymptote is achieved, therefore dictating the scale at which spatial correlation disappears. Specifically, points at a distance A_E apart retain only about a third (precisely, 36.79 %) of their maximum spatial correlation $C_E/(Co+C_E)$.

Kriging was initiated upon completion of the fitted semivariogram. A 61x56 grid of sites (240 seconds in longitude and 220 seconds in latitude, with origin (-284620 W, 120650 N), with 4 second gaps) was chosen for prediction. Kriging provides the BLUP, Best (=minimum variance) Linear Unbiased Predictor, of the measurement that would be obtained at each site under the structure of the stochastic process. For ordinary kriging, each prediction is a weighted average of the values obtained at sampled sites, with weights heaviest at nearby sites as dictated by the fitted semivariogram. The theory also provides standard errors for the predictions. Kriged values can then be represented graphically by standard contouring or grayscale graphical routines (Figure 3).

GIS

A National Wetlands Inventory (NWI) map of the Murrells Inlet Estuary was used as a base layer. The NWI map, in ARC/INFO digital format, was provided by the South Carolina Land Resources Commission. Habitat classes were aggregated to water bodies, lowland and upland areas for purposes of illustration.

The South Carolina Wildlife and Marine Resources Department conducted an Intertidal Oyster Survey for the Murrells Inlet Estuary during 1982. Each reef was located on a 7.5' topo quad, and the oyster population associated with each reef was classified according to strata characteristics (Table 1). The strata differ with respect to reef structure, substratum characteristics, presence or absence of clams, depth of shell matrix, density of live oysters and proportion of shell to live oysters. All oyster reefs examined during this survey are included in Figure 1. Ancillary data (bottom type, size of oysters, shell matrix, depth, elevation, strata, water depth and reef area) for all reefs in the intertidal zone were collected during the survey. The reefs were then given unique identification codes, which were used to identify the reef polygons in the ARC/INFO data layers.

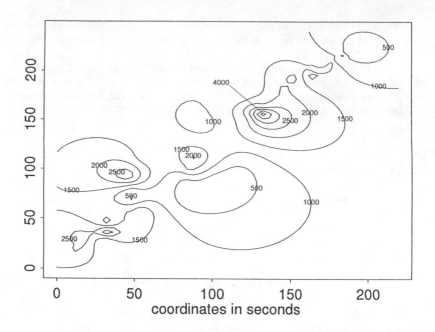

Figure 3. Contour map of kriged oyster settlement values.

The South Carolina Department of Health and Environmental Control has classified the water (creeks and bays) in Murrells Inlet as being non-polluted (62% of total area), conditionally polluted (18%), or polluted (20%) (refer to Figure 4). Conditionally polluted areas were degraded during periods of heavy freshwater runoff and were considered non-polluted during drier periods. Polluted waters are always closed to commercial and recreational fishing activities.

The kriged output matrix from S-Plus was converted to ASCII format. The kriged output was a 220 by 240 second grid, with cell sizes of 4 seconds, relative to a reference point that was located at the southwestern most point. The ASCII file was read into the GRID module of ARC/INFO using the ASCIIGRID command. The grid was resampled and projected in UTM coordinates to overlay on the NWI basemap. The overlay was contoured and all areas outside the creek network were masked to show only predicted oyster recruitment patterns in the creeks.

RESULTS AND DISCUSSION

Oyster recruitment in Murrells Inlet varied throughout the estuary and ranged from 0 to 5076 oysters per m², with an average of 1465 oysters per m². Recruitment was directly related to reef type, substrate type, surrounding water quality, and proximity of reefs to marinas and public fishing grounds. GIS spatial overlay procedures indicated that oyster reefs were located throughout the three water quality zones: non-polluted (69 % of total reef area), polluted

(24%), and conditionally polluted (7%). All reefs in conditionally polluted waters were located in one area adjacent to a storm drain for the Murrells Inlet watershed (Figure 4).

STRATA	DESCRIPTION
"A"	12,093 bushels of live oysters per acre. Greatest yield per acre of densely clustered live oysters. Exhibits little exposed dead shell or mud and the shell matrix is not visible.
"B"	2,608 bushels of live oysters per acre. Characterized by having no vertical clusters in the standing crop. Found mostly in the lower intertidal zone, oysters are frequently single. Located on heavily shelled grounds with thin shell matrices.
"C"	1,957 bushels of live oysters per acre. Characterized by vertical clusters with spatial separation. Substrate is usually mud with little or no surrounding shell. Spatial separation between clusters ranges from a distance equal to the height of an individual cluster to approximately one meter.
"D"	295 bushels of live oysters per acre. Characterized by scattered live oysters usually integrated with large quantities of "washed" or dead shell. Found in the lower intertidal zone on hard substrate. Hard clams are found sympatrically in this area.
"E"	7,277 bushels of live oysters per acre. Characterized by over-growth. Oysters are tightly clustered, totally covering the substrate. Usually found at the highest oyster growing elevation and is further characterized by small oysters with sharp, thin shells
"F"	4,021 bushels of live oysters per acre. Characterized by mostly vertical clusters of oysters. Similar to "C" strata, except the substrate consists of shells with few horizontal live oysters and very little mud.
"F1"	1,926 bushels of live oysters per acre. Characterized by small, vertical clusters evenly dispersed within a substrate of small, single horizontally oriented oysters. Very little exposed mud is associated with "F1" strata.
"G"	5,199 bushels of live oysters per acre. Characterized predominantly by vertical, clustered oysters. Spatial separation between clusters is equal to or less than height of the standing crop. Substrate habitat is mud with little or no shells or single live oysters.
"M"	Less than 20 bushels of live oysters per acre. Characterized by scattered live oysters, which are generally small and show negligible aggregation. Surrounded by a highly permeable mud substrate.
"P"	Near minimum density of oysters in intertidal zone. Characterized as recently harvested areas with very few market size oysters remaining. Considered to be productive since present condition is due to harvesting and they will propagate to the next highest category by natural or artificial recovery.

Table 1. Intertidal oyster strata descriptions.

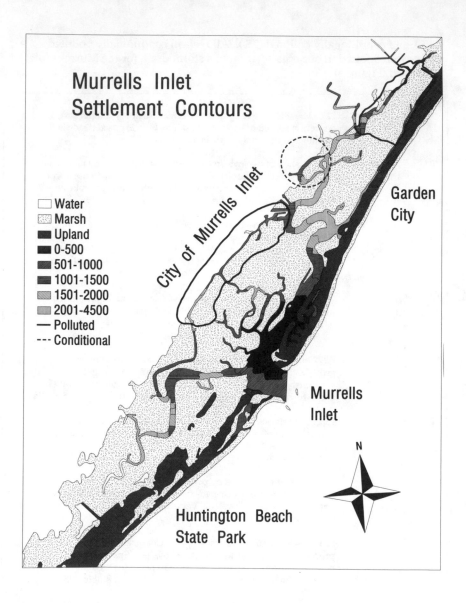

Figure 4. Map of Murrells Inlet Estuary showing pollution zones and kriged contours of oyster recruitment.

Initial results indicated that average recruitment was highest in conditionally polluted (1641/m²) and non-polluted waters (1526/m²; Table 2).

Polluted waters were characterized by lower recruitment (1212/m²) and smaller average size of oysters (Table 2).

Pollution Zone	% Total area	Oyster recruits/m²	Mean length (mm)	Mean width (mm)
Conditional	18	1641	27.1	17.4
Non-polluted	62	1526	27.0	18.9
Polluted	20	1212	25.4	16.8

Table 2. Recruitment and mean size of oysters on reefs in different pollution zones in Murrells Inlet, SC

Average recruitment was highest on the strata "D" reefs (1859/m²) and lowest on strata "G" (967/m²) and "P" (928/m²) reefs (Table 3). Similarly, the smallest average size (length and width) was recorded for oyster recruits in strata "G" and "P" reefs.

Highest average recruitment occurred in the conditionally polluted area which was dominated by strata "D" reefs. However, higher recruitment occurred on strata "D" reefs in non-polluted areas (Table 4). The highest average recruitment (2986 oysters per m²) and the largest average size per recruit (length 35.22 mm, width 21.6 mm) was observed on strata "M" oyster reefs in polluted waters. These reefs were located behind a residential/restaurant district of Murrells Inlet. With exception of the strata "M" oyster reefs, all other strata exhibited higher recruitment and larger average size of recruits in non-polluted waters than in polluted waters (Table 4).

Strata	Oyster recruits/m²	Mean length (mm)	Mean Width (mm)
"B"	1510	26.6	18.6
"C"	1322	30.1	20.7
"D"	1859	28.1	18.6
"F"	1113	25.7	17.2
"G"	967	20.9	14.4
"M"	1783	27.7	18.9
"P"	928	20.3	13.8

Table 3. Recruitment and mean size of oysters on different reef strata in Murrells Inlet, SC.

Results of the kriging showed three major areas of high recruitment and two areas of low recruitment (Figure 4). High recruitment was predicted for an area dominated by strata "B" reefs that are utilized as lease lots for commercial oyster fishermen. These areas are monitored to prevent overfishing and are in areas where the water quality is good and there is little boat traffic. The second area was located in the protected, pristine waters surrounded by Huntington

Beach State Park. A third high recruitment area was located in a polluted zone adjacent to a residential section of Murrells Inlet. The apparent high recruitment in this area was attributed to high oyster densities at a single sampling location and may not be representative of the overall health of the oyster populations in this region of the estuary.

Low recruitment was predicted for two areas, one located near a densely populated trailer park at the north end of Murrells Inlet. This area is closed to fishermen year round and is classified as a polluted zone. High concentrations of pesticides have been measured in adult oysters collected in this portion of the estuary (A. Fortner, personal communication). The second low recruitment area is located in the public fishing grounds. This area is generally depleted of live adult oysters yearly. Overharvesting (removal of substrate and adults) and high boat traffic may account for the observed low recruitment.

Strata	Pollution Zone	Oyster recruits/m²	Mean length (mm)	Mean width (mm)
"B"	non-polluted	1525	26.6	18.6
"B"	polluted	1219	25.5	18.6
"C"	non-polluted	1381	30.4	20.9
"C"	polluted	562	21.5	13.1
"D"	conditional	1747	27.5	17.5
"D"	non-polluted	2000	27.7	19.4
"D"	polluted	1645	32.2	19.7
"F"	conditional	1392	25.8	17.2
"F"	non-polluted	1035	26.1	17.9
"F"	polluted	1020	25.4	16.8
"G"	polluted	967	20.9	14.4
"M"	non-polluted	1182	18.2	15.5
"M"	polluted	2986	35.2	21.6
"P"	non-polluted	928	20.3	13.8

Table 4. Recruitment and mean size of oysters on reef strata located in different pollution zones in Murrells Inlet, SC.

CONCLUSIONS

The combination of innovative sampling methods, spatial statistics for point-sampled data, GPS, and GIS provided an effective research tool for analyzing ecological patterns at the landscape scale. Observed oyster population recruitment patterns could be related to probable causative factors (e.g., overharvesting, water quality, and adjacent land use). Continued technological developments and integration of these techniques into ecological research and resource management programs can provide the basis for environmental decision support systems which can be utilized to address important questions at relevant spatial scales.

ACKNOWLEDGEMENTS

Funding for this project was provided by NSF grant BSR-8514326 and NOAA grant NA90AA-D-SG672. The South Carolina Coastal Council provided funding for the intertidal oyster survey. Digitization and editing of the basemap and data layers were performed by Sharon J. Lawrie. Jim Monck, Mike Yianopoulos and Ray Haggerty completed the intertidal oyster resource assessment in the study area. F. Danny Spoon assisted in the field and laboratory components of this study. ARC/INFO is a trademark of Environmental Systems Research Institute, Inc. S-PLUS is a registered trademark of Statistical Sciences, Inc. This paper is Contribution Number 955 of the Belle W. Baruch Institute for Marine Biology and Coastal Research.

REFERENCES

Broadley, T.A., 1989. The Remote Setting of Oyster Larvae: Lecture/Laboratory Handbook, IEC Collaborative Marine Research and Development Limited, Victoria, British Columbia, 61 p.

Burrough, P.A., 1987. Spatial aspects of ecological data, In, Data Analysis in Community and Landscape Ecology. R.H.G. Jongman, C.J.F. ter Braak, and O.F.R. van Tongeren (eds.), Pudoc, Wageningen the Netherlands, 213-251.

Callahan, J.T., 1984. Long-term ecological research, Bioscience, 34:363-367.

Cressie, N., 1985. Fitting variogram models by weighted least squares, Mathematical Geology, 17:563-586.

Cressie, N.A.C., 1991. Statistics for Spatial Data, Wiley, New York.

Franklin, J.F., C.S. Bledsoe and J.T. Callahan, 1990. Contributions of the long-term ecological research program, Bioscience, 40:509-523.

Hidu, H, and H.H. Haskin, 1971. Settling of the American oyster related to environmental factors and larval behavior, Proceedings of the National Shellfish Association, 61:35-49.

IGBP, 1988. The International Geosphere-Biosphere Programme: A Study of Global Change, IGBP Secretariat, Stockholm, Sweden, 200 p.

Journel A.J. and C.J. Huijbregts, 1978. Mining Geostatistics, Academic Press, New York.

Kelly, D., R.P. Cote, B. Nicholls, and P. Ricketts, 1987. Developing a strategic assessment and planning framework for the marine environment, Journal of Environmental Management, 25:219-230.

Michener, W.K. and P.D. Kenny, 1991. Spatial and temporal patterns of *Crassostrea virginica* (Gmelin) recruitment: relationship to scale and substratum, Journal of Experimental Marine Biology and Ecology, 154:97-121.

Michener, W.K., 1992. Global positioning system activities in the long-term ecological research program, GIS World, 5:58-63.

Rao, G.R., 1992. Programs for kriging in Matlab and Splus, M.S. Thesis, Department of Statistics, University of South Carolina, Columbia, SC, 29063.

Ripley, B.D., 1981. Spatial Statistics, Wiley, New York.

Schaeffer, D.J., E.E. Herricks, and H.W. Kerster, 1988. Ecosystem health: I. Measuring ecosystem health, Environmental Management, 12:445-455.

A KNOWLEDGE-BASED GIS METHOD FOR VISUAL IMPACT ASSESSMENT IN TRANSMISSION LINE SITING

Ralph Miller
Duke Power Company
Charlotte, NC   28201-1010
(704)   382-6741

Wei-Ning XIANG
Department of Geography and Earth Sciences
University of North Carolina at Charlotte
Charlotte, NC 28223
(704)   547-4247

ABSTRACT

Visual impact assessment is an important part of siting new electric transmission lines. Two of its characteristics are multicriteria and fuzziness. First of all, it is data-intensive and involves many factors such as view distance to the proposed towers, the amount of fore-ground screening, and the number of visible towers. Secondly, it is usually conducted by professionals with imprecise knowledge. Instead of numbers, they usually use fuzzy expressions. Presented in this paper is a knowledge-based GIS model to streamline visual impact assessment. It integrates CAD with an expert system shell and GIS. The knowledge base contains decision rules based on such factors as the amount of fore-ground screening, the number of visible towers, view distance to the proposed towers, the amount of visible right-of-way clearing and the scenic quality of the viewshed. Instead of numbers, all the rules are in natural language. The burden of data processing and graphical presentation is handled by the use of GIS. The model is designed for field use and run on a laptop computer. The model has been tested through a case study in Cherokee County, North Carolina. It however has the potential to be used in other areas.

INTRODUCTION

The last several decades saw a dramatic increase in public knowledge and concern over environmental issues. The electric power industry has been scrutinized by educated and informed citizenry more than ever before. One of the issues of growing concern in the 1990's is the location and construction of new electric transmission lines. Concerns regarding electric and magnetic fields and effects on property values cause this to be a very emotional topic, especially with those property owners closest to the proposed alignment. Another aspect is the visibility of the towers themselves and what effect this has on the visual landscape. The subject of landscape aesthetics, i.e., perceived beauty, has interested landscape architects and other land planners for many decades and influenced artists for centuries.

Scenic resources were included in national environmental legislation, such as the Wild and Scenic Rivers Act and the National Environmental Policy Act, as public assets that should be protected the same as more commonly identified resources, namely soil, water, and air. The last several decades has seen the discipline of visual resource assessment move from academic theory toward practical applications. The United States Forest Service implemented a visual management system in the mid-seventies which categorizes national forest lands into certain zones based on such factors as character type/variety classes (physical features of the land), human sensitivity levels for scenic quality (view distance and use volume or use duration criteria), and visual quality objectives or goals set forth by the Forest Service (from total preservation to retention to maximum modification). Areas are mapped either manually or with computers with the different factors appearing on separate overlays

577

on a base map. A composite map is then drawn with parcels divided and labeled according to the data on the overlays. The visual management objectives are combined with other resource objectives (i.e. timber, recreation, water) to produce a master plan for all the National Forest lands. Proposed human activities that could alter the visual landscape are reviewed using this master plan. The actual visual impact of a proposed project such as a clear cut logging operation or a transmission line is still a very subjective decision by human "experts" experienced in this field.

Visual impact assessment is usually conducted by professionals with imprecise knowledge. Often this knowledge has been developed through a combination of formal training and years of practical experience. Instead of numbers, they use fuzzy expressions such as "very high," "quite low," and "almost invisible." There is no standardized way to evaluate potential impacts on visual quality. The decision or reasoning path followed is not usually documented in a way to be easily understood by persons not experienced in this area.

For transmission line siting, a method to determine the relative levels of visual impact of proposed alignments was developed by Duke Power Company and has been in use for about four years. Pulling together information from a variety of knowledgeable sources, the system consists of a form which is filled out by an analyst in the field. Factors such as the number of towers visible or visible above the horizon line, the view distance, and the amount of right-of-way clearing visible are scored and weighted along with an evaluation of the amount of fore-ground screening and the scenic quality of the viewshed. Hand calculations and several matrices yield a final resultant impact. This form is completed for every viewpoint deemed important in corridor evaluation - residences, highways, overlooks, and lakes being the most common.

The implementation of this method has, however, only been supported by computer technology in a peripheral way. A CAD type software (PC NewPerspectives)* is currently used to generate 3-dimensional terrain models from each viewpoint studied, which simulate the proposed transmission structures in true perspective and scale to the existing landforms. The computer generated terrain models must be carried in the field as paper plots. Time becomes important when numerous viewpoints must be studied and even more important when doing preliminary studies before any public announcement of a project. This situation is further complicated by the fact that all the charts and calculations are completed manually.

The method presented here provides a promising solution. It integrates a CAD software which is currently used with an expert system shell and a Geographic Information System (GIS). The burden of data processing and graphical presentation is handled by the use of GIS. The knowledge base contains decision rules based on such factors as the amount of fore-ground screening, the number of towers visible above the horizon line, the number of visible towers, view distance to the proposed towers, the amount of visible right-of-way clearing and the scenic quality of the viewshed. Instead of numbers, all the rules are in natural language. The model is designed for field use and run on a laptop computer. The model has been tested through a case study in Cherokee County, North Carolina.

METHODOLOGY

The System

The system presented here consists of three components (Figure 1). The first part is a GIS. Within the study area one of the layers of data is a map showing all the critical viewpoints that will need to be analyzed. Each point is geographically referenced to

---

* PC NewPerspectives is a trademark of Visual Simulations, Inc., Hubbard, Oregon, USA.

correlate with all the other layers in the data base. Attribute information is also entered for every viewpoint, such as an ID number and description.

The second part of the system is a CAD software called PC New Perspectives (NEWPER), which generates the three-dimensional terrain models with the proposed towers and right-of-way graphically illustrated. These figures are necessary to evaluate the number of towers and amount of corridor clearing visible from each viewpoint. Each drawing is saved as a digital slide.

The final component is the knowledge base. Using an expert system shell, decision rules are written which follow the logical thought process and calculations that appear on the visual impact form developed by Duke Power Company. The shell, VP-Expert,* can display the digital slides captured by the NEWPER program, perform the visual analysis, and save the results to a file to be transferred back to the GIS data base for geographical display and further data processing as part of the total siting study.

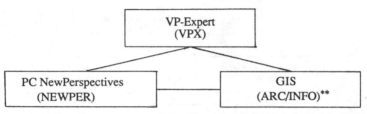

Figure 1. The Components of the Knowledge-Based GIS Model for Visual Impact Assessment

Implementation of the System

The implementation of the system consists of three steps (Figure 2).

The first step is data base and rule base preparation. Site specific terrain models are built with the NEWPER program. Viewpoint locations are input into the GIS data base and the inference rules are entered into the knowledge base of the expert system shell.

The second step, the visual impact assessment, begins with the GIS program. A map is displayed with all the critical viewpoints identified with ID numbers. This gives the analyst a geographic perspective of the various points. After picking a site on which to do the visual impact assessment, an ASCII file containing the point ID is transferred into the expert system shell. The rule base first displays a terrain model of the viewpoint, which has been saved as a PCX slide in the NEWPER program and loaded in the same directory as the knowledge base. The analyst can view the slide to determine the number of towers visible and other parameters which he will need to know. Once the slide is removed the system goes back to the expert system shell to start the consultation (i.e. inference). A series of questions follow with the analyst either typing in brief responses or highlighting multiple choice answers. These questions deal with the number of towers visible and whether they are visible above the horizon, the amount of foreground screening, the amount of right-of-way clearing visible, and the scenic quality of the viewshed. At the conclusion of the consultation, a rating for visual impact is determined. The range of values goes from very high

* VP-Expert is a trademark of WordTech Systems, Inc., Orinda, CA, USA.

** ARC/INFO is a trademark of Environmental Systems Research Institute, Inc., Redlands, CA, USA.

impact to no visual impact with six categories in between. Once this is finished the system starts over with the GIS map being displayed.

The final step, presentation and interpretation of the results, can be done in several ways. Maps showing each viewpoint location along with the associated impact rating can be displayed in the GIS system. Also, the expert system shell can produce graphics and text explaining the process of the inference. Finally, the user's notes, entered during the consultation, and the impact ratings are saved in a report file which can be printed for later reference.

I.   Preparation

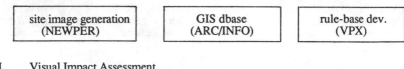

II.   Visual Impact Assessment

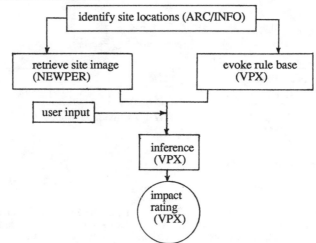

III.   Presentation and Interpretation of the Results

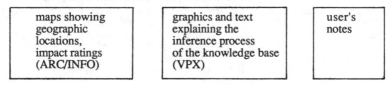

Figure 2.   Three Steps of System Implementation

THE CASE STUDY

Duke Power Company is completing a siting study for a proposed 161 kV transmission line in Cherokee County, North Carolina. (Figure 3) The line would run approximately 16 miles, 9 miles of which will probably follow an existing line right-of-way with wooden pole structures being replaced with steel towers. The remaining 7 miles would be a totally new corridor. This area of North Carolina is extremely mountainous and the scenic quality is very high. Therefore, Duke recognizes that the visual implications of larger structures on the existing line and new structures for the remainder of the line are very important to the citizens of the area and to the public travelling on the many scenic highways. The task of the siting team is to include the full range of environmental constraints, from high to low, into a suitability map. Visual resource analysis yields one of the dozen or so data base layers that are incorporated into the final composite map. Visibility was assessed from along all the major highways in the study area and combined with the landcover data to produce a data layer showing visibility from major roads with the screening affects of trees and topography taken into account. Several alternate corridors were identified on the suitability map and, after further study, a route with the least environmental impacts was chosen.

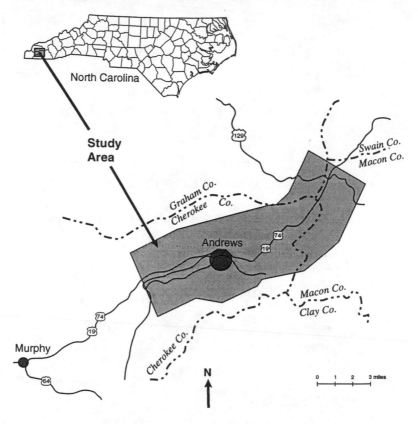

Figure 3. The Study Area

Database and Knowledge Base Development

Numerous residences and one church were identified in close proximity to the selected route, mostly near the existing right-of-way portion of the alignment. Terrain models from each residence were built using the NEWPER software and proposed tower locations were digitized off USGS 1:24000 topography maps from preliminary engineering data. An example is shown as Figure 4. These perspective views were saved as slides for later display during the actual assessment phase. The viewpoints were digitized into ARC-INFO from USGS topography maps for display and georeferencing purposes. The rules written in natural language were coded into the knowledge base of the expert system shell (VP-Expert). Table 1 lists a sampling of the rules.

581

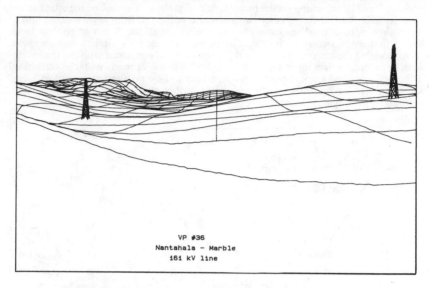

VP #36
Nantahala - Marble
161 kV line

Figure 4. Terrain Image Generated by NEWPER

TABLE 1. Examples of Rules in the Knowledge Base

RULE 1
IF   Distance = less_than_1/2_mile AND
            Screening = open
THEN   Vis–rating = high;

RULE 2
IF   Distance = less_than_1/2_mile AND
            Screening = light
THEN   Vis–rating = medium_high;
•
•
•

RULE 28
IF   Towers_vis_sky = five_or_more
THEN   Score1 = 40;
•
•
•

RULE 33
IF   Vis_rating = low AND
            Factor = moderate
THEN   Impact = very_low;

RULE 34
IF   Vis_rating = low AND
            Factor = low_moderate
THEN   Impact = no_visual_impact;

582

Using the procedure outlined in Figure 2, eighty-two residences and one church were studied. A sampling of their scores is listed in Table 2.

Presentation and Interpretation of the Results

Table 2 presents the visual impact ratings of some of the 83 viewpoints analyzed in the case study.

TABLE 2. Examples of the Viewpoints Analyzed

VP Number	Impact	Type	Location	Comments
1	moderate	residence	Topton Hill, behind Church	--
9	low-moderate	residence	very close to existing R/W	can see 5 existing structures
14	moderate	residence	under construction	can see 4 existing structures
15	very low	residence	--	top half of towers only
24	low	church	Hwy. 19/74	can see 3 existing structures
27	low	residence	Brittian Terrace	8 houses
38	no impact	residence	--	along drive, quarter mile

None of the viewpoints scored as very high or high impact, while only one ranked in the moderate-high category. Eighteen scored as moderate and fifteen scored as low-moderate. Seventeen viewpoints ranked low and twelve ranked very low. Nineteen locations were assessed as having no visual impact from the proposed transmission line.

Figure 5 shows a sample screen captured during a consultation which shows a text display explaining the process of inference.

```
less_than_1|2_mile        half_to_1_mile          one_to_1_1|2_miles
1_1|2_to_2_miles          2_to_2_1|2_miles        more_than_2_1|2_mile

Choose the amount of foreground screening?
open                      light                   moderate
heavy

How many visible towers are skylined?
none                      one_or_two              three_or_four
five_or_more

------[ RULES ]------                 ------[ FACTS ]------
Vis_rating = medium_high CNF 100      VP_number = 12 CNF 100
Finding Score1                        VP_type = residence CNF 100
Testing 25                            Location = hwy. 19 CNF 100
RULE 25 IF                            Comments = small pine screen CNF 100
Towers_vis_sky = none                 Distance = less_than_1|2_mile CNF 100
THEN
Score1 = 0 CNF 100                    Screening = light CNF 100
Finding Towers_vis_sky                Vis_rating = medium_high CNF 100
```

Figure 5. Screen Captured From VPX Explaining the Inference Process

CONCLUSIONS

Using a knowledge-based GIS method for visual impact assessment in transmission line siting has several advantages. First, all the rules and logic are written in natural language. This fits well into the field of visual analysis which uses fuzzy, non-precise terms. Second is the time factor. The program can be loaded on a laptop computer for field work. Through a series of questions and multiple choice responses it will, in a matter of seconds, do the analysis and calculations and give the visual impact rating. The method previously used involved filling out many forms and doing calculations by hand. Third, the digital terrain models are displayed by the expert system before each viewpoint consultation. This saves office time, eliminating the need to plot hard copies of all the viewpoints being studied, and also saves field work time, because the analyst does not have to be continually unrolling and rolling paper plots. The connection with GIS yields yet another improvement. The GIS database is automatically updated as the field assessment is done. Therefore, this information is ready if needed for additional analysis with the other data layers in the project area.

The model has been successfully tested through a case study in Cherokee County, North Carolina. The results were promising and appear to have high potential for widespead application.

REFERENCES

Duke Power Company. 1992, "Nantahala - Marble 161KV Line, Transmission Facility Siting and Environmental Report," Nantahala Power and Light Company, Franklin, N.C.

Priestley T. 1984, "Aesthetic Considerations and Electric Utilities: An Introductory Guide to the Literature," Electric Power Research Institute, Project 2069-4.

The Landscape Architecture Program, College of Architecture and Urban Studies, Virginia Polytechnic Institute and State University. 1989, "Cane River-Nagel Station Transmission Line Project."

U.S. Department of Agriculture. 1974, "National Forest Landscape Management Volume 2," U.S. Department of Agriculture, Forest Service, Agriculture Handbook Number 462.

Wagner, R. H. 1971, "Environment and Man," W. W. Norton & Company, Inc., New York, pp. 404-408.

BEYOND METHOD, BEYOND ETHICS: INTEGRATING SOCIAL THEORY INTO GIS AND GIS INTO SOCIAL THEORY

Roger P. Miller
Department of Geography
University of Minnesota
Minneapolis, Minnesota 55455

ABSTRACT

Those who work in technical fields such as GIS often assume that disciplinary debates centering on social theoretical questions are irrelevant, and focus instead on issues of data integrity, system structure and performance, and improving presentation of data to end users. Questions of how the results of GIS may be used and misused, ways in which choices of variables for inclusion or exclusion may affect the kinds of questions that can be asked, and how reliance on technologically sophisticated analysis systems privileges certain individuals while excluding others from decision-making processes are ignored. An awareness of the major debates in the social sciences, centering on theories of deconstruction, post-modernism, and the agency/structure debate have serious implications for GIS practitioners. The influence need not be one-way, however. Informed use of GIS can make significant contributions to the development of social theory, as well.

INTRODUCTION

Most who work with geographic information systems (GIS), when they consider the situation at all, think of themselves as technical analysts, blessedly removed from the pointless theoretical debates that seem to be modern-day equivalents of the Scholastic controversy over the number of angels that can dance on the head of a pin. More formally put, the current empirical and positivist basis of GIS has led to a general dismissal of the idea that the theoretical debates rocking the social sciences have any relevance for "practitioners" or "applied geographers." However, such a blanket rejection appears increasingly unwise and untenable.

RECENT DEBATES IN SOCIAL THEORY

To understand why, we first must examine the broad nature of these theoretical debates, which tend to be lumped together under the three general rubrics of **deconstruction, post-modernism**, and the **duality of structure and agency**, each of which corresponds to a major strand of the current debates.

- Deconstruction is based on the idea that there are viable **alternatives to positivism**, and insists that "reality" is socially constructed, rather than "naively given." The goal of analysis is to carefully "deconstruct" the discourses through which socially-constructed reality is constituted. This involves more than merely looking at the words used to convey meaning. Deconstructivists examine the "unspoken implications" that

585

attach to words and images, the contexts in which words are uttered and images presented, and how the ensemble of meaning is created in the mind of the reader or analyst. What the speaker or author intended is of secondary importance in deconstructivist analysis. Criticism is elevated to a higher position than authorship, since meaning is created by the audience.

- Post-modernism adds an emphasis on **diversity and intersubjectivity,** reflecting the variety of newly voiced experiences that now compete in the vacated space once occupied by the unitary, universalizing and authoritative voice of positivist science. Post-modernism tries to redress the long-standing inequity that has existed because "white, Eurocentric male" discourse has been "privileged," to use a favorite term. Post-modernist analysis is an outgrowth of deconstructivism in that it not only acknowledges that reality is socially constructed, but also that different social groups have different realities, all of which can make claims to validity.

- The agency/structure debate grows out of this marked enthusiasm for questioning "received authority" and existing structures of domination and control. If the legitimacy of social authority is no longer assured, what keeps people acting according to social rules, customs and mores? How much knowledge and influence do individual actors wield? How much understanding do they have of the social actions in which they participate? Do they act of their own volition, under conditions of relatively complete knowledge, or are they the puppets and pawns of hidden structural forces, whose true nature lies hidden behind ideological duplicity?

Theorists such as Anthony Giddens have posited a model that loosely follows Marx's dictum, that "[people] make history, but not under conditions of their own choosing" (Giddens, 1984). This is a position that attempts to mediate between the idea that we are controlled by social structures and institutions, and the opposite notion that social structures and institutions exist only insofar as individuals freely act to create them (the so-called "Social Contract"). Giddens makes an analogy with language. The existence of linguistic structure enables speech and communication, but the speech acts of individuals replicate, continue and change the very structures that make such acts possible. Over time, languages change as their structures are modified. And when those who use a language disappear, so does the language. Social structures and institutions provide the context for human activities, but are likewise legitimated, reproduced and changed by those very activities they make possible. And far from being unthinking automatons, individuals have considerable knowledge about the social rules within which they operate, even though they can only change those rules incrementally, if at all.

Faced with these theoretical developments, how do GIS professionals, both developers and practitioners, respond?

To the claim that reality is socially constructed, most state what seems to them obvious: "What's there, is there." Once again, to put this more formally, they are saying that empiricist positivism is the only correct approach -- the "given" one based on common sense and general experience. Most GIS professionals would be pleased to learn that the majority of scientists would support their position, and in fact would claim that the striking advances that have contributed to the very high quality of life in the industrially advanced nations of the West are directly attributable to "the scientific method." However, strict reliance on empirical facts is much harder when we are dealing with social constructs and situations than it is when we are reporting on the locations of geophysical features, lot lines or utilities.

When presented with the competing claims of groups who have not generally been heard in scientific and policy debates, including the voices of women, racial and ethnic minorities, post-colonial citizens of Third World countries, and those with different sexual orientations, GIS practitioners often throw up their hands and exclaim, "Facts are facts, regardless of who holds them." The role of GIS is simply to present those facts in a value-neutral fashion, they say, and practitioners need strive only for accuracy and presentational clarity. However, *your* facts and *my* facts may be quite different, because the ends to which knowledge is put very much affect how that knowledge is constructed. Given that much of our knowledge (or data) is partial, indeterminate, or provisional, interpretation always must play an important role. Then the question becomes, *whose* interpretation is best served by a given form or type of data? Such claims cannot be adjudicated *a priori*, but necessarily involve political processes of confrontation, contestation and compromise (Olsson, 1991).

When faced with such opposition, GIS practitioners may find themselves falling back upon claims that, "Experts know best." As professionals, they avoid the realm of values, and concentrate on technical issues of data collection, organization and presentation, eliminating the need to deal with issues of power relationships and domination. They simply provide the best possible information to those who are responsible for making decisions, be they members of the public, elected officials, or technical experts. Politics, as such, really doesn't play a role in this process.

Clearly, however, experts know that which they're trained to know. By ignoring issues of social theory and the social construction of reality, they remove themselves from the possibility of understanding a very important aspect of social interaction -- how the relative positions of actors in a given situation affect the ways in which information is presented and received. Having been, for the most part, in the advantageous position of holding power, GIS practitioners and specialists have not had to confront the ways in which different forms of presentation, and unequal power relationships, affect communication and the decision-making process. Many of the situations in which GIS data are used are complex and politically controversial; to pretend that the technically sophisticated presentations associated with GIS analysis don't affect the reception of these data is naive.

RELAXING THE ASSUMPTION OF OBJECTIVITY IN GIS

What would happen if we started thinking of GIS theory and practice as other than value-neutral and bias free?

- We would see a shift in emphasis similar to that in the socio-theoretical realm. This involves a decreasing concern with **praxis** (questions of "how do we do this?"), and greater concentration on issues of **ontology** ("what exists?") and **epistemology** ("how do we know what we know?").

As in other technical areas, there already exists a formal language and a large body of literature dealing with these issues. We tend to take for granted that the categories for which we obtain, organize and present data exist, but this is manifestly not the case. As we increasingly use GIS analysis to address issues of social efficiency and the distribution of scarce resources, we necessarily deal in categories which resist precise definition. In our society, are the "disadvantaged" those with incomes less than 75% of median income, those with incomes less than 50% of median income, or those with incomes below the "poverty line." What *is* the poverty line, and how do we determine its exact level? We can make precise operational definitions based on objective criteria, but this should not blind us to the fact that these definitions are ultimately arbitrary, especially when we deal with categories such as race, class and other social concepts.

- A concern with ontology and epistemology leads naturally into a fourth major branch of philosophy -- **ethics**. Such a development means that GIS practitioners and theorists are not merely technical functionaries, but cognizant, socially-aware actors. In other words, GIS analysts have a responsibility to consider the ultimate disposition of their efforts, rather than functioning blindly in obedience to orders. As soon as we acknowledge that we function within a political process by providing certain forms of information, we must also take responsibility for the ends to which that information is put. This acknowledgment can be both empowering and frightening.

SOCIAL THEORY IN GIS

There are numerous ways in which GIS practitioners and theorists can incorporate notions from the recent debates in social theory outlined above. Three examples are given here; many more could be advanced.

- A first approach involves an explicit acknowledgment of a major ontological and epistemological problem. Although we are often trying to model very complex and indeterminate situations, we tend to treat all data as highly determined. We need to develop ways to incorporate data that are less than precise into our analyses, while making it clear that such data are conditional. A number of approaches could be developed, including the use of fuzzy set theory, stochastic modeling, Markov-chain analysis, and other methods. Some of these issues are being addressed on the presentational side by those working on problems of map generalization, but we still have a long way to go in deciding how to incorporate behavioral, social, and economic data into the databases associated with our geographic information systems. Indeterminacy often more accurately reflects the social nature of the situations we model and analyze than do the exact presentations we frequently deliver.

- A second example is based on the fact that we often exclude data from sources that do not conform to our notions of measurement accuracy, replicability or lack of bias. Such stringency has done little to ensure that our data are, in fact, accurate or useful. For instance, for social and economic information we are overly dependent on existing large data sources, particularly the US Census of Population and Housing, as well as Federally-mandated manufacturing and commercial censuses. The availability of such resources often has the unintended consequence of limiting the questions we can or do ask. If data of the correct sort are not available in the census, we often refuse to ask questions that would require us to perform expensive and time-consuming surveys to provide the answers.

  As an illustration of this phenomenon, it is interesting to compare 19th century Dunn and Bradstreet reports with those of today. In the 19th century, subjective assessments of trustworthiness of principals in various businesses were commonly included in reports to customers. This information was obviously subject to biases based on ethnic, religious, class, or other grounds. Nevertheless, despite its biases, such information was essential for the utility of the Dunn and Bradstreet system at a time when there were few institutional guarantees protecting investors. Today, Dunn & Bradstreet data tend to be based on "objective" measures of corporate performance, including capital to earnings ratios, investment performance histories, and productivity ratios. This reflects our increasing dependence on "objective" measures, which, however, still fail to provide complete protection from risk.

  When we depend on existing sources of information, the questions that we can ask are limited by the data that have been collected, often for purposes quite different from our own. Rather than excluding imperfect data, we should utilize them in ways that acknowledge the biases they contain, while attempting to relate those biases to alternative viewpoints of those who created or collected the data. Too often, we find that we have excluded data because they don't conform to *our* biases, not because they are bias-free!

- We pay a great deal of attention to issues of presentation, while ignoring the fact that information provision is an explicitly political process. By concentrating on methods of presentation, we are implicitly acknowledging the fact that our data are not value-neutral. We know that certain forms of presentation are more useful, convincing or intimidating than others. The very fact that "experts" are presenting data processed by computer adds weight to the results, regardless of the quality of the initial data on which the results are based.

This is an area in which we need to think clearly about the inherent power relationships that arise in technical and scientific analyses. We might consider Jurgen Habermas' notions of barriers to true communication acts (Habermas, 1984). This involves recognition of the role played by unequal power, unequal access to information, and the subtle and not-so-subtle cues that are inherent in various modes of discourse. We could also think about the role that information plays in legitimating authority to act in our society, a role in which GIS plays an increasingly powerful part. We should be cognizant of our power, and be careful not to abuse it ourselves, or let others misuse it.

GIS IN SOCIAL THEORY

These three examples provide only a brief overview of the ways in which an active engagement with issues in the theoretical wings of social science disciplines can enrich GIS practice and theory. Must the enrichment be one-way? Can GIS not only utilize current social theory, but actually contribute to the debates in such areas?

In some cases, emphasizing the disjuncture between modes of presentation and quality of information would help in this process. This is probably the major contribution that GIS practitioners can make in areas such as facilities management, geophysical resource monitoring, and other more technical, empirically-based areas in which GIS is used. In other areas, however, especially those that involve large-scale social, political, or economic processes, GIS can help researchers, analysts and policy makers understand how theory itself should be modified and reconstructed. GIS information can provide a good "reality check" for the sometimes ethereal formulations of academicians and politicians!

For instance, in constructing theories of urban residential patterns and geographic mobility, researchers have tried to determine the differential effects of socioeconomic status, race and ethnicity, and family status. The best tests of such theories have been based, in effect, on GIS models, using statistical analytical techniques such as factorial ecology or principal components analysis (Shevky and Bell, 1955; Rees, 1971).

In the 1960's, as academicians and policy-makers grappled with questions about the historical effects of race on the social mobility of African-Americans, an early form of GIS technology was utilized. Researchers were testing several competing explanations for the high levels of residential segregation and the low levels of social mobility experienced by African-Americans. Were they simply the "last of the immigrants," or had they been discriminated against systematically over a much longer historical period? Research at several large urban social history projects concluded that Black Americans had been systematically discriminated against in ways that were not typical for other immigrant groups, such as the Irish, Germans, Italians, Eastern European Jews, and others. Residential locations of households were compared across different groups, and African-Americans were found to have experienced far more residential segregation and less access to jobs than other groups (Hershberg et al., 1981).

In carrying out the analysis, however, researchers found that they had to utilize the information contained in large databases such as the manuscript schedules of the US Census, city directories and special censuses in creative ways. Engagement with issues of how one determined ethnicity or race from the census led to a rethinking of what constituted race or ethnicity in general. For instance, was the ethnicity of a person with grandparents born in Bohemia, but one parent born in Silesia and another born in Pennsylvania, German, German-Polish, or American? Did Polish ancestry mean different things in Poland and in Chicago? Dealing with such questions in the databases that were part of these early social history GIS endeavors led to the development of new ideas about ethnicity and its effects on social mobility and community formation. More importantly, a dialectical process of theory formation, database construction, and analysis of the categories of data occurred. The process involved an active dialog between those constructing the databases, and those creating the social theory, to the benefit of both.

GIS can contribute to the development of social theory in an even more important way. Researchers are finally beginning to acknowledge the role of space in the construction of social theory. Human activities are concretely situated in both time and space. Heretofore, much greater emphasis has been placed on the role of time, privileging history over geography, and creating a curiously aspatial notion of how societies function. Economic, political and social models have been seriously deficient because they have been viewed as universal constructs, ignoring important spatial variations in the distribution of resources, attitudes and opportunities. Post-modern theories, however, explicitly acknowledge spatial variability as an important component of social theory, while analysts grappling with the structure/agency debate insist on the importance of the spatial "situatedness" of human activities. Both developments provide opportunities for contributions based on GIS analysis.

Let's take theories about the economic transformation of contemporary society first. There is a major debate between structuralists and postmodernists who are trying to interpret current developments in the US economy. David Harvey and Ed Soja, for instance, have interpreted the recent transformation of the American economy in very different ways (Harvey, 1989; Soja, 1989). In each case, however, the crux of the argument rests on assumptions about the behavior of firms and the relocation of economic activities for which scanty data exist. Is a true economic transformation occurring, in which our understanding of the behavior of firms according to economic location theory no longer hold, or are we seeing some combination of economic restructuring and geographic relocation which is a continuation of trends that began shortly after the Second World War?

None of the principals in the debate has yet undertaken the multiscalar empirical research that would help answer these questions, and to do so clearly requires the use of GIS analysis. However, GIS experts need to familiarize themselves with the terms of the socio-theoretical debate if they are to assist in the creation of geographical information systems that will be useful in this area. The mere replication and updating of the formats and data that sufficed for economic analyses based on Keynesian economic theory simply will not work in a situation in which basic economic premises and assumptions no longer seem valid. A host of new considerations need to be factored into our analytical models, including interactions among local, regional, national and global economies; labor force mobility, deskilling and job insecurity; outsourcing and subcontracting of production; competition for public subsidies at the local, state and federal levels; and changing production models based on "just-in-time" manufacturing, which eliminate stockpiling and warehousing of components. If GIS practitioners are unaware of the changing nature of the economy their systems will be utilized to help analyze, it is doubtful they can design them to be useful into the 21st century.

GIS can also make a major contribution in helping to model the spatial context of human activity in theoretical debates about structure versus agency. Theorists from Anthony Giddens to Allan Pred have utilized the time-geographic research of Torsten Hägerstrand to illustrate the movement of individuals through time and space. Hägerstrand's model traces individuals through temporally and geographically located stations as they join other individuals and utilize specific resources to accomplish tasks. Although primarily a heuristic model, several researchers have used time-geographic methodology to examine situations related to social theoretical issues, especially the structure/agency debate (Miller and Gerger, 1985; Pred, 1986). Research has been limited by the difficulty of creating adequate GIS models to test large-scale empirical situations. To utilize GIS in this fashion, it is necessary to include fine-grained information on times associated with a wide variety of social phenomena which we currently tend to describe in purely geographical terms. GIS holds considerable promise in this field precisely because its data structures, coupled with appropriate stochastic or Markov-chain techniques, can handle the interdependencies that characterize time-geographic analysis.

As databases of information become more sophisticated, and incorporate data more sensitive to the issues that have recently gained prominence in social debates, GIS will be able to contribute significantly to the development of social theory. It is our responsibility as GIS practitioners to actively participate in such debates, rather than passively sit on the sidelines, ignorant of the implications of our work.

REFERENCES

Giddens, A., 1984. *The Constitution of Society: Outline of the theory of structuration.* Berkeley: University of California Press.

Habermas, J., 1984. *The Theory of Communicative Action.* Translated by T. McCarthy. Boston : Beacon Press.

Harvey, D., 1989. *The Condition of Postmodernity : An enquiry into the origins of cultural change .* Cambridge, Mass: Blackwells.

Hershberg, T., A. N. Burstein, E. P. Ericksen, S. W. Greenberg, and W. L. Yancey, "A Tale of Three Cities: Blacks, Immigrants, and Opportunity in Philadelphia, 1850-1880, 1930, 1970," in T. Hershberg, ed., *Philadelphia: Work, space, family, and group experience in the nineteenth century*. New York: Oxford University Press.

Miller, R. and T. Gerger, 1985. *Social Change in 19th-century Swedish Agrarian Society*. Stockholm: Almqvist and Wiksell International.

Olsson, G., 1991. *Lines of Power/Limits of Language*. Minneapolis: University of Minnesota Press.

Pred, A. R., 1986. *Place, Practice and Structure: Social and spatial transformation in southern Sweden, 1750-1850*. Totowa, N.J. : Barnes & Noble.

Rees, P. H., 1971. "Factorial Ecology: An Extended Definition, Survey, and Critique of the Field," *Economic Geography* 47 (Supplement), pp. 220-233.

Shevky, E. and W. Bell, 1955. *Social Area Analysis*. Stanford: Stanford University Press.

Soja, E. W., 1989. *Postmodern Geographies: the reassertion of space in critical social theory*. New York: Verso.

Capabilities Needed in Spatial Decision Support Systems

George Moon
george@cvgis.prime.com
Mark Ashworth
mark@cvgis.prime.com
Computervision GIS
45 Vogell Road, 7th floor
Richmond Hill, Ontario
Canada, L4B 3P6

Abstract

Capturing spatial digital data has been a key focus during the 1980's. There now are enough digital data available and capture procedures in place that the key to the 1990's is how quickly can the digital data be used to solve real pressing problems. One approach is to use decision support systems.

Building a spatial decision support system (SDSS) on a three level framework will be discussed. The capability of the fundamental level uses a toolbox approach. At the next level, the capabilities of the tool kit must be combined in a cohesive manner to form the SDSS builder software. The SDSS builder enables the decision maker to generate specific SDSS applications. As new and improved tools are added to the toolbox level, the SDSS builder software must make them available to the decision maker. Existing specific SDSS applications can be updated to take advantage of the changes.

This is ongoing work being done using the Application Tool Box in SYSTEM 9.

Introduction

There is enough good quality digital data available that the key to successful use of Geographic Information Systems, or more generally Spatial Information Systems, in the 1990's is how quickly and effective can the digital data be used to solve real pressing problems. One approach is to use decision support system methodologies.

This paper will briefly reference what is meant by decision support in a spatial context. The building of a spatial decision support system (SDSS) on a three level framework will be discussed. Finally examples of how one may approach loosely defined problems with a SDSS will be given.

Spatial Decision Support System

Problem solving with spatial information systems (SIS) often involves neatly ordered problems. Decision support systems can assist in solving many problems that stretch beyond a neatly packaged SIS system.

Armstrong and Densham (1990) discuss spatial decision support systems (SDSS) in terms of six characteristics. These characteristics are loosely given as:

1. limited problem domain
2. use of different data types
3. ability to model
4. the availability of graphic displays for reporting
5. flexibility in the facilities provided for decision making
6. being able to extend the capabilities of the system.

The papers by Ashworth (1988, 1990) describe spatial support tools available in a SIS that can effectively be used to build spatial decision support systems. These tools are also extended and used by the development team at Computervision to build user applications. One such application is described by Zhang and Wilkinson (1990). The basis for another application that builds topology from a set of user defined 'rules' and which has been built by using and extending the tools is described by Moon et al. (1991). Yet another example was presented by Wong (1992).

Spatial Support Tools

The spatial support tools described by Ashworth (1988, 1990) have the characteristics outlined above. The following section highlights some of the points that are important for the support tools.

1. Limited problem domain - a limited problem domain implies that the user needs to be able to have available and to organize cleanly the tools necessary to do the job. Unnecessary options or functionality needs to be eliminated to ease the task. The framework for the Application Tool Box (ATB) used to build certain limited domain applications is based on providing a set of independently accessible processing functions that address different SIS or data management operations. These operations are available on their own or as servers to which clients can send requests.

2. Use of different data types - three main spatial data types, point data, vector data, and polygon data are supported. Raster and triangular networks are also supported. Data can also be read from different data sources and passed to different data sources as part of the set of tools. This is important as will be discussed below. One significant point about the use of different data types is that the tools hide the data type from the user. The user does not have to know whether they are dealing with point, vector or polygon data. The operations work on any of these types. See Berrill and Moon (1992) for more details.

3. Ability to model - what if questions can be explored using a rich set of processing functions that allow the user to work with object pointers that always references the current active spatial data managed by the SIS database system or copies of spatial data that could be used for things such as temporal exploration over discrete intervals or to make available for modeling, data from diverse sources.

4. The availability of graphic displays for reporting - features (objects) within the system know how to display themselves freeing the

user from this task. All the expected graphic capabilities for two way interactions are available either through command line, menu interaction or an API.

5. Flexibility in the facilities provided for decision making - a three level framework is supported. The tools available or their derivatives are designed to be used by operators, specialists and programmers. The operator would be expected to use a highly interactive application where the operator executes well defined commands. The specialist may package highly interactive applications for operators or may be a more sophisticated operator working on much less structured problems. The programmer would use the API to build applications for decision making.

6. Being able to extend the capabilities of the system - the programmer mentioned above can use the ATB facilities to add their own specialized routines into the ATB.

By making reference to previous work it thus can be argued that the facilities necessary for a spatial decision support system exist in current GIS systems. What is explored in more detail is to what extent unstructured problems can be solved. By exploring the authors experience with the system they use, certain capabilities are shown.

Unstructured Problem Solving

The brief outline and the references given above are used to demonstrate that it is possible to build spatial decision support systems to handle orderly problems. That is, problems that can be broken into the sequence of problem analysis, database assembly, analysis and reporting. Orderly problems can allow exploratory work to be done but as outlined in the paper by Abel et al. (1992) there are classes of problems where an orderly approach is not sufficient.

These are problems that have such variability or complexity that a high degree of judgment or experimentation is necessary before solutions are found. They also may be problems that are very loosely defined so that the very definition of the problem and its solution evolves.

Abel et al. (1992) argues that for loosely structured problem solving the decision maker needs easy access to the full set of resources needed to develop a solution. He discusses and presents work on two key issues. One is the user interfaces and the other is the data management facilities.

Prototype work on an object oriented interface and on data management facilities needed to manage widely varying data is presented by Abel et al. (1992). The need to manage widely varying data that may be distributed throughout a heterogeneous network is also discussed by Allam et al. (1990) and others.

The rest of the paper will show an example of how a system was put together from the above mentioned application tools to allow the management and analysis of data from a variety of sources and to present to decision makers facilities where they can collectively interact in the problem solving domain.

The Simulation

The example consists of a system to monitor and to update in real time a utility network. The model is where data is being shared between different departments, and facilities need to be available for decision support to monitor an electrical network and analyse various situations.

The data sources are:

Electrical network data stored and managed in a SYSTEM 9 database.

Base map data stored and managed in Intergraph files.

Resource data stored and managed in ESRI Arc/INFO files.

Raster images both as picture attributes and as raster data types stored in a SYSTEM 9 database.

Plant information managed and stored in a Computervision CADDS file.

Corporate information store.

A number of users are able to monitor information sources and get updated information on their graphic screen. Each user's graphic screen shows the same spatial information. Updates to the information from any source including updates from one of the users, will be seen by all the users. Each of the users has a set of tools available to assist in making decisions that may be necessary.

On each screen are a base map and a representation of the utility network. There is no need to manage the base map information in a single vendor's system. To the decision maker though, the base map information is a valuable reference and appears as completely integrated information available for query or analysis.

Resource data is available in a similar way to users that need the information for purposes of aiding in decision making. Again there is no need to force the data to be managed in any particular vendor system. The tools that are available for doing spatial analysis work on the resource data, the base map data and the utility data. The decision maker is not concerned about the source of the data nor its maintenance. The data is managed in a homogeneous analysis environment.

The raster data type and the raster attributes are used to provide schematic information and detailed pictures of a site or scanned documentation that may be necessary to review to assist the decision maker. Since the raster data type and raster attributes are part of the data model, queries, including spatial queries can be made on the information.

The CADDS system is used to design and maintain detailed plant information. A 3D visualization system is available for walking through plant designs, looking at DTM information and using a 3D locator to query the scene.

The Corporate information store contains customer information which is managed by the IS department. An integrated link to the spatial data is used to preserve database integrity.

Diagram metrically the available facilities look to a decision maker like figure 1.

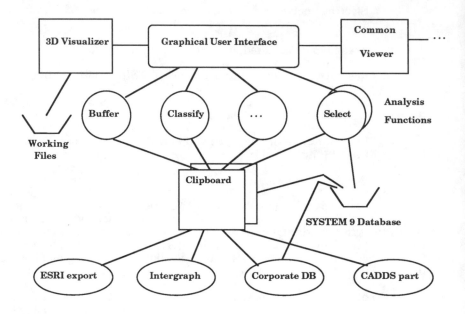

Figure 1. Representation of User's Available Facilities

The Results

A set of slides is presented that show the result from on decision example where a generator site has been damaged and there is a toxic leak. The decision maker is notified of the problem. Given the location of the failed site, all owners within 1000 meters of the leak must be notified. A list of telephone numbers is obtained from the corporate information database based on the spatial analysis. The nearest emergency vehicle is dispatched. The schematic for the generator builder is brought up and the 3D detail of the builder is examined.

Another set of slides shows some speculative analysis for electrical line siting.

Conclusions

This paper has presented briefly the type of facilities needed in a GIS/SIS to support the decisions that will be made using the digital data available in the 1990's. Existing SIS exhibit many of the features necessary for SDSS. Examples to illustrate how far these systems can already be used show that decision makers can indeed expect to manage and answer difficult questions of the future.

Giving the decision maker a rich set of facilities, allowing the facilities to work with data from different data sources, providing the means to distribute the processing within the user's system and providing common, concurrent display capabilities, opens up many possibilities for timely decisions.

It is believed that the future decision makers will have to use information from many different sources. While some systems are capable of handling very rich data varieties within their database management facilities, the ones that do not demand all data to be stored and manage within their system will have obvious appeal to a broader group of users whose job it is to make complicated decisions based on their best knowledge and information.

Acknowledgements

The authors would like to acknowledge the support of Computervision during the preparation of this paper, which is published with their permission.

References

Abel, D.J. et al. (1992) Support in Spatial Information Systems for Unstructured Problem-Solving, Proceedings of the 5th International Symposium on Spatial Data Handling, Charleston, South Carolina, U.S.A, Aug. 3-Aug. 7, 1992, pp. 434-443.

Allam, M.M. et al. (1990) Integrating Corporate Databases in a Heterogeneous GIS Environment, Proceedings of GIS/LIS'90, Anaheim, California, U.S.A., Nov. 7-Nov. 10, 1990, pp. 887-903.

Armstrong, M.P. and Densham, P.J. (1990) Database Organisation Strategies for Spatial Decision Support Systems, International Journal of Geographical Information Systems, No. 4, 1990, pp. 3-20.

Ashworth, M.A. (1988) Concurrent Display Process in a Spatial Analysis System, Proceedings of GIS/LIS'88, San Antonio, Texas, U.S.A., Nov. 30-Dec. 2, 1988, pp. 132-140.

Ashworth, M.A. (1990) Performing Speculative Analysis in the GIS Environment, Proceedings of GIS/LIS'90, Anaheim, California, U.S.A.,

Nov. 7-Nov. 10, 1990, pp. 588-597.

Berrill, A. and Moon G.C. (1992) An Object Oriented Approach to an Integrated GIS System, Proceedings of GIS/LIS'92, San Jose, California, U.S.A., Nov. 7-Nov.10, 1992.

Moon, G.C. et al. (1991) Maintaining Data Integrity in a GIS Environment, Proceedings of the Canadian Conference on GIS, Ottawa, Ontario, Canada, Mar. 18-Mar. 21, 1991, pp. 80-90.

Wong, T. (1992) Beyond Dynamic Segmentation, Presented at the URISA Conference, Washington, D.C., U.S.A., July 12-July 16, 1992.

Zhang, G and Wilkinson, S. Simultaneous Multi-Layer Spatial Overlay, Proceedings of GIS/LIS'90, Anaheim, California, U.S.A., Nov. 7- Nov. 10, 1990, pp. 506-515.

TOWARDS BETTER COMMUNICATION IN COUNTY GOVERNMENT

Robert F. Morris
Hillsborough County
Engineering Services, GIS Section
P. O. Box 1110
Tampa, FL 33601

ABSTRACT

Hillsborough County, Florida, like other counties, is actually composed of several constitutional offices competing for the same tax dollars. Three offices have established four geographic information systems departments (GIS) to address the needs of planning, infrastructure and parcel mapping. To be truly successful these systems must communicate directly with each other to access the most current information. As these systems are operated by different constitutional offices, no hardware connections have been established. To facilitate the communications between systems and develop overall goals and plans that would benefit the citizens of the county, an informal GIS User's Group has been formed. This group meets every month with members from each GIS system and other interested agencies to resolve problems and share information on systems and projects. A steering group focuses on issues dealing with the county-wide needs of the GIS systems. Four working groups address specific issues: Database, Education, Hardware, and Projects. The User's Group has already been successful since its recent inception: participants have come from all constitutional offices sharing information and reducing duplication of effort; establishing database standards for sharing information; and educating other departments in the uses and benefits of the GIS systems.

INTRODUCTION

During the past five years, Hillsborough County, Engineering Services, under the County Administrator, has developed a complete parcel level 1:2400 basemap. Hillsborough County covers over 1,000 sq. miles and 36 townships. Approximately 180,000 parcels were mapped, which includes unincorporated Hillsborough County and the cities of Temple Terrace and Plant City. This does not include the City of Tampa, which is the largest of the three cities within the county. The GIS system used is Genamap from Genasys, Inc. running UNIX on HP9000 series using two 370's and one 835 with Oracle from Oracle, Corp. as the relational data base management system (RDBMS).

As this program advanced, other agencies decided to enhance their programs with the purchase of a GIS system. Planning and Development Management, also under the County Administrator, focuses on water/waste water utilities and infrastructure mapping. The Property Appraiser's Office established a system to maintain the parcels and will build the parcel map for the City of Tampa. These two systems are similar to Engineering's with the exception of using Informix as their RDBMS. The City-County Planning

Commission has established a GIS system using Genamap, HP9000 and Oracle. All of these systems are utilizing the Engineering basemap.

As these systems have grown, more agencies now see the potential and have started requesting that their projects utilize the GIS systems. However, as there are always more requests than time, smaller agencies are starting to think about developing their own PC-based GIS system (PCGIS). Creating numerous systems without standards creates many more problems that these agencies often do not realize.

Due to the competing nature of our system of government coupled with the decreasing tax base, cooperation is not always as good as it could be. Until a coordinating council is formed, a GIS User's Group has been formed to try to address some of the issues that face the GIS systems and their use within the county. The User's Group tries to bring together those individuals that are responsible for the work and share information that will be of mutual benefit.

GIS USER'S GROUP

The User's Group is an unofficial, voluntary effort that has a steering group and four working groups. General meetings are open to any agency and as a result bring together individuals that work on GIS or related databases. Agencies that normally do not work with each other have been able to meet, and as a result, many have found information that they were in need of or were creating with great effort. Some agencies are looking for information and support that is not just an over-simplified vendor solution when building a GIS system. Members are able to share experiences, whether good or bad, and learn what data is available for their use.

Meetings are held monthly to present details about on-going projects, inform the members of the working groups efforts and to bring forward other areas of concern. The general membership has representatives from:

Agency	Department
City of Tampa	Housing and Development Coordination
Clerk of the Court	MIS Department
County Administrator	Building Permits
	Community Services
	Dept. of Information Technology
	Engineering Services
	Environment Protection Commission
	Parks
	Planning and Development
	Public Utilities
	Real Estate
	Road Department
	Emergency Dispatch Operations

Agency	Department
	Emergency Operations Center
	Fire Department
	911 Emergency Operations
Health Department	
Planning Commission	
Property Appraisers	
School Board	
Sheriff's Office	
Supervisor of Elections	

Steering Group

A steering group meets monthly with one representative from each of the four GIS systems, one person from the Building Permits Department, and one from the City of Tampa. The purpose of this group is to try to identify problems and suggest solutions to items that are hindering the success of GIS within the county. If there are issues that need technical review, then they are given to a working group to pursue. Each working group has as members those individuals with special interest or skills relative to the group and one person from the steering group. This keeps the steering group informed as to the direction of the working groups and assists the working groups with a GIS perspective which many members do not have.

WORKING GROUPS

Database Working Group

During the past 10 to 15 years, computer technology has dramatically changed the way many agencies perform their fundamental duties. Each agency or department purchased computers and developed databases to suit the needs of their own department. With the advent of GIS, it is no longer feasible for each department to ad hoc their systems to the extent that currently exists. For example, street name and address is common to almost every department but each uses a different record format, spelling convention, and abbreviation. Requests made using street based information can not be easily matched, making the task difficult and not cost effective.

The database group is developing standards that will allow all agencies to match on street name and address. This street file standard will define the naming conventions for abbreviations, directionals and prefixes in addition to the record field sizes for each record. The 991 Emergency Operations is responsible for street names and address assignments and would maintain the street file on a local area network (LAN). This file could be searched by any agency for spelling, addresses, changes, deletions and additions. By creating a standard in conjunction with the database on a LAN each agency can maintain up-to-date street information automatically, thus eliminating the need for duplicate data entry efforts. As each agency adopts the standard and conforms, the uses for and viability of the GIS system increase dramatically.

The database group is also attempting to help answer the question: "What data and/or maps does your department have?" This is a very hard question to answer unless one knows of the data personally. As more projects are completed and more users are added, the ability to quickly search for maps and data created by other departments will be very important. To help identify data that may be useful to other agencies, the database group has initiated a database catalog that will be maintained on a LAN. Each member of the User's Group has completed preliminary summaries of existing databases. This survey has two purposes: 1) it shows commonality across databases assisting with standardization and 2) is the basis for a catalog indicating accuracy, scale, and update cycle.

Education Working Group

While many agencies are now aware of GIS, they do not always know how to successfully utilize GIS in their projects. Often departments insist that their 1:12000 or 1:24000 scale maps be digitized and then expect the data to perfectly align with the 1:2400 basemap. Or they may develop a plan without knowing what other resources and methods that may be easier or better suited could be applied to their project. Basically, it is thought that there exists on the computer a button that will get their project done. They don't realize that it is not just one, but many buttons! The education group was formed to address these issues.

As part of its education efforts, the group has provided tours of the various GIS systems. While some have a basic knowledge of GIS, very few have seen it at work. The tours are given in small groups at each of the four systems, thus showcasing each department's system and accomplishments. Also, it is planned to tour facilities of local businesses using GIS to learn of other applications and what additional services might be provided. In addition to the tours, this group is developing an executive seminar to address the use and limitations of GIS within the county. It is hoped that this will give upper management a better understanding of the needs of GIS. In the future, it is thought that the education group could help provide training on the system or provide a forum to learn specific skills developed on other systems.

Hardware Working Group

The hardware group was formed to find a solution to the lack of communication between the systems, which is a big obstacle to an effective county-wide GIS. Currently, data layers and database extractions are put onto an optical disk and then copied onto the other systems. The use of this "sneaker-net" causes the data to be continually outdated and results in each system having to store redundant information. The group is exploring how to make the connections between constitutional offices and how to work with the exchange and maintenance of data. As the cost of GIS systems decreases and the computing power increases, small agencies want their own in-house GIS system. The hardware group is testing and evaluating several PCGIS solutions that might allow more users of the database. The goal is to extend the use of GIS to include mapping, planning and regulatory applications, in real-time, within the county. Because many departments are not familiar with GIS technology and its associated benefits and pitfalls, this group is trying to determine what is a good solution given the existing hardware and software. Creating a list of equipment known to provide a suitable GIS environment platform will greatly assist other departments in making cost-effective decisions.

Projects Working Group

Some projects require a large effort in a short time-frame or should be done but are of no immediate value to one GIS system. The project group was formed to develop teams to tackle problems relying on individuals from many departments with various backgrounds to complete specific projects. For example, trying to make the US Census' TIGER/Line file fit the basemap involves a major effort and is only marginally useful to any one department but overall is of great value. Other examples include trying to build a simplified interface that allows outside departments to browse the map database and make decisions based on accurate, up-to-date information.

Some GIS systems are only used to create maps, replacing a paper-based task with a new electronic tool. The final product is a plotted map for presentation and/or planning use. Few systems are used in real-time to regulate using information from multiple agencies. The power of GIS to integrate divergent databases into an effective regulatory tool is one of the most exciting areas and one which requires much thought and planning. Supporting other departments in the use of GIS will be a major task that involves more than just connecting across a network.

As an example, the building department needs to have a street address with each permit issued. If one needs to be assigned, then a request is faxed to the 911 Emergency Operations where maps are checked and an address assigned and returned by fax. Once addressed, other agencies need to know of the transaction to provide vital services such as fire and police protection. With the basemap completed, the address checking and assignment could be done directly from the GIS system in real-time. While this has not yet been fully implemented, many departments are working to achieve this goal. Another example is the use of GIS to approve building permits, checking for compliance with the State-mandated Comprehensive Plan and its concurrence review process. By having the parcels, sewer, water, roads, zoning, schools and other required infrastructure mapped, the system could be used to assess if new construction can be granted at the proposed site. If the application is denied, it would quickly show the missing pieces.

CONCLUSION

To use GIS systems for real-time regulation related tasks requires the cooperation of many departments and agencies with a common understanding of the issues. The User's Group provides a forum in which these goals can be openly discussed and solutions can be found. The User's Group has been able to demonstrate that GIS is not something to be feared, but a tool to assist each agency in its mandated tasks. By getting the involvement of many agencies, the suggestions made by the User's Group have more impact and acceptance.

The User's Group can not replace management in making decisions, but it can help management to focus on those issues which are most pressing. For example, the success of the User's Group may hasten the establishment of a GIS Coordinating Council for the planning and management of county-wide GIS systems. Thus, a GIS User's Group can be a highly effective tool to improve intra-department communication and cooperation, enhancing county-wide productivity.

INTEGRATION OF GPS TECHNOLOGY AND DIGITAL IMAGE PROCESSING TO CREATE A WILDLIFE MANAGEMENT DATABASE

Jennifer Nealon
Salem State College
Department of Geography
352 Lafayette Street
Salem, MA 01970

ABSTRACT

This paper is the synopsis of Project Thor from an educational perspective. It shows the ability of undergraduate students to work with a conservation group seeking a low cost GIS. The procedures for building this GIS are laid out in a step by step fashion.

INTRODUCTION AND SETTING

Project Thor is part of a five year, cooperative, research project conducted by Salem State College at the Massachusetts Audubon's Ipswich Wildlife Sanctuary. The object of the project is to employ Geographic Information Systems (GIS), Digital Image Processing (DIP), and Global Positioning Systems (GPS) to identify, classify, map and model vegetation units at various scales. The goal of the project is to derive a database for wildlife development and management. Research and data collection is compiled by undergraduate students in the Geography Department's Cartography program under the direction of Professor William Hamilton.

The research site, shown in figure one, is a wildlife sanctuary consisting of approximately 2,500 acres located on the Ipswich River in Northeastern Massachusetts. The Sanctuary has been divided into 206 upland, wetland and aquatic management units.* The hydrology of the area is complex. The wet meadow is marked Bunker Meadow on the image, shrub and wooded swamps are highlighted with dark borders, while the floodplain forests follow the river areas of the Ipswich. All of these areas are managed with aid of an introduced drainage system.

PRODUCTS REQUIRED

The two principal tools in use for data collection and manipulation are the IDRISI software package and GPS technology. The former is a raster based GIS package for

---

*Marc Lapin, Ipswich River Wildlife Sanctuary Ecological Management Plan, (Submitted to the Mass Audubon Society 18 September 1990), IN-1.

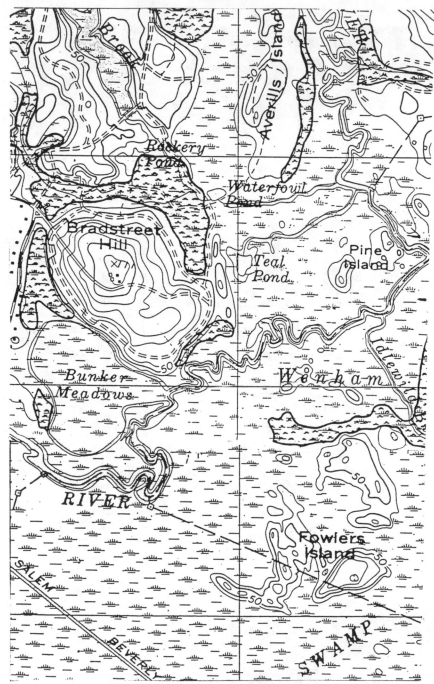

Source: U.S.G.S. Massachusetts Georgetown and Salem
Topographic Quadrangles, photoreviesd 1979.

the PC. In terms of this project the software was used
primarily for its image processing facilities. The
program's capabilities include "supervised and unsupervised classifications, color space transformations,
principal components analysis and a variety of filter
operations"[**].

GPS technologies utilized include a base station, Trimble
Navigation's Pathfinder Basic Software and Basic Hand-Held
GPS Receiver with an outboard antenna. A GPS is a receiver
that collects positional data from satellites. The base
station is made up of GPS hardware and software at a known
location. In this case the base station antenna was
surveyed on to the roof of Meier Hall at the Salem State
College campus. By using data collected at the base
station it is possible to determine the time and magnitude
of errors incurred in the field. The Pathfinder software
calculates differentials to make the necessary corrections.
The two to five meter accuracy the calculations provide is
particularly important in this instance because the
vegetation types and land use classifications change
quickly within the Sanctuary. Had this accuracy not been
available, problems may have been introduced by erroneous
spectral signature assignment caused by locational
inaccuracy. The outboard antenna, when attached to an
adjustable rod, overcomes the difficulty of obtaining
points in areas that are undesirable (i.e. poison ivy patch
and swamps), and under areas with thick canopy.

UNSUPERVISED CLASSIFICATION

Infrared photos, air photos and the pre-existing Sanctuary
management plan were used to create the unsupervised
classification the composite image. A LandSAT image
covering the study site was loaded into the IDRISI program
to create the image. The LandSAT Thematic Mapper scene is
comprised of seven bands in the visible, reflected
infrared, and thermal infrared. A June scene was chosen
because it includes vernal pools and full leaf cover. The
fore mentioned maps and photos made it possible to equate
spectral signatures on the composite image with known and
discernable areas on the ground.

Figure two is one of the images used for the unsupervised
classification and is composed of bands seven, five and
four. Reflection from the water is lost in this image but
when compared to figure one it is possible to pick out the
delineations of Bunker Meadows, Bradstreet Hill and
Averills Island. Crude delineations have been added to the
image to give orientation and aid in the identification of
Pine Island, the meadow and the hill.

[**]John C. Cartwright, "IDRISI - Spatial Analysis at a
Modest Price", GIS World, 4 (December 1991): 96.

FIGURE TWO
UNSUPERVISED CLASSIFICATION
TM BANDS 7, 5, 4

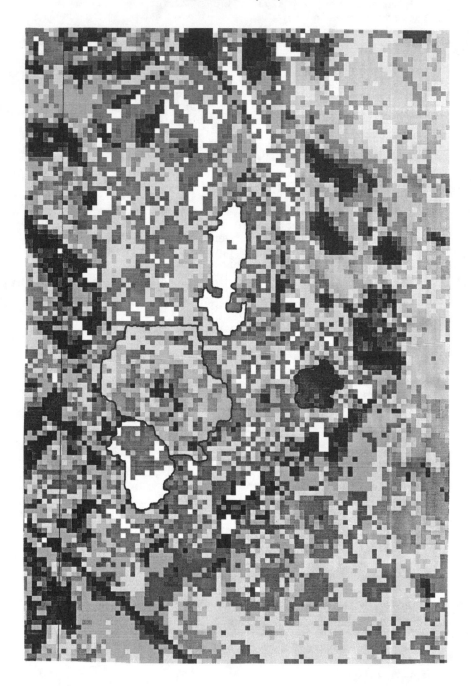

SUPERVISED CLASSIFICATION

This classification was created by ground truthing of the unsupervised. For high resolution and accuracy, ten training sites were assigned to each of the previously determined land uses. The training sites or locations for ground verification were assigned randomly within each land use. The approximate latitude and longitude of each site was determined by means of a digitizer with a map. This process allowed the sites to be easily located on the ground with the GPS.

The first land uses verified in the supervised classification were homogeneous tree stands. By confirming the locations of these stands it is possible to immediately assign each of the stands a spectral signature. The ability to tie a signature to a specific tree type aids in the signature identification of more complex stands.

The supervised classification is already in operation at the Sanctuary. It is being used to track invasive plant species that pose a potential danger to the existing ecologic balance. Buckthorn is an example of such a species. By using the GPS to locate prominent thickets of Buckthorn it is possible to determine this spectral signature. Not only does this provide an accurate map of current Buckthorn locations on the LandSAT image but it can also be used for modeling.

Once the spectral signature of the plant is determined soils maps are overlaid on the LandSAT image. Hydrology and canopy cover information are also mapped over the LandSAT image. Once the preferred conditions of Buckthorn are determined and mapped, it is possible to divide the areas into categories of intruded and unaffected. Preventive steps can now be taken to protect the un- affected areas with that have desirable conditions for Buckthorn.

OBSTACLES

The predicaments encountered in this project came in two forms. The first was GPS interference and selective availability. Interference with signals sent by the satellite can be blamed on a multitude of causes but most commonly is related to atmospheric and ionospheric delays.[***] This type of obstruction can be rectified with the differentials calculated from the base station. The selective availability (SA) is not as easily remedied. SA affects the positional accuracy and velocity components of

---

[***]Jeff Hurn, GPS: A Guide to the Next Utility, (Trimble Navigation, 1989): 47.

non-authorized GPS receivers. The military reserves these accuracies for authorized users in governmental agencies. "SA is implemented by altering satellite position and/or clock data broadcast in the navigation message"****. The only way to overcome the burden of SA in this project was to revert to the older and more time consuming methods of pace-and-compass, and transit use.

The second obstacle in the project occurs in signature assignment. Two prominent vegetation types in the Sanctuary, conifer and herbaceous wetland, have the same or similar spectral signature. This caused confusion until the predicament was realized. The two were differentiated with topographic location and proximity to other vegetation types.

FUTURE RESEARCH

In conjunction with Project Thor data is being collected and research is being conducted to enhance the current database and increase its modeling abilities. Weather and hydrology data are being recorded for flood and drought management models. Wetland and meadow chemistry is being monitored for analysis. Existing soil maps are being made more accurate in content with chemistry testing while the boundaries of the map are corrected with GPS. The focus of the project will turn to an October scene to enhance the study of temporal change. This new information will be used to expand the width and breadth of the relational database that Project Thor has created.

CONCLUSION

Project Thor, as the foundation for the larger five year project is close to being complete. It has succeeded in exposing students to a diverse range of methodologies and protocols. Project work has incorporated lab use and field work with classroom learned theory. In this aspect Project Thor is a success.

****LT.Col. Joseph Wysocki, "GPS and Selective Availability - The Military Perspective", GPS World (July/August 1991): 40.

ACKNOWLEDGEMENTS

Research for this project was expedited by the students of the Salem State College Geography Department, Jeff Logan, Douglas Briedwell and Tim Smith.

IDRISI software is registered product and/or trademark of the Graduate School of Geography, Clark University.

GPS Pathfinder Basic System and Trimble Basic Hand-Held GPS Receiver are registered products and/or trademarks of Trimble Navigation LTD.

THE SPATIAL DATA TRANSFER STANDARD (FIPS 173):
A MANAGEMENT OVERVIEW

Kathryn Neff
U.S. Geological Survey
526 National Center
Reston, VA 22092

ABSTRACT

The Spatial Data Transfer Standard (SDTS) was approved by the
Department of Commerce as Federal Information Processing Standard
(FIPS) 173 on July 29, 1992. As a FIPS, the SDTS will serve as the
national spatial data transfer mechanism for all Federal agencies and
will be available for use by State and local governments, the private
sector, and research organizations. FIPS 173 will transfer digital
spatial data sets between different computer systems, making data
sharing practicable. This standard is of significant interest to
users and producers of digital spatial data because of the potential
for increased access to and sharing of spatial data, the reduction of
information loss in data exchange, the elimination of the duplication
of data acquisition, and the increase in the quality and integrity of
spatial data.

The success of FIPS 173 will depend on its acceptance by users of
spatial data and by vendors of spatial information systems.
Comprehensive workshops are being conducted, and the tools and
procedures necessary to support FIPS 173 implementations are being
developed. The U.S. Geological Survey, as the FIPS 173 maintenance
authority, is committed to involving the spatial data community in
various activities to promote acceptance of FIPS 173 and to providing
case examples of prototype FIPS 173 implementations.   Only by
participating in these activities will the members of the spatial data
community understand the role and impact of this standard.

INTRODUCTION

Recent advances in geographic information system technologies and
digital cartography have increased the demand for digital spatial
data. Unfortunately, existing hardware and software capabilities and
the lack of data exchange standards have inhibited the transfer of
spatial data between data producers and users. The SDTS was designed
to facilitate data transfer between dissimilar spatial data bases.
Implementation of SDTS, or FIPS 173, will increase access to and
sharing of spatial data, reduce the cost of developing data bases, and
improve the quality and integrity of spatial data and related
documentation.   In addition, FIPS 173 will reduce duplication of
effort in data production and maintenance and will make a national
spatial data infrastructure feasible.

STATUS

In April 1991, after nearly 10 years of development and testing, the
SDTS was issued by the National Institute of Standards and Technology
(NIST) as a proposed FIPS. Following a 90-day formal public review
and comment period, the Technical Review Board (TRB) overseeing the
development of the SDTS met in October 1991 to arbitrate the review

comments. The document was then edited according to decisions made by the TRB. The edited SDTS was forwarded to the Department of Commerce for processing as a FIPS in February 1992; approval was granted on July 29, 1992. The FIPS 173 implementation will be effective February 15, 1993; use of FIPS 173 is mandatory for Federal agencies 1 year from this date.

FIPS 173 will serve as the national spatial data transfer mechanism for all Federal agencies and will be available for use by State and local governments, the private sector, and research and academic organizations. The success of any standard, such as FIPS 173, depends on its acceptance by the user community. Therefore, the U.S. Geological Survey (USGS), as the designated FIPS 173 maintenance authority, is committed to providing implementation support to the greatest extent possible to increase access to and use of the standard.

IMPLEMENTATION SUPPORT

The USGS has identified several key program elements necessary to promote acceptance of FIPS 173. The first element, FIPS approval of the SDTS, is complete. FIPS 173 has now entered a 5-year maintenance cycle, at the end of which it will be possible to modify FIPS 173 to meet the changing demands of the user community. Because of its modular design, FIPS 173 can be changed as the requirements for its use change. Additional approvals will be sought from the American National Standards Institute and the International Standards Organization during 1993 in an effort to broaden access to the FIPS 173 among the commercial and international communities.

Profile development is an important element for the successful implementation of FIPS 173. A profile is a clearly defined and limited subset of a standard that is designed for use with a specific type of data. FIPS 173 contains a full range of capabilities and options designed to handle a wide spectrum of possible geographic and cartographic data structures and content. Because handling this range of options is such a difficult task for encoding and decoding software, the best way to implement FIPS 173 is to define a profile with few, if any, options. Software can then be designed to handle just these options. Regardless of which options are specified for a given profile, all profiles will share important common characteristics.

The USGS plans to coordinate the development of profiles with the user community to ensure maximum consistency among all FIPS 173 profiles. The first of these profiles, the Topological Vector Profile (TVP), is in a review and testing period that will end late in 1992. The intent is to have the prototype TVP rigorously tested to ensure that it appropriately handles vector data and then to forward the profile to the NIST for FIPS approval as an amendment to FIPS 173. The USGS Digital Line Graph data will be available from the National Digital Cartographic Data Base in the TVP once it is approved as an amendment to FIPS 173. Recently, the USGS has started to develop a prototype raster profile; this effort is expected to continue through 1993. The design of a raster profile will follow a similar sequence of events as those involved in the development of the TVP - developing a draft profile and test data sets, conducting a test and demonstration period to evaluate the completeness and robustness of the draft profile, and finalizing the profile based on test results. It is expected that this raster profile will be limited to georeferenced data, sampled uniformly and in a geodetic or cartographic coordinate system, as opposed to raw sensor data. The requirement for additional profiles, such as CAD/CAM and graphics profiles, will also be evaluated in the future.

User guides are critically needed to increase the knowledge and understanding of FIPS 173 within the community. FIPS 173 describes content, structure, and format; it is not an easy document to comprehend. To address the complexity of the document, user guides need to be developed for FIPS 173, for the various profiles being defined, and for the software tools being developed. The USGS will coordinate the development of these user guides over the next few years.

Software development is an integral part of FIPS 173 implementation. Software tools, such as encoding, decoding, and display tools, must be developed. The vendor community is expected to assume a large part of this responsibility. The USGS is designing a spatial data transfer processor to support FIPS 173 transfers of its own digital spatial data, such as Digital Line Graphs, Digital Elevation Models, and Digital Orthophotos. The USGS is also developing a suite of public domain software tools designed to support the encoding and decoding of logically compliant FIPS 173 data in and out of the required ISO 8211/FIPS 123 physical file implementation. This software will be available to the vendor community to develop turnkey systems conforming to FIPS 173.

One part of FIPS 173 presents a standard model for a spatial features data dictionary as well as a list of terms and definitions for entities and attributes. This feature and attribute glossary provides a foundation for standardizing spatial features. The glossary now contains only a limited set of hydrographic and topographic features. Because this glossary is not complete, conformance is optional in the prototype TVP; however, conformance to the model is mandatory. For this part of FIPS 173 to be useful, additional terms and definitions must be included for other categories of data, such as cadastral, geodetic, and geologic, and the set of hydrographic and topographic features must be expanded. The NIST intends to establish a Spatial Features Register, designed to facilitate this effort. Information from the Federal community will be coordinated through the data category subcommittees of the Federal Geographic Data Committee (FGDC); however, the USGS intends to solicit information from the non-Federal spatial data community as well. A strategic plan to maintain this part of the FIPS 173, using the NIST Spatial Features Register, is being developed. Because the register will allow users to dynamically update the glossary, this part of FIPS 173 will evolve over time.

The USGS will continue to conduct FIPS 173 workshops and other presentations to educate the spatial data community and to promote the use of FIPS 173. Implementation presentations are planned for the major professional organizations, such as the Association of American Geographers, the American Congress on Surveying and Mapping, the American Society for Photogrammetry and Remote Sensing, Automated

Mapping/Facilities Management International, the Institute for Land Information, and the Urban and Regional Information Systems Association. In addition, the USGS, the NIST, and the Standards Working Group of the FGDC plan to sponsor implementation workshops.

The final program element necessary to promote acceptance of FIPS 173 is program coordination. This coordination involves developing support activities within the USGS, facilitating similar activities outside the USGS, and interfacing with related standards development activities in the spatial data community, both nationally and internationally.

CONCLUSION

The Department of Commerce's approval of the SDTS as a FIPS is a major milestone for the spatial data community. Although the USGS is committed to coordinating a wide range of activities designed to promote acceptance of FIPS 173, all members of this community must contribute to these efforts to ensure the success of FIPS 173. For additional information concerning the SDTS, or FIPS 173, or how to participate in these activities, please contact:

SDTS Task Force
U.S. Geological Survey
526 National Center
Reston, VA 22092
(703) 648-4566
(703) 648-4591
FAX (703) 648-5542

GEOGRAPHIC INFORMATION SYSTEMS FOR MODELING BISON IMPACT ON KONZA PRAIRIE, KANSAS

M. Duane Nellis, Professor and Head
Department of Geography, Kansas State University
Manhattan, KS 66506-0801

Jon D. Bathgate
TGS Technology, Inc.
2625 Redwing Road, Suite 312
Ft. Collins, CO 80526

John M. Briggs
Division of Biology, Kansas State University
Manhattan, Kansas 66506

ABSTRACT

The Konza Prairie Natural Research Area, Kansas is representative of the Flint Hills, one of the few extensive areas of tallgrass prairie remaining in North America. Konza Prairie is managed to preserve the site as a premier example of tallgrass prairie and to provide an array of burning and grazing (especially bison) treatments to facilitate research on the effects of burning and grazing on plant composition, primary production, consumer density and diversity, nutrient dynamics, and hydrology. Since 1988, habitat selection by bison (_Bison bison_) as influenced by spring burning of grasslands have been studied on Konza. The bison herd was introduced onto 500 hectares of Konza Prairie in October, 1987, and presently consists of 120 individuals. Utilizing a geographic information system, spatial data on land cover condition, topography, fire frequency, soils and density of bison were analyzed to determine present habitat selection pattern and to predict future spatial diffusion patterns onto newly available watersheds. Through development of selectivity indices for various spatial components and integration of these indices through a geographic information system it was possible to model spatial patterns of bison impact on the tallgrass prairie environment and to predict future diffusion patterns.

INTRODUCTION

Human impact is severe and widespread over every biogeographical realm and no ecosystem is spared and no geographic locale unaffected (Morain 1984). It is impossible to reverse the effect of human impact. However, steps have been initiated to restore and preserve some ecosystems and their inhabitants.

One of the purposes in ecosystem restoration is the actual restoration process itself. Studying a natural, self-sustaining ecosystem may reveal principles concerning the functioning of a stable system (Hulbert 1972). With an extensive knowledge of the ecosystem dynamics, effective

618

management plans can be developed (Carins and Crawford 1991). A model to predict patterns of bison dispersion would help in management decisions regarding the reintroduction of American bison to the prairie ecosystem.

The pristine tallgrass prairie ecosystem is now at risk as unique prairie grasses and wildflowers are threatened by human impact. Human impact has also caused the larger wild animals, herbivores and predators, to abandon the prairie (Kolata 1984).

The largest remnant of tallgrass prairie that escaped the plow is represented by the Flint Hills on the southwestern edge of the biome (Figure 1). Most of the Flint Hills escaped the plow because the land was too steep and rocky to be converted into an agroecosystem. Within the tallgrass prairie there are three major dominant warm season tallgrasses; big bluestem (<u>Andropogon gerardi</u>), Indian grass (<u>Sorghastram nutans</u>), and switchgrass (<u>Panicum virgatum</u>).

Fire and grazing are an integral part of the tallgrass prairie ecosystem processes that maintain productivity of tallgrass prairie (Knapp and Seastedt 1986). Beyond fire, to approximate the pristine condition, the prairie ecosystem must also contain the native ungulates, specifically bison (<u>Bison bison</u>).

The purpose of this paper was to examine <u>Bison bison</u> grazing preferences in order to develop a model to predict patterns of bison dispersion. This model could be applied when reintroducing the animal to the prairie ecosystem.

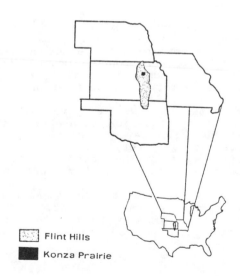

Flint Hills
Konza Prairie

Figure 1. Konza Prairie in the Kansas Flint Hills

STUDY AREA

The study area for this project is the Konza Prairie (Figure 1), located in the Flint Hills of northeast Kansas. Konza prairie is 3,488 hectares of virgin tallgrass prairie (Kolata 1984).

Fire has been reintroduced to the Konza prairie in hopes of simulating presettlement fire frequency. In October 1987, bison were reintroduced onto 469 hectares. The bison are allowed to roam free within the area comprised of five watersheds (Vinton 1990). The herd has grown to 120 animals and in the Summer 1992, an additional five watersheds were opened.

HABITAT MODELING AND SPATIAL ANALYSIS

The difficulty in analyzing ecological processes, such as foraging patterns and dispersal, has often caused spatial dynamics to be ignored. However, prediction of landscape-level phenomena, such as wildlife habitat, requires the development of spatial analytical models (Turner and Gardner 1991).

Geographic Information Systems utility in habitat modeling has been greatly increased through the use of remote sensing technology (DeWulf et al 1988; Turner and Gardner 1991). Inputting remotely sensed data into a GIS will allow for a more thorough analysis of spatial and temporal characteristics of habitats (Hodgson et al. 1987). The linkage of models with GISs and remote sensing technologies has begun, and functional models are being constructed (Turner and Gardner 1991).

Field observation were taken twice a week from April 7, 1991, until January 31, 1992. The first observation of the week was taken in the morning and the second was taken in the afternoon. The bison were found, located on the proper aerial photograph, and recorded by grid coordinates (30 meter cells) and the number of bison in each grid cell.

In the KSU remote sensing laboratory the GIS layers, the dependent (monthly bison location layers) and independent (soil, surficial geology, percent slope, slope aspect, vegetation, distance to water, and burn frequency of watersheds) variables, were developed within IDRISI, a raster based Geographic Analysis System (Eastman 1990).

Image processing of 1990 Landsat Thematic Mapper (TM) data was used to create the vegetation layer. A Transformed Vegetation Index (TVI) of the Landsat image was then classified into eight classes to form the vegetative cover layer.

Model Design

The first step in the model construction was to induce class relationships with bison distribution for every attribute layer during each month. This was accomplished by producing selectivity indices for all attribute layers

620

during each month. The selectivity index values were used to weigh each class within an attribute.

After the class relationships had been modeled, the next step was to model the attribute relationships. The relationships were determined by measuring each attribute's association with the observed monthly bison distribution. The associations were measured using the Cramer's V statistic, which is a measure of association for nominal variables. The Cramer's V statistics were utilized to rank the attributes. The attributes, whose classes were weighed, were then given a weight according to their rank. The predicted bison distribution image was produced by overlaying, with an additional method, these weighted attributes.

Attribute Association Trends

Burn treatment had the highest average association. This variable had the highest association, or was tied for the highest, during April, May June, and August. Soil type had the second highest average association, just slightly below burn treatment. It had the highest association, or was tied for the highest, during July, August, September, November, and December. Distance to permanent water sites had the third highest average association. Vegetation had the next highest average association, followed by slope aspect and finally percent slope.

It can therefore be concluded that burn treatments were the most important attribute in determining bison distribution during spring and part of the summer. During late summer, fall, and early winter, soil type was the most important attribute. Distance to water was an important attribute mostly during late fall and winter. Vegetation had a lower association (possibly because of the discrepancies within the layer). It appears that slope aspect and percent slope were not very important in determining bison distribution.

Attribute Selection Trends

After the watersheds were burned, bison tended to prefer the burned watersheds during spring and summer because the biomass vigor of warm season grasses are more lush in the course of the recovery. From spring until early summer they preferred the annual and 20-year burned watersheds. From mid to late summer, they preferred the unburned watershed that was accidently burned that spring in a wildfire. By late summer the bison have heavily grazed the burned watersheds so in fall they begin to graze on the ample forage in the unburned watersheds. During fall and winter, bison highly selected the unburned watershed with a 4-year burn frequency and slightly and moderately selected the 20-year burned watershed.

Bison were also observed having a nonrandom usage of soils. During late winter and spring they selected the Benfield-Florence (upland) soils. They had a high preference for alluvial (lowland) soils during the heat of the summer, so they could be closer to water. During fall and early winter, they preferred the Dwight-Irwin (ridge top) soils.

A more even use of the whole area in regard to distance to permanent water sites was observed during spring to mid-summer. However, there was definite trends during late summer, as bison preferred to be near water sites. In fall and winter, however, bison selected areas farther away from the water sites. This is because during this time, the bison preferred upland and ridge top soils, which are located farther from the permanent water sites.

CONCLUSION

This study has shown the need for Geographic Information Systems to adopt spatial analytical modeling. Without this capability, the power and modeling utility of GIS is greatly limited.

Even though the prediction model was developed from low level statistics, general locational attribute trends were observed. It is not known, however, if these trends are constant. Long-term research is needed to test this. Such knowledge could be applied to develop sound management plans. These plans would be of value when reintroducing bison to tallgrass prairie ecosystems, and they may be the determining factor in establishing a national tallgrass prairie preserve.

ACKNOWLEDGEMENT

The authors wish to thank the National Science Foundation for supporting this research through grant NSFBSR-9011662. We also wish to thank the Division of Biology for use of the Konza Natural Research Area for this research.

REFERENCES CITED

Carins, J. and T. Crawford. 1991. Integrated Environmental Management. Chelsea, MI: Lewis Publishers.

DeWulf, R. D., R. E. Goosens, J. R. MacKinnon, and Wu Shen-Cai. 1988. "Remote Sensing for Wildlife Management: Gian Panda Habitat from Landsat MSS Images: GEOCARTO International. 3(1):41-50.

Eastman, J. R. 1990. IDRISI: A Grid-Based Geographic Analysis System. Worcester, MA: Clark University.

Hodgson, M. E., J. R. Jensen, H. E. Mackey, and M. C. Coulter. 1987. "Remote Sensing of Wetland habitat: A Wood Stork Example" photogrammetric Engineering and Remote Sensing. 53(8):1075-1080.

Hulbert, L. 1972. "Management of Konza Prairie to Approximate Pre-White-Man Fire Influences", Third Midwest Prairie Conference Proceedings, Kansas State University, Manhattan, Kansas.

Knapp, A. K. and T. R. Seastedt. 1986. "Detritus Accumulations Limits Productivity of Tallgrass Prairie," Bioscience 36(10):662-668.

Kolata, G. 1984. "Managing the Inland Sea," Science 224:703-4.

Morain, S. 1984. Systematic and Regional Biogeography. NY: Van Nostrand Reinhold Company.

Turner, M. G. and R. H. Gardner, eds. 1991. Quantitative Methods in Landscape Ecology. NY: Springer Verlag.

Vinton, M. 1990. "Bison Grazing Patterns and Plant Response on Kansas Tallgrass Prairie." Master Thesis. Kansas State University, Manhattan, Kansas.

PRACTICAL EXPERIENCES OF USING OBJECT-ORIENTATION TO IMPLEMENT A GIS

Richard G. Newell
Smallworld Systems Ltd.
Cambridge, England

ABSTRACT

This paper describes some of the benefits and experiences gained over the last 3 years of designing and implementing a major generic GIS platform. A number of the concepts of object-orientation such as encapsulation and inheritance, as well as other object related concepts such as object-centred data models are described, together with their benefits in building a GIS. In particular the use of an object-oriented environment to integrate disparate systems is described. The context in which these experiences have been gained is a system which initially has been targeted at AM/FM, local government, retail and environmental applications and is now installed at over 60 sites in Europe. The paper concludes with an assertion that a system of equivalent power and flexibility would have taken an order of magnitude more effort had object-oriented techniques not been used. Further, the ability to customise the system for specific applications takes far less effort than in conventional systems.

INTRODUCTION

Many modern GIS's are described as being object-oriented, as their software structure, or the structure of the data that they handle is different from the systems that came before. In this paper, we would not wish to argue where to draw the line as to when a system may deserve the accolade "object-oriented", but we will describe various aspects of object-orientation in the context of where we have found them useful in building real production systems.

The technology that preceded object-orientation in database applications such as GIS was relational, and then any system whose logical data structure was based on tables earned the title "relational". There were degrees of "relationalness" depending upon how many of Codd's rules were satisfied by the system.

There does not seem to be such a precisely defined set of rules to describe an object-oriented system, but there would appear to be degrees of "object-orientedness", depending upon how many of the concepts of object-orientation are supported. It is common place these days to hear the word object-oriented to describe any of the following concepts:

- A kind of programming language and its associated programming methodology.
- Object-Oriented Analysis: a paradigm for analysis (see Booch 1991).
- A database that can hold things more complex than fields, such as complex structures, images and multi-media representations. Sometimes these are held in BLOBS - Binary Large Objects.
- A GIS database that focuses on real world objects, at a higher level of abstraction than polygons, links and nodes.
- A graphical user interface (GUI) based upon icons

In our experience, all of the above bring considerable advantages to the implementation of a GIS, particularly the first which provides a framework for implementing the others. A number of authors have described various approaches to the application of object-oriented design principles to GIS (e.g. Egenhofer 1989, Oxborrow 1989, Worboys 1989, Aybet 1992)

Just as in the case of the relational model, taking the first step into tables gave considerable advantages, so the first step towards object-orientation also provides advantages.

What is the first step into object-orientation for GIS? May one propose that it is moving away from the CAD concepts of tiles (sheets), layers and graphical features to centre ones attentions on real world objects, their attributes, relationships and behaviours. This is in fact the stage that many of today's object-oriented systems have got to. It does bring considerable advantages over the previous approaches that were based on CAD principles, but it does not incorporate any of the principles of object-oriented programming. Thus there would appear to be three levels of object-oriented system as follows:

- object centred data models
- encapsulation of behaviour
- inheritance between classes

Previous authors, ourselves included, have protested that the first level does not deserve the title "object-oriented", but as so many people are now using the term it seems pointless to resist. Here, based upon our experience, we will focus on the benefits and relevance of these concepts.

Our experience is based on building a generic software platform, Smallworld GIS, and implementing it in customer sites throughout Europe during the last two years. In a previous paper (Chance et al 1990), we published our philosophy and can now speak with the benefit of 2 more years of hindsight.

ARCHITECTURE OF SMALLWORLD GIS

Smallworld GIS is built using an interactive object-oriented programming language, Smallworld Magik. Magik takes the roles of both the macro language and systems language in more conventional systems. The language is used for far more than the occasional writing of a macro or for minor customisation: most of the system is written in it.

The Magik environment is underpinned by a virtual machine, written in C, which contains a full and extensible set of primitives for handling graphics and interaction, database, and remote access to alien processes, such as external database systems and analysis systems.

All persistent data is front ended in the object-oriented environment by a "virtual database", an object that understands queries and provides a seamless interface to all databases. The virtual database allows applications code to be independent of the specific database.

The system itself contains a version-managed datastore for handling multiple users involved in long transactions (Easterfield et al 1990).

The Magik language has a syntax which is procedural in style with familiar and readable constructs for conditionals, expressions, loops, and procedure calls. The message expression is formed by a concatenation of the object class and the message. The semantics have primarily been inspired by Smalltalk, Lisp and Clu. In particular the ability to define iterators and iterator methods (borrowed from Clu) is particularly powerful.

Another powerful facility is the ability to define dependencies between objects so that a change in one object results in messages being sent to dependent objects to update themselves. For example object editors update themselves automatically in response to a graphics hit. Another use for this is in the implementation of derived fields in a database, for example a length field may be updated automatically every time a geometry centre-line is changed. This is equivalent to using triggers in SQL.

New classes can be defined by a special procedure which creates the first instance (the exemplar) of the class. All further instances are cloned from the exemplar.

Magik is weakly typed. It is now our firm belief that the small loss in performance and the absence of compile-time type checking available in strongly typed systems is a price worth paying for the considerable additional flexibility gained by run-time message evaluation in a polymorphic system.

Magik supports both encapsulation and multiple inheritance, thus it supports the building of object-oriented systems comprehensively.

Of equal importance to development productivity is the interactive development environment in which new classes and their methods can be defined dynamically on-line.

The system can be developed incrementally, rarely is it necessary to rebuild the whole system. Classes and their methods can be explored on line, including the inheritance hierarchy and dependency hierarchy of any class.

The development methodology is based more on a sequence of prototypes which can be evaluated, resulting in a production system.

OBJECT-CENTRED DATA MODELS

Object-centred systems are normally based on a data model out of the entity-relationship mould, in which objects are modelled by their attributes and their relationships to other objects. Such models are easily accommodated in the relational model. Objects may be related to spatial entities such as areas, chains or points.

A particular problem arises in representing these spatial objects in a relational structure in that they are of variable size and also they need to be indexed in 2 dimensions (2 dimensional range queries) which requires spatial indexing techniques (Newell et al 1991). The solution is either to go to a different model, or to compromise the relational model and hide the internal structures behind an acceptable interface. This is a good example of where encapsulation can be used.

In our work, the fundamental persistent storage is tabular, but using the principle of encapsulation (see below) the table/record structure is made to look like an object data structure. At the lowest level, a table looks like an indexed collection, a record looks like a slotted object and a field is a slot. Tables understand messages about relational algebra. Any program written that works for an indexed collection will work for a database table also.

In our system, a spatial characteristic of an object, such as a "building outline", is treated as an attribute or slot. Other systems may model the object by saying a building "is a" polygon.

By focussing on the objects, it is straight forward to have one object with several alternative representations. For example, a building may be represented by both an outline and a property seed point. It is also straight forward for different objects to share the same geometry. Models which are focussed on the geometry normally allow a limited degree of sharing such as two polygons sharing a common dividing boundary. More complex sharing such as a boundary being shared by a building, land parcel and parish is either implemented by duplicating the geometry or by heavy customisation.

In our system, a linear object may be represented by a chain comprising multiple links. Two or more roads with a shared section can easily be represented in such a model.

In summary, an object-centred data model has the following benefits:

- One object may possess multiple alternative geometries, which may be useful for generalisation.
- One geometry, such as a node, link or polygon may be shared by any number of objects.
- Explicit topology for objects from many themes (classes) can be modelled, one is not limited to single theme coverages.
- It is easier to provide generic support for a much richer datamodel than is available in the geometry centred data model.

ENCAPSULATION AND POLYMORPHISM

Encapsulation insulates the programmer from the internal structures of his objects. This is radically different from the relational approach, in which all of the semantics of the model are provided by the programs which operate on the data. Encapsulation means that programs can be written which are independent of the detailed internal implementation of an object. Polymorphism allows functions and operators to be defined and used on different data types.

One of the benefits of encapsulation is that changing the internal implementation of a class has no effect on the rest of the system. We have found that we can make fundamental changes to the system data models without any changes being necessary at all to other parts of the system. The chances are that if there are excessive side effects to a change then there is something wrong with the system design - and indeed we have encountered this problem on more than one occasion. Methods that are private to a class, as opposed to those that are public need to be clearly identified.

Polymorphism combined with encapsulation is extremely powerful. New classes can be defined, and regardless of their internal representation, provided the new class is given all of the required behaviour, no change is needed to the rest of the system in order to accommodate it. For example, a new class needs standard methods, many of which can be inherited from generic classes (see inheritance below).

Examples of common inherited methods on an object such as a house are as follows:

house.draw(a_graphics_system)
the house has the behaviour to draw itself in the supplied graphics system. The message draw essentially asks the house for its geometric properties which it then uses to draw on the screen or plotter.

house.delete
the house knows how to delete itself

house.goto
centre the house in the middle of the screen. Such a facility may be used after finding the house via a query. The **goto** message provides a very general gazetteer capability using any object instance to locate the screen display.

house.select(a_predicate)
the result of this is an object which contains the results of a query, namely a set of houses.

This last example, is a very good example of the power of encapsulation, since it provides an easy way to implement lazy evaluation of the set that results from the query. In our system, the internal representation of such a result is to hold the predicate unevaluated. Thus if the result of the above is held in an object called **house_set,** then further selections can be invoked as follows:

house_set.select(another_predicate)

Internally, the system builds up a complex predicate, which only gets evaluated, and possibly optimised, when it is necessary to do so.

This approach can be generalised to more complex analysis and transformation functions. For example, to generate a contour map from a grid, representing a density function, which has been derived by filtering a set of incidents (e.g. held in an accident_table which had been selected from a database of incidents could be achieved as follows:

```
a_predicate  <<  predicate.new(:age,:gt,65)
accident_selection  <<  accident_table.select(a_predicate)
grid_model  <<  accident_selection.filter(a_filter)
contours  <<  grid_model.contour(a_height_iterator)
```

In our implementation of the geometric primitives, we have included a set of behaviours between pairs of geometric objects which may interact topologically. Thus a water distribution main centre-line "knows" what to do if it is inserted and its ends collide with another distribution main. A land parcel boundary "knows" what to do when it coincides with another land-parcel boundary. Thus while the user may create geometry as if it were in a CAD system, the geometric primitives "know" what to do in order to modify the internal structures when geometric coincidences and overlaps occur.

This behaviour is driven by a set of user-defined rules which define what happens when pairs of geometries from the same topologically related group (or "manifold") interact.

A summary of the main benefits of encapsulation with polymorphism is as follows:

- new classes with the same signature (behaviour pattern) can be added to the system and then take their full part in the system without any other modifications being made.
- programs can be written which are independent of the type of data upon which they operate. This helps reuse.

- changes to the internal implementation of a class can be localised, provided the interface to the class (the public methods) is unchanged. This helps to reduce maintenance costs.
- external systems databases can be encapsulated within an agent object so that programs written within the GIS are defended from the differences between different external systems.

In addition to these, a major benefit that we have experienced is the ease with which interfaces to external systems such as DBMS's, analysis packages and specific user's programs written in conventional languages, can be represented in the GIS by an agent object which encapsulates the behaviour of the external system. Within the GIS environment, neither the programmer's interface, nor the user interface is compromised. The object-oriented environment provides a seamless integration across external systems.

INHERITANCE

Inheritance is an extremely powerful aspect of an object-oriented programming system. It is a mechanism for factoring out behaviour and structure common to different classes. This means that code is reused to a far greater extent than it is in a procedural system. Smallworld Magik supports inheritance of both structure and behaviour. Both single and multiple inheritance are supported so that new object classes can inherit data structures and behaviours from multiple parent classes.

Smallworld Magik supports the concept of mixins. A mixin is a repository for common behaviours from which other dependent classes can inherit. Mixins are a major help for reuse of code, this is especially useful when multiple inheritance is used.

One of the most common applications of inheritance is in modifying or extending an object editor. As a consequence of creating a new object class, the system generates an editor for instances of that class automatically. The programmer may then add to or modify the editor without replicating any of the unchanged behaviour.

Another common application is in the use of application templates. A generic water model for example can be specialised for specific customers by subclassing. Thus any changes made to the generic template will always manifest themselves automatically in each derivative system.

The GIS has standard network and area classes, so specific applications such as water, gas and planning inherit from these classes without replication.

The main benefits of inheritance are:

- code reuse is maximised so that new systems are produced more quickly and systems are easier to maintain.
- changes made to a parent class are automatically propagated to all classes that have inherited from it.
- because code is not replicated, it is far easier to maintain consistency in both the internals of the system and in the user interface.
- the ability to customise and develop the system for specific applications is far quicker than with conventional programming.

OBJECT-ORIENTED PROGRAMMING

Programmers familiar with the procedural approach take time to adapt to thinking in terms of objects with behaviour as opposed to functions that act on data. The learning curve for programmers new to object-orientation to adapt to the new way of thinking, where one is for ever making new objects that never seem to do anything, is typically between 3 and 6 months; a few never get the hang of it. Whereas in the procedural approach, typical procedures can be quite long, a typical method in an object-oriented program is quite small, often less than 5 lines and the vast majority less than 20. Long methods are actually doing something procedural.

It is a common reaction for newcomers to worry about when does anything happen and where does it happen. Instead of a procedural script which may be represented by a flow chart or program flow diagram, we have a large collection of disconnected components, which only become connected via message passing at the time an action is commenced. This is done by sending a message to the first object in the action. As messages travel from object to object, the system determines which methods should be invoked. In our system, which is weakly typed, this occurs at run-time.

Anything that can be done in an object-oriented system can also be done in a procedural system, especially if the procedural system supports polymorphism. It is just that it is generally more difficult, and some things are extremely difficult. It is almost always the case that object-oriented programs are very much shorter than the equivalent procedural programs.

We would not claim that object-oriented design is easy, it sometimes takes a lot of deep thought and clear thinking to design the correct inheritance hierarchy. There are very few programmers on the market (at least in Europe) that have an extensive experience of object-oriented programming, and the majority have none beyond a classroom exercise in Smalltalk. The educational establishments are still not paying enough attention to the teaching of object-oriented methods and so it is a long hard grind for an organisation to climb the learning curve. We can also advocate that for certain tasks, a procedural approach is more appropriate.

CONCLUSION

Following nearly 3 years experience of implementing a generic GIS using an interactive object-oriented programming environment, we are left in no doubt as to the power of the approach. Even without object-orientation, we can recommend the use of object-centred data models and complete and powerful interactive front end languages for the implementation of major systems such as a GIS.

Object-oriented programming concepts such as polymorphism and inheritance make it far easier to write general purpose programs which can be heavily reused in different applications and contexts. This aspect of reuse yields much shorter lead times and reduced costs in producing benchmarks and new applications.

Perhaps the most convincing argument for the power of object-orientation is that we have managed to produce a successful generic GIS platform using about 12 man years of development effort. Our previous experiences of producing large but less complex systems in the past leads us to believe that it would have taken an order of magnitude more effort to produce an equivalent system using conventional techniques.

REFERENCES

Aybet, J (1992): Object-Oriented GIS: What Does it Mean to GIS Users, EGIS '92 Conference Proceedings, Munich, March 1992.

Booch, Grady (1991) Object-Oriented Design with Applications, Benjamin/Cummings Publishing Company Inc., ISBN 0-8053-0091-0.

Chance, A., Newell, R.G. & Theriault, D.G. (1990). An Object-Oriented GIS - Issues and Solutions, EGIS '90 Conference Proceedings, Amsterdam, April 1990.

Egenhofer, M.J., and Frank, A.U. (1989), Object-Oriented Modelling in GIS: Inheritance and Propagation. Auto-Carto 9, Baltimore, April 1989.

Newell, R.G., Easterfield, M. and Theriault, D.G. (1991), Integration of Spatial Objects in a GIS: Auto-Carto 10, Baltimore, March 1991, Vol. 6, pp408-415.

Oxborrow, E. and Kemp, Z. (1989), An Object-Oriented Approach to the Management of Geographical Data. Conference Proceedings: Managing Geographical Information Systems and Databases, Lancaster University, September 1989.

Worboys, M., Hearnshaw, H. and Maguire, D. (1989) The IFO Object-Oriented Data Model. Conference Proceedings: Managing Geographical Information Systems and Databases, Lancaster University, September 1989.

USING DIGITALLY COMPILED 1:10 000 DATA

FOR 1:50 000 AND 1:250 000 DATASETS

Mary Ogilvie
New Brunswick Geographic Information Corporation
P.O. Box 6000, 985 College Hill Road
Fredericton, New Brunswick, Canada
E3B 5H1

Rao Irrinki
Department of Natural Resources and Energy
P.O. Box 6000, HJF Forestry Complex, Regent Street
Fredericton, New Brunswick, Canada
E3B 5H1

ABSTRACT

This paper looks at the data sources being used in the province of
New Brunswick, Canada, to produce a comprehensive set of digital
topographic databases. It will review the specifications for the aerial
photography and digital compilation, and then review the procedures used in
generalization and filtering of this data to produce provincial coverage at not
only the 1:10,000 scale, but also in datasets suitable for work at 1:50,000 and
1:250,000. Resources used in manipulating this data would be noted.

INTRODUCTION

The province of New Brunswick is in Eastern Canada. It shares a border with
the State of Maine, and is about the same size. There has been a long history
of integrated surveying and mapping in the province, beginning with the
Atlantic Provinces Surveying and Mapping program (APSAMP) in the 1960's,
through the LRIS program in the 1970's and 1980's to the activities carried out
by the New Brunswick Geographic Information Corporation, created in 1990.
These various programs put in place a comprehensive system of survey
control, topographic base mapping, and property mapping for the entire
province.

The administrative nature of the province makes it easy to handle geographic
information on a provincial basis -- since the 1960's, counties have not been
an important administrative or political unit, and most administrative
functions, and survey and mapping functions, have been carried out on a
provincial basis.

This paper focuses on *one* of the benefits of having a comprehensive 1:10,000
database for the province - that this database can be exploited to provide
digital data not only to users of 1:10,000 data, but also to those who prefer to
work in scales of 1:50,000 and 1:250,000. It is not a paper that explores the
nuances of generalization theory.

630

THE GEOGRAPHIC INFORMATION FRAMEWORK

The Geographic Information Framework for the province of New Brunswick is essentially in place. A network of 20,000 survey control monuments has been established and maintained throughout the province since 1977. Topographic base mapping in orthophoto format at the 1:10,000 scale was completed in 1983, and topographic base mapping for the urban areas was completed at the same time. The urban maps are line maps and are at larger scales than the orthophoto series. A digital line map series from new aerial photography at the 1:10,000 scale was begun in 1983 and is scheduled for completion in March 1994. Property mapping, which consists of a hard copy map, overlayed on the topographic series, showing the boundaries of all parcels and the property identifier, has been completed since 1986. It has been maintained since that time, as has the attribute data files containing ownership information linked to the property maps. In 1991 work was begun to convert all these hardcopy property maps to digital form. That will be complete in March 1994 as well.

An automated assessment system has been in place since 1980. This system, recently upgraded, does automatic calculation of assessed values based on market trends and input of information from inspection visits. The files in this system are linked to the property mapping, and provide the province with the tax role and notification of assessed values.

This geographic data is distributed to the public through 5 regional offices which function not only as the centers for distribution of data, but also as the registry and mapping offices for the province. Distributed access to some of the data is achieved through subscriptions to microfiche systems. Distributed access to some of the data through electronic terminals is likely to happen within the next 12 months.

The New Brunswick Geographic Information Corporation is the agency responsible for all this information and activity. As its name implies it is a Crown Corporation. This feature requires the Corporation to become self-sufficient by 1995, but does allow it to keep revenues generated from the sale of data, and to invest these revenues in additional products or services to its clients.

STATUS OF MEDIUM AND SMALL SCALE MAPPING

As must be obvious from the preceding comments the collection of topographic data for use at the 1:10 000 scale is considered a provincial responsibility in Canada. Recommendations from a Task Force in the 1980's on federal mapping responsibilities made it quite clear that although there was a role for federal agencies in the production of 1:1 000 000, 1:500 000, 1:250 000, 1:100 000, and 1:50 000, that federal agencies should not use valuable resources trying to provide mapping at scales larger than 1:50 000. Following adoption of this recommendation the Canadian government no longer produces data at scales of 1:25 000 or larger.

Complete coverage of the province at 1:50 000 is available from the federal government in hard copy maps. Only 25 % of the province is covered in federal digital data at this scale. The federal government has instituted a program to create the 1:50 000 digital database for the country in cooperation with the provinces, and New Brunswick has had a limited amount of data collected under this program.

Federal coverage at scales of 1:250 000 is also available in hard copy. These hard copy maps were scanned, and the project to structure them was sponsored jointly by the federal government and the province, but the source data is probably 30 years old.

THE IMPETUS FOR UNDERTAKING THE SMALLER SCALES

<u>Why produce this data provincially?</u>
Figure 1 shows the province in 10 000, 50 000 and 250 000 windows.

FIGURE 1
SHEET LAYOUT - 1:250 000, 1:50 000, 1:10 000

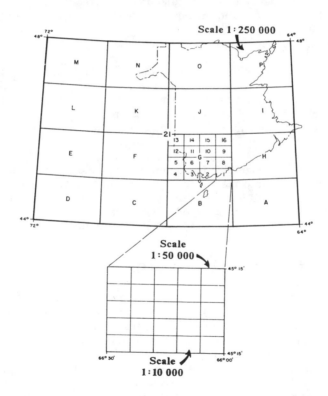

There was a requirement from the resource departments for digital data in 50 000 windows. The Mineral Resources Branch of the Department of Natural Resources and Energy was particularly keen to have access to data in these windows to use as a base for its geological data. They were keen to have not only digital data in these windows, but also to have data that was as current as the 1:10 000 data being collected by the province. There was also a desire to have data in 250 000 windows for smaller scale portrayal of geological information. This data was not available from the federal government.

Why use the 1:10 000 data?
Several years ago the Mineral Resources Branch of the Department of Natural Resources and Energy began some experimentation with merging 25 of the 1:10 000 files into a 1:50 000 window to determine the difficulties involved. This work demonstrated that the approach was certainly feasible, but needed some software improvements to make it efficient.

Given the economic situation in the province, the alternative of funding separate programs for producing 50 000 data and 250 000 data was out of the question. And even if funding could have been found to produce additional series, the costs of maintaining three separate series would have been extravagant.

In any event, major users want upwardly compatible data - data which can have the same content, same level of timeliness, etc, no matter what scale is being used.

THE PROCESSING INVOLVED

This section will describe the process used to create the 1:10 000 database, the software used in this data collection, and and the generalization and filtering processes done to the 1:10 000 dataset in order to create the 1:50 000 dataset and the 1:250 000 dataset.

The original 1:10 000 data was collected by digital stereo compilation from aerial photography flown specifically for the 1:10 000 mapping project. The work was done by LRIS, in Summerside, Prince Edward Island, using conventional stereoplotters connected to CARIS workstations. The positional horizontal accuracy is +/- 1.5 m to +/- 2.5 m. Vertical accuracy is +/- 2.5 m. 90 % of all well-defined features fall within these positional accuracies.

The features of CARIS that make it attractive for base mapping applications are a compact data structure, and an ability to transfer data relatively easily to other systems such as ARC Info and Auto CAD. The CARIS data structure uses source ID's, themes, and features. These make it fairly simple to define a new series based on a sub-set of an original series, and to selectively retrieve themes and handle them appropriately.

User surveys were conducted among the major potential government department users to determine what themes they required in smaller scale data. They were also asked what resolution they felt was right for the various scales. From this information decisions about content at these scales were made, and macros written to pull out the data for the appropriate scale.

In addition some additional software was developed to help deal with the placement of text on the screen and any hard copy outputs. This software removed multiple occurrences of the same name, and assisted in the placement and orientation of the label on the map face or screen.

Filtering of the data was also done automatically. No complex algorithms were employed here - the intent was only to further compact the data to make it quicker to display and plot , and less hungry for disk space. Generally the rules employed in the filtering have the effect of creating datasets that are about the same density as the 1:10 000.

The larger series, the 1:10 000 series, will be completed by March 31, 1994. By May of that year, the 1:50 000 series will be complete, and by June the 1:250 000 series will be done.

Figure 2 shows the steps in the procedures, with some estimate of the cost for completing this for the province.

FIGURE 2
STEPS IN COMPILING THE THREE SERIES

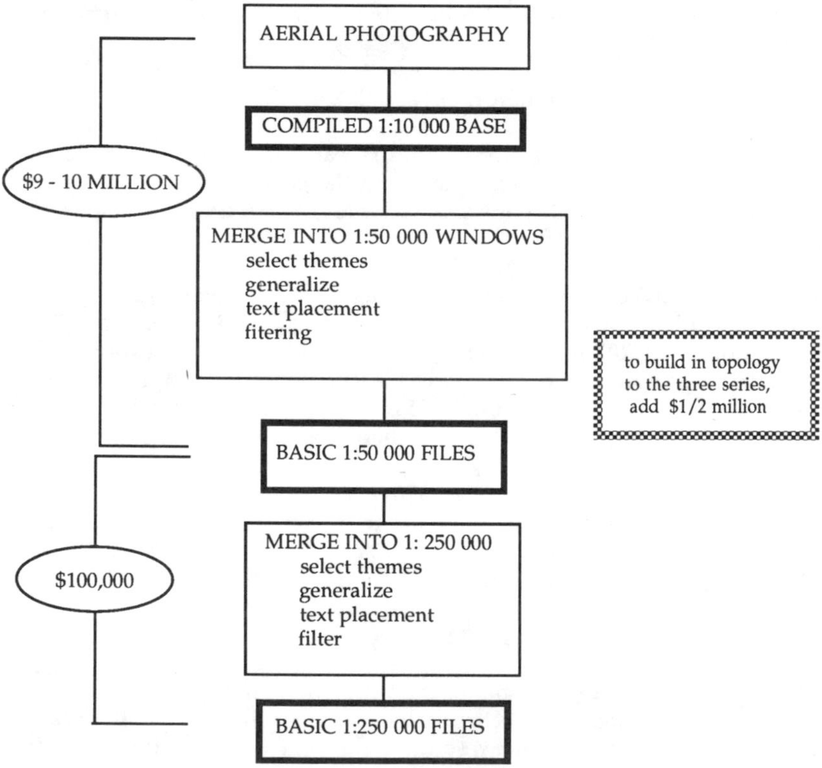

CONCLUDING REMARKS

It is clear from our experience that it is feasible to create datasets at scales of 1:50 000 and 1:250 000 from data collected for use at 1:10 000 scales. In terms of cost of production, cost of maintenance, and convenience for the user, it is the best way to produce these datasets. Most of the procedural issues with respect to creating the new series have been sorted out. What we will be dealing with next are the procedural issues associated with maintaining all series from the same update information.

IMAGE PROCESSING OF GRAPHS USING GIS FUNCTIONS AND PROCEDURES

Micha Pazner, Assistant Professor
Department of Geography
The University of Western Ontario
London, Ontario N6A 5C2
(519) 661 2111 x4501

Wenhong Liu, Ph.D. Student
Department of Geography
The University of Western Ontario
London, Ontario N6A 5C2
(519) 661 2111 x4501

ABSTRACT

The paper explores the application of GIS image processing functions
and procedures to non-spatial data variables that are graphed and
represented in pixel (image) format. The question we are asking is: can
certain GIS tools lead to new types of data graphs? As opposed to
familiar existing graph forms (e.g.: scatter plot, bar chart, pie chart,
etc.) which are line-based "vector" graphs, the new graphs are area or
"raster" based. We term this concept GRASP (GRAph Space
Processing).

THE GRAPH SPACE PROCESSING (GRASP) CONCEPT

The paper presents the GRASP concept through an example of how GIS
image processing functions and procedures can be applied to bivariate
data in graph form. The example is drawn from the area of digital
image classification of remote sensing data. We show how pixel based
image data can be used to create, manipulate and analyse *images of bivariate scatter patterns*. The visual interpretation of the patterns in
multivariate space is used to guide the classification process.
Specifically we examine two pixel based scatter graphs which we term
ScatterGrid and *ScatterSpread*. The paper provides a description of our
initial experiments. The results are being assessed to establish whether
the new graph forms and methods can be used to identify irregular
shaped and more natural class boundaries that are superior to the
regular elliptic equiprobability contoura of the maximum likelihood
classifier model. In this exploratory data analysis exercise, raster
pattern interpretation replaces the use of statiscal measures.

The significance of this work is in extending the applicability of unique
GIS operations and procedures to areas of information science that lie
outside of the relatively specialized realm of spatial data handling. The
work is expected to lead to contributions in the area of inforgraphics by
aspects of GIS whose development took place in a spatial data context.

636

THE 2-BAND *SCATTERGRID* OF REMOTE SENSED DATA

We now give an example of how GIS image processing functions and procedures can be applied to bivariate data in graph form. Our example happens to be drawn from the area of digital image classification of remote sensing data. The data used is Landsat TM satellite data of London, Ontario, collected on June 19, 1985. Spectral band 2 (Green) and 4 (Near Infrared) were selected. This reflectance data is represented by digital numbers in the 0-255 range. It is common to display this data with black for zero, turning gradually to gray and white with ascending value.

Remote sensing image classification is used for transforming reflectance information in satellite images into land cover feature information in the form of thematic maps. The classification stage is preceeded by sampling and analysing the data. Graphic and statistical analysis of training sites are used to guide the classification process, which is followed by testing of the accuracy of the results. As part of the statistical analysis of training sites the data are routinely put in graph form, including bivariate scatter plots, 3-axis scatter plots for comparing 3 spectral bands, and co-spectral plots. In these scatter plots the reflectance value range in each band is represented by an axis. The purpose is to identify clusters of coinciding value sets which represent distinct land cover features. These may appear as planes in a 2-D scatter plot, or as "clouds" in a 3-axis graph. One common limitation of this representation is that often a black point or dot marks a co-occurence, rather than a number specifying the frequency of that particular numeric combination, or a visual cue of magnitude such as a gray pattern or color sequences. Richards (1986) refers to these clusters in multi-spectral space as information classes. We note that the traditional scatter plot, much like other graphs, is a point and line based graphic rather than an area based representation using pixels. We now show how area and pixel based pictorial representations can be used to create, manipulate by applying image processing techniques, and analyse scatter patterns in a multivariate space. Specifically we examine two pixel based scatter graphs which we term *ScatterGrid* and *ScatterSpread.*

PREPARING THE DATA: CREATING THE *SCATTERGRID* GRAPH

To begin with, training areas must be selected in the image. This is a standard initial step in supervised remote sensing image classification. Figure 1 presents a previously published diagram (Lillesand and Kiefer, 1987) of the classification procedure. In general, at least two training areas should be chosen for each land type or "class". For convenience, training areas are often rectangular sets of pixels that are aligned with the grid's X and Y axis. A typical training area will contain a statistically significant number of pixels, eg. 10, 20, 50, 100, or more—as the case may predicate based on such factors as spectral homogeneity, confidence of positive class identification, and other factors. For instance, small samples of 9 pixels (arranged in a 3x3 array) may be perfectly adequate for a Deep Ocean class. On the other hand, more spectrally variable vegetation classes may require larger samples, such as a 8x12 array, for a more optimal characterization of their spectral signatures. In our example, training areas were selected for the following classes: water, forest, urban, natural vegetation. and bare soils.

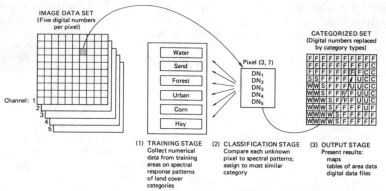

Figure 1: Basic steps in supervised classification.
Source: T.M. Lillesand & R.W. Kiefer.

The training areas were located on 2 of the LANDSAT TM images: Band 2 in the Green portion of the visible spectrum and Band 4 in the Near Infra Red (NIR) portion of the electro magnetic spectrum. Three or four training areas, chosen throughout the scene, were selected for each class. In our experiment we used the MAP II Version 1.5 Map Processor program (Pazner, Kirby and Thies) on the Macintosh. The band 2 training rectangles were copied and pasted in a systematic way to a blank map. The same process was repeated for band 4, taking care that the training areas were: (i) taken from identical locations in the band 2 and 4 images, and (ii) were pasted in the same new locations on the 2 new blank maps.

Once the two training images (for band 2 and band 4 data) were completed, and contained spatially registered sets of training areas for each of the classes, the data was in a form that was conducive to generating a scatter plot, or in our case, a ScatterGrid graph. It is possible to represent the 2-band multispectral space as a 256x256 data matrix image. A ScatterGrid is effectively created by assigning the frequency (or count) of all the spatially co-occurring pairs of reflectance values in the two training images, to the appropriate entries in the ScatterGrid array. We note that it is possible to generate a scatter grid using the full satellite images, rather than just the training sets. The resulting scatter grid 'cloud' may not reveal any cluster structuring, other than perhaps a cluster of values for water and another cluster for all the rest. However, this scatter grid is used to generate a "friction" map, as is explained later on.

MAP II 's Combine operation was used to generate a new unified training area map. Combine looks at the values of each of the gray-scale maps of band 2 and band 4 on a cell-by-cell basis, and records all of the

638

existing combinations of reflectance values. In the resulting map, the pixel values are simple serial numbers of combinations, while a text field in the legend contains the ordered pair of input values which generated the particular combination ID number. An additional "count" field in the legend provides the number of occurences throughout the entire area of the maps of that particular combination. The information that needs to be recorded in a scatter graph includes only the information in the count and text fields of the map's legend. The text field combination locates the scatter diagram coordinate, and the count entry is the Z value in the graph for that coordinate. Whereas the map depicts cartographic space, the scatter diagram depicts a bivariate feature space, in our case this is a two-band spectral space. In order to generate a ScatterGrid graph, we used a small separate program called "ScatterGrid" which was created for the purpose of this experiment. The resulting ScatterGrid raster image was then imported back to the MAP II program.

As a computer image, the ScatterGrid graph is now a picture which can be colorized for improved visualization. Figure 2 shows a graytone version of the scatter grid graph for band 2 and band 4 created by the ScatterGrid program. In this image, the [X, Y] coordinate of each "scatter pixel" is determined by the brightness values of a satellite image pixel in two bands, 2 and 4 in our case. The value of each pixel in the scatter grid is the number of satellite image pixels that share a specific combination of the two brightness values in band 2 and band 4. The most frequent co-occurences appear as white. This graph illustrates the clustering tendencies and the variability of the co-spectral properties found within each cover class, as represented by the training area samples.

Figure 2: ScatterGrid—A scatter plot in raster GIS image format

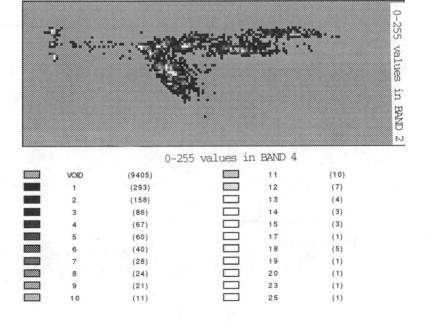

0-255 values in BAND 4

	VOID	(9405)		11	(10)
	1	(293)		12	(7)
	2	(158)		13	(4)
	3	(86)		14	(3)
	4	(67)		15	(3)
	5	(60)		17	(1)
	6	(40)		18	(5)
	7	(28)		19	(1)
	8	(24)		20	(1)
	9	(21)		23	(1)
	10	(11)		25	(1)

639

In Figure 2 above, one scatter pixel represents a set of one or more satellite image pixels with the same two-band digital value combination. It is now assumed that there is a core pixel for each information-class cluster. The other pixels belonging to a cluster are most densely distributed around the core pixel. Based on this assumption, the scatter grid space can be partitioned into information-class areas according to the clustering pattern around core pixels. The boundaries of these areas can be used as class decision thresholds by an image classification program. The output of the classification program would be a thematic image where class types replace the reflectance value in a given spectral band in the original satellite image. We now provide a model or procedure composed of a series of GIS image processing functions that can be used to partition the area of the scatter grid into information classes. The class threshold values can then be used to specify the decision thresholds for the classification program.

1. Identifying a Core Pixel for Each Information Class

The cores may be visually identified from the training site data in the scatter graph. To better highlight these cores visually, a digital image processing filtering function, such as an averaging moving box filter, can be applied to the scattergrid graph. The result of this operation is a new image in which each pixel value is set to the mean value of its immediate neighbors. The Scan filter operation in MAP II was used for this purpose. In the resulting map, the pixel with the highest value in each group is identified as the core for that group. Cores were identified for all nine clusters. These core pixels are isolated (using the Recode operation) and extracted onto a separate layer.

Figure 3: Using a filter to identify cluster cores

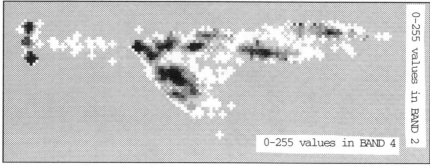

2. Creating Distance Buffers Around Cores

In the scattergrid graph, pixels other than core pixels are assigned to a group of pixels according to their proximity to the core of each land-cover type. The proximity of two locations in the cartographic plane is assumed to be the length of a straight line. This can be achieved by performing a buffer spreading operation ("Spread" in MAP II). Spread creates a buffer map outlining distances to the cores.

However, as the distance increases, the density of pixels around the core decreases at different rates. Some pixel groupings are clustered more tightly than others. It would be desireable to take pixel density into account when calculating distance buffers. The buffer should sweep outward over a hypothetical surface that undulates with changes in density. The higher the count of a two-band brightness value combination that occurs on a location (with a specific coordinate in the scattergrid), the less difficulty is associated with "spreading" through that location. The difficulty of traversal can be represented by a friction map layer used in conjunction with the Spread operation. (Pazner et. al. 1992 [Tutorial]: pp. 88). We next describe how a friction layer, such as is shown in Figure 4 below, is produced.

Figure 4: A Friction map in bi-variate space

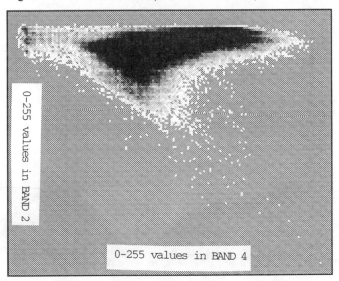

First, a scatter grid using the full satellite images, rather than just the training sets, is created. The friction map is then generated from this scatter graph, by assigning a new value to each pixel. Blank pixel locations are set to a sufficiently large value that will block spreading. Pixels with non-blank values will be assigned the reciprocal of their value (which represents the number of pixels with the same two band combination). In practise the friction value is multiplied by a constant in order to convert fractions into integers that can be handled by the program.

Figure 5 is the buffer map generated from spreading over a friction map. The value of each location is the distance from that point (ie. the center of the pixel) to the nearest core. All pixels can now be associated visually with different information classes according to their proximity to a specific core. Colorization of the spread result can aid substantially in the interpretation of this map, which we term a *ScatterSpread* graph.

Figure 5: ScatterSpread graph—A Spread operation with friction on ScatterGrid

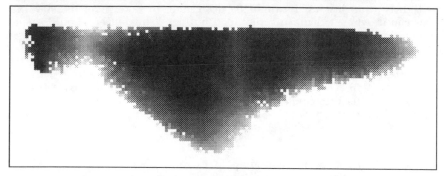

3. Partitioning the Area Into Different Information Classes

Based on this information, it is possible to delineate boundaries between the different information classes. Sets of rectangles of different size and proportion can be fitted into each information class (Figure 6). The use of rectangles is well suited to the grid based representation of the scatterspread graph. The upper-left and lower-right coordinates for each rectangle provide the lower and upper threshold values for a parallelepiped classifier.

Figure 6: Possible parallelepiped class boundaries

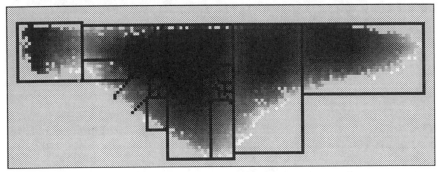

4. Parallelepiped Classification

The Parallel operation functions as a parallelepiped classifier. These classifiers are based on a simple, one class at a time, multiband value-range set intersection (ie. conjunction) operation. The sets are defined as a collection of X and Y values ranges in several spectral bands. If a particular pixel satisfies all the specified ranges, it is assigned to a particular class. Actual algorithms vary in terms of support of multiple classes, versus one class at each pass, and in detecting set conflicts that would lead to errors of comission.

Is is emphasized that image classifiers operate in multispectral <u>image space</u>. Spectral space representations such as the ScatterGrid and ScatterSpread can be used to assist in determining optimal value ranges that will be specified in the classification function. This is the purpose in delineating the rectangles (in spectral space), as described in (3) above.

We note that current classification procedures do not generally rely on information in graphs in grid form; instead, they rely on line graphs such as scatter plots and statistics in numeric and/or tabular form. The advantage of raster based graphs is the ability to manipulate the information further, by way of image processing functions and visualization, in order to extract additional implicit information. For example, using a raster graph we were able to apply a spreading (buffering) operation with friction to derive the ScatterSpread graph. This graph was conducive to delineating the rectangles used by the classifier.

In our example, Parallel is used to isolate pixels that satisfy the set intersection of specific value ranges in the band 2 and band 4 images. Parallel assigns a user-specified value to these pixels. The result from the operation is a new map layer identifying a specific class or land type.

5. Creating a Thematic Map

All the land type maps derived one at a time from Parallel, can be overlayed on a new single map using the Combine operation. This thematic map is the end product of the image classification procedure. The map shows the spatial locations and distribution, pattern, etc., of the various classes. A portion of the resulting thematic map is shown in Figure 7.

CONCLUSION

We have just began experimenting with the GRASP notion. It is clear that, as with all exploratory data analysis tools and techniques, there is potential for use and danger of abuse. We are still in the process of refining the technique and in the midst of a comparative classification accuracy study to determine the strengths and weaknesses of the statistical versus the image processing classifiers. We feel that applied correctly, image processing of graphs has the potential to provide insight of the let-the-data-speak-for-itself type which is not available through mathematical or statistical approaches. Indeed a family of software has emerged in the marketplace that rather than generate graphs from statistics (scientific data visualization), generates statistics from scanned graphs (graph quantization and graph pattern recognition). The message is that there is implicit information in traditional line graphs that can be extracted through image processing methods.

Figure 7: The Classified Image Result

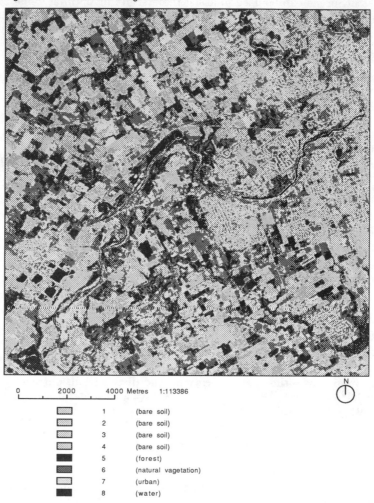

1	(bare soil)
2	(bare soil)
3	(bare soil)
4	(bare soil)
5	(forest)
6	(natural vagetation)
7	(urban)
8	(water)

According to Tufte, time series graphs and the yet more abstract multivariate data graphs evolved from spatial graphs—maps. Is it possible by the same token to evolve multivariate data graph processing from spatial data processing techniques? It appears that GIS procedures may hold more promise than single functions. For instance, the Clump family of GIS functions, while very useful, are not so special; they are merely a type of function knowns as "region growers" in image processing. Nevertheless, there are a few individual functions which are more specific to GIS. Can, for example, a viewshed or intervisibility function be used to highlight local trends in mutivariate space as observed from a particular location in that space? Sets of image processing functions used in GIS joined together to form higher level procedures, such as in the example given in this paper, may contribute unique GRASP methods and perhaps eventually give birth to new forms of data graphs.

ACKNOWLEDGEMENTS

We thank Mr. Yair Medini for his work on the ScatterGrid program and K. Chris Kirby for his outstanding technical and programming support. Thanks also to Professor Cheryl Pearce for providing remote sensing data and lending her expertise in image classification of the study area. Work on this project is conducted at The UWO GIS Laboratory which was established through a generous grant from the Academic Development Fund of The University of Western Ontario.

REFERENCES

Lillesand Thomas M. and Ralph W. Kiefer 1987, Remote Sensing and Image Interpretation, Second Edition, John J. Wiley & Sons Inc., New York .

Pazner M., K. C. Kirby, and N. Thies 1992, MAP II Map Processor Version 1.5, John Wiley & Sons, Inc., New York and ThinkSpace, Inc., Ontario.

Richards J.A. 1986, Remote Sensing Digital Image Analysis: An Introduction, Springer Verlag, Berlin.

Tufte, Edward R. 1983, The Visual Display of Quantitative Information, Graphics Press, Connecticut.

A NONPARAMETRIC TEST FOR PATTERN DETECTION AND ITS USE IN GIS

P.A. Rogerson
Department of Geography
University at Buffalo
Buffalo, NY 14260*

ABSTRACT

Methods of spatial analysis are increasingly being integrated within GIS. Methods for pattern detection in spatial data represent a capability for GIS that is particularly desirable. In this paper, I develop and evaluate a nonparametric test for pattern detection in spatial data, and discuss the use of the method within GIS. Nonparametric techniques for pattern detection are well developed for one-dimensional data. Such techniques are attractive because they make no assumptions regarding the distribution of the underlying data. In this paper, I extend a test of Hewitt et al. for detecting seasonal trends in monthly data to the case of spatial data. I use Monte Carlo methods to (a) derive the null distributions of the statistics, and (b) assess the statistical power of the test.

INTRODUCTION

The advent of GIS has provided us with the opportunity to explore, describe, portray, and analyze spatial data in many new ways. The search for pattern in spatial data represents one of the most fundamental of users' quests. Detecting patterns in crime, arson, and disease represent examples of important and meaningful problems that require timely attention and accurate analysis. And, while there is indeed a voluminous literature on techniques for pattern recognition, there seems at present a clear lack of facilities, either within or outside of GIS, for exploring data to find potentially interesting patterns and for testing spatial data to identify subregions that depart statistically from the remainder of the region.

This paper addresses the specific task of identifying spatial outliers when (a) numeric data are available on at least an ordinal scale, (b) data are arrayed on a grid, and (c) there is no prior hypothesis regarding the distribution of the underlying data, thus prohibiting the use of a parametric test. The test is motivated through a generalization of the test originally developed by Hewitt et al. (1971) to detect seasonality in monthly data.

---

*1992-93 Address: Center for Advanced Study in the Behavioral Sciences, 202 Junipero Boulevard, Stanford, CA 94305.

Hewitt et al. developed a nonparametric test for detecting seasonality in monthly data by employing Monte Carlo methods to determine the null distribution for the maximum of the rank sums (where rank 12 is assigned to the highest value, and 1 to the lowest) calculated for successive six-month periods. For example, suppose monthly data are ranked with the following results for January through December, respectively: 6, 3, 2, 4, 9, 10, 1, 8, 7, 12, 5, 11. The maximum rank sum for any six-month period (including those six-month periods that include December and January) is 49, which is the sum of the ranks for August through January. The question arises whether the high ranks have somehow clumped together, leading to a maximum rank sum that indicates a statistically significant seasonal effect. Alternatively, it may be that the maximum rank sum is not particularly high under the null hypothesis of no seasonal effect. Hewitt et al. used 5000 Monte Carlo trials where ranks were randomly assigned to the twelve months of the year. From this experiment they found the cumulative distribution of the rank sum under the null hypothesis of no seasonal effect by calculating the proportion of trials where the maximum rank sum exceeded specific maximum rank-sum values. Table 1 depicts their results. We conclude from the table that in our hypothetical example, the "observed" maximum rank sum of 49 is consistent with the null hypothesis. We find from the table that the critical rank sum needed to reject the null hypothesis at the approximate $\alpha=.05$ is 55.

Largest Rank Sum	Probability of obtaining this value or higher under the null hypothesis
57	.0134
56	.0248
55	.0464
54	.0766
53	.1260
52	.1914
51	.2908
50	.3826
49	.4958
48	.6086
47	.7258
46	.8310
45	.9138
44	.9614
43	.9904
42	.9986
41	1.0000

Table 1
Cumulative Distribution of Maximum Rank-Sum Statistic
Under Null Hypothesis

Source: Hewitt et al. (1971).

The test is motivated at least in part by the failure of the usual chi-square test, where observed and expected frequencies are compared, to display any power in those cases where there is a cluster of months that display only moderate differences between observed and expected values (of the same sign). Even when the chi-square test does signal a significant result, it still falls to the researcher to determine where the seasonality in fact lies.

EXTENSION OF HEWITT'S TEST TO SPATIAL DATA

It is also possible to use the chi-square test when spatial frequency data are being used. But as in the one-dimensional case, the chi-square test will not be sensitive enough to pick up those cases where there are only modest differences in observed and expected values, but where these modest differences are clustered together in space. In this section, I describe how Hewitt's test may be extended to the case of data collected for a two-dimensional grid.

We first define a window of grid cells; we will be interested in the sum of the ranks within the window. The test statistic is simply the maximum sum of the ranks found within the window after the window has been moved over the entire surface of the grid.

To illustrate the procedure, a 2 x 2 window was moved over four different grid sizes. The grids were of size 4 x 4, 5 x 5, 6 x 6, and 7 x 7. To determine the null distribution of the test statistics, integers representing ranks from 1 to 16, 25, 36, and 49, were randomly assigned to the respective grids. A Monte Carlo simulation of 10,000 trials yielded the results in Table 2.

	4 x 4		5 x 5		6 x 6		7 x 7
Rank Sum	CDF	Rank Sum	CDF	Rank Sum	CDF	Rank Sum	CDF
44	.3656	81	.8249	121	.8572	170	.9053
45	.4525	82	.8600	122	.8811	171	.9189
46	.5377	83	.8917	123	.9008	172	.9328
47	.6184	84	.9163	124	.9193	173	.9415
48	.6966	85	.9369	125	.9352	174	.9520
49	.7656	86	.9536	126	.9483	175	.9604
50	.8248	87	.9653	127	.9602	176	.9682
51	.8732	88	.9767	128	.9692	177	.9733
52	.9152	89	.9845	129	.9778	178	.9783
53	.9422	90	.9907	130	.9843	179	.9835
54	.9652	91	.9938	131	.9881	180	.9862
55	.9792	92	.9966	132	.9928	181	.9889
56	.9890	93	.9976	133	.9951	182	.9920
57	.9947	94	1.0000	134	.9973	183	.9940
58	1.0000			135	.9986	184	.9960
				136	.9997	185	.9969
				137	.9999	186	.9985
				138	1.0000	187	.9989
						188	.9992
						189	.9997
						190	1.0000

Table 2
Cumulative Distribution of Rank Sum Statistic
Under Null Distribution

To illustrate the power of the test, random numbers were drawn from a uniform (0,1) distribution and entered into the grid. Then an increment, delta, was added to each of four cells located near the center of the grid. Table 3 demonstrates the ability of the rank sum statistic to reject the false hypothesis of no spatial effects. Entries in the table represent the proportion of times that the null hypothesis was rejected, and were calculated upon the basis of 1000 Monte Carlo trials. As expected, the power of the test rises with increasing values of delta. The table reveals that the power of the test is adequate only when the delta values reach about 0.7 or higher. Thus the area of interest must have a raised incidence or increment that is slightly higher than the mean (0.5) before it is detected with regularity.

There is also a noticeable tendency for the power to decline with increasing grid size for a given value of delta. This implies that it is harder to detect a given level of raised incidence on a large grid than it is on a small grid. Interestingly, the decline appears more marked for $\alpha=.05$ then for $\alpha=.01$.

	4 x 4 Alpha:		5 x 5 Alpha:		6 x 6 Alpha:		7 x 7 Alpha:	
	.01	.05	.01	.05	.01	.05	.01	.05
Delta:								
.1	.034	.135	.022	.086	.026	.080	.014	.067
.2	.041	.134	.040	.125	.046	.123	.035	.098
.3	.101	.269	.080	.205	.094	.200	.067	.151
.4	.151	.407	.134	.311	.154	.308	.122	.242
.5	.262	.560	.262	.521	.272	.470	.230	.418
.6	.477	.787	.424	.700	.430	.683	.397	.639
.7	.643	.899	.632	.864	.668	.873	.586	.844
.8	.862	.984	.880	.983	.910	.976	.884	.983
.9	.991	.998	.986	.999	.996	1.000	.990	1.000

Table 3
Power of Rank Sum Statistic

It is more interesting to contrast the power of the statistic with the power associated with alternative tests. Both the chi-square goodness-of-fit test and the maximum rank-sum test were carried out on a 5 x 5 grid with a sample of 500 assigned via a probability rule to the cells of the grid. Grid cells had equal probabilities of having observations assigned to them, with the exception that a 2 x 2 subregion in the middle of the grid had a probability of assignment that was increased by a factor of delta. Table 4 summarizes the results. As in Table 3, power increases with both delta and α. The maximum rank-sum statistic is seen to display substantially greater power than the more usual chi-square statistic.

Delta:	Maximum Rank Sum $\alpha=.05$	Chi-Square $\alpha=.05$	Maximum Rank Sum $\alpha=.01$	Chi-Square $\alpha.01$
1.250	.272	.134	.107	.037
1.375	.455	.281	.231	.111
1.500	.706	.541	.464	.309
1.750	.941	.900	.828	.779

Table 4
Comparative Power of Chi-Square and Rank Sum Statistics

DISCUSSION

The nonparametric test presented here may prove useful in those situations where ordinal data are available on a square grid and where detection of a subregion that differs significantly from the remainder of the region is desired. Perhaps more importantly, the test suggests a general procedure that may be employed in situations that do not meet these somewhat restrictive conditions. Specifically, the procedure of randomly assigning ranks to regions and generating Monte Carlo samples to find the null distribution may be used for grids and windows of any size, though computation time increases quite rapidly with grid size. More challenging would be to extend the test to regional configurations not on a grid -- one could presumably address individual problems by defining contiguity matrices.

A fundamental complication when going from one dimension to two is that there emerge several possibilities for generalization, and it is not apparent that one is any better than another. Should one use a 2 x 2 window, or a 1 x 4 window? How big should the window be relative to the grid? These choices must ultimately rest upon the nature of the inquiry.

The test described in this paper could also be extended to cases where one edge of the region is "wrapped" around to another, just as December is wrapped around to January in the Hewitt et al. test. Such an extension might be appropriate, for example, in cases where the outlying grid cells represent suburbs and the inner ones a central city. Wrapping cells around as if they were on a torus would take explicit account of the potential for the outlying suburbs in different regions of the grid to be similar to one another.

The test seems particularly well-suited to raster-based GIS. Even an exploratory capability including the rank-sum within a given window, and an ability to move the window around the grid would give the user a good feel for subregions that were potential outliers.

The present analysis is most encouraging because the illustrative results demonstrate that the power of the test is considerably greater than the ordinary chi-square goodness-of-fit test. Finally, Freedman (1979) has shown that a Kolmogorov-Smirnov test outperforms the Hewitt et al. test with seasonal data, indicating that attempts to generalize the Kolmogorov-Smirnov test to spatial data may also prove fruitful.

ACKNOWLEDGMENTS

The author gratefully acknowledges the assistance of Presidential Young Investigator Award SES8553055 from the National Science Foundation. This work was carried out as part of the GIS and Spatial Analysis Initiative of the National Center for Geographic Information and Analysis.

REFERENCES

Freedman, L. S. 1979. The Use of a Kolmogorov-Smirnov Type Statistic in Testing Hypotheses About Seasonal Variation. <u>Journal of Epidemiology and Community Health</u>, Vol. 33, pp. 223-228.

Hewitt, D, Milner, J., Csima, A., and Pakula, A. 1971. On Edwards' Criterion of Seasonality and a Non-Parametric Alternative. <u>British Journal of Preventive and Social Medicine</u>, Vol. 25, pp. 174-176.

PARALLEL TERRAIN FEATURE EXTRACTION

Demetrius-Kl. Rokos
Marc P. Armstrong

Department of Geography
The University of Iowa
Iowa City, IA 52242

ABSTRACT

Spatial models often must handle large datasets to capture the complex interactions of physical systems. This leads to computational intensity. Fortunately, many geographical problems can be decomposed into relatively independent parts that can be processed simultaneously. This paper describes a terrain feature extraction algorithm that operates in a parallel computing environment. The morphology of a drainage basin is derived using information provided by a DEM. The time efficiency of the algorithm is investigated, as well as the relative speedup accomplished over a sequential version of the algorithm.

INTRODUCTION

As researchers have incorporated increasingly detailed and realistic descriptions of physical geographic processes in their models, data volumes and computational complexity have increased commensurately. The physical limit of processing speed (Polychronopoulos, 1988) and the cost of high performance traditional computers, however, have limited both the ability of researchers to engage in near-real-time modeling and the size of problems that can be examined. Parallel programming provides a solution to these constraints by integrating and harnessing the power of multiple processors to produce relatively inexpensive high performance computer environments. Before algorithms can be efficiently implemented in parallel architectures, however, they must be decomposed into relatively independent segments. Fortunately many spatial algorithms treat geographic space as an aggregation of interacting spatial units, thus facilitating their decomposition. This paper describes the decomposition and parallel implementation of a terrain feature extraction algorithm (Bennett and Armstrong, 1989). The algorithm is designed to operate on a digital elevation model (DEM) to determine the hydrologic category (e.g. channel, divide, slope break) of each DEM cell and then to inductively construct an initial approximation of the hydrologic feature networks.

FEATURE EXTRACTION ALGORITHMS

The detailed description and delineation of the subcatchments and hillslopes of a drainage basin are essential parts of many drainage basin studies. Because of their well-defined boundary conditions and explicit internal connectivity, subcatchments and hillslopes provide an "unambiguous template for structuring models in geomorphology, hydrology, landscape ecology and other fields" (Band, 1989a:151). Two basic topographic features are used to derive this information: drainage and divide networks. Drainage networks provide a spatial framework for integrating spatial elements of the land surface and a structure for dissecting space into functional areas (Jarvis, 1977). The drainage network also defines the main flow pattern of water, sendiment, nutrients and pollutants in a watershed. The divide network, on the other hand, defines the extent and the shape of the runoff contributing area for each channel segment and acts as the boundary of relatively independent watershed partitions. As Band (1986) noted, encoding these feature networks may be quite tedious and time consuming for any but the smallest data sets which "has provided a strong restriction on the scale and the complexity of the watershed research that is generally attempted" (Band, 1989a:151). Researchers, therefore, have sought ways to automate this process.

Peucker and Douglas (1975) designed three feature extraction algorithms that are the basis of much of the consequent work in the area. The first used the elevation information in a 3 by 3 elevation window to classify the center cell of the window into categories (peak, pit, pass, ridge, ravine, slope, break and flat) using the following method:

1) Calculate elevation differences between the center DEM cell and its neighbors
2) Describe each feature type as a unique combination of the following characteristics: difference of the sum of all positive elevation differences with the sum of all negative elevation differences, number of sign changes and number of points between two sign changes.
3) Classify the center cell into a feature type by matching its characteristics with the characteristics of one of the prototypes created for the different feature types.

In the second algorithm the center cell is classified as a ridge if it is higher or a channel if it is lower than all four of its row and column neighbors. The third algorithm uses the basic principle of a "one-step move": In the neighborhood of every cell, the highest point in the window is flagged. All remaining cells are classified as channel cells. Similarly, if the lowest cell in every elevation window is marked, all remaining cells are classified as ridges.

In 1983 Mark proposed a process-based approach in which he assumes that "a drainage network represents those points at which runoff is sufficiently concentrated that fluvial processes dominate over slope processes" (Mark, 1983:289) and that the center cell in a 3 by 3 elevation window will drain to the neighbor having the steepest downslope towards it. Considering that every cell produces a unit quantity of runoff, every cell will drain this quantity plus the runoff of all the cells draining to it, if any. By using a predefined threshold over which runoff could be considered concentrated, drainage network cells were defined as those cells that drain more water than the threshold. A more detailed implementation of the algorithm can be found in O'Callaghan and Mark (1984).

Marks *et al.* (1984) identified and delineated drainage basins and sub-basins using two measures: slope and exposure. Slope was defined as the angle between the terrain and the horizontal and exposure as the direction of slope. The goal of the algorithm is to identify which, if any, of the neighboring cells belong in the same basin as the center cell. A neighboring cell is in the same basin as the center cell, if its exposure faces toward the center cell within a quantization level of 8 divisions of the circle. Slope is considered only in almost flat terrain, when the exposure cannot be defined. A similar algorithm by Morris and Heerdegen (1988) also used drainage direction to derive the drainage network. Their algorithm went a step further since it used the drainage network to recursively define the boundaries of the drainage basin.

Jenson's (1984) algorithm assumes that if an elevation profile is V-shaped, then water will tend to concentrate there and form part of the drainage network. After identifying the drainage cells in a basin, the algorithm connects them using the following method: When two drainage cells are adjacent and the lower one belongs to a certain channel, then the other cell also belongs to the same channel. Finally, all non-drainage cells are assigned to the basin of a drainage network. This is accomplished by iteratively assigning labels to non-drainage cells that drain to (are uphill and adjacent to) already labeled cells.

In each of these algorithms, we described only the rules that are used to identify features. Following this step, each algorithm attempts to construct the topology of the feature networks by connecting adjacent cells that perform similar hydrologic functions. This step, however, often fails to produce the expected feature networks. Very often the original data sets include errors which appear in the form of pits or sinks, which rarely occur naturally (O'Callaghan and Mark, 1983). These artificial sinks cause the interruption or distortion of existing drainage lines or the introduction of erroneous non-existent drainage lines. There has been considerable discussion in the literature about techniques that either identify and remove erroneous pits from data sets (e.g. Marks *et al.*, 1984) or correct errors in the derived drainage lines (e.g. Band 1986; Morris and Heerdegen, 1988).

Another issue that feature extraction algorithms should address is scale. First, the degree of detail that an algorithm uses to represent feature networks should vary with respect to research objectives. Furthermore, there are no consistent criteria in the literature for the classification of a valley as a part of a drainage network or a ridge as part of a divide network. Since drainage networks are dynamically defined as the set of points where fluvial processes dominate over slope processes (Mark, 1983), their extent may be altered by changes in precipitation input and land cover. To deal with scale and dynamic feature definition issues, some feature extraction algorithms employ thresholds that enable the scale and the detail of feature networks to be controlled. For example, Mark's (1983) algorithm classified a DEM cell as a drainage cell if the area contributing runoff to that point was larger than a predefined threshold.

In the next section we describe an alternative approach that addresses these shortcomings. This alternative, yet still computationally intensive, approach: 1) uses inductive logic to identify features and assemble them into networks, 2) allows control of the level of detail in feature network representation, and 3) can derive feature networks even in flat areas.

AN INDUCTIVE BIT-MAPPED CLASSIFICATION SCHEME FOR TERRAIN FEATURE EXTRACTION

The algorithm presented here uses a "rule based system driven by inductive logic and physical law" to extract hydrologically significant features from digital elevation models (Bennett and Armstrong, 1989:60). Induction encompasses "all inferential processes that expand knowledge in the face of uncertainty" (Holland *et al.*, 1989:1) and can be used when there are no objective and indisputable criteria to extract hydrologic features from digital elevation models. It should be noted that the results of the inductive process cannot be guaranteed to be correct. To improve the reliability of inductive predictions, domain specific knowledge is used and the predictions are fed back to the system for re-evaluation and improvement.

This algorithm was developed to extend previous work in 3 ways (Bennett and Armstrong, 1989:60): a) identify points at which the rate of flow is modified by relief (slope breaks), b) construct a topologically complete data structure that closely describes the topology of the basin, and c) subdivide the terrain into homogeneous areas with respect to slope and aspect. The algorithm consists of four steps:

1. The DEM cells are classified into hydrologic categories from which an initial approximation of the basin's morphology is derived.
2. The results are refined to improve the topographic model.
3. The fragmentation of the derived features is resolved by identifying and eliminating erroneous pits and connecting divides through passes.
4. The topological model of the basin is completed by connecting drainage and divide networks.

In this paper we will focus on the first step of this algorithm. In a 3 by 3 elevation window the algorithm records the two-dimensional shape of the four directed, symmetrical transects that cross the center cell. This information is used to classify each transect into one of the following hydrologic categories: 1) ridge, 2) valley, 3) slope break, or 4) flat. The shape of each transect is stored in a six element boolean array (Table 1) that is compared with prototypes generated for each hydrologic category. Each transect is placed in the category with which it has the greatest similarity. This method was chosen because it has three advantages (Bennett and Armstrong, 1992):

1. Characteristics of features are captured in a simple manner avoiding redundancy.
2. Classification can proceed even if data are incomplete, erroneous, or inconclusive.
3. Classification is a simple pattern matching operation.

Bit Mapping:

1	true if center point is topographically significant
2	true if first point of transect is higher than mid point
3	true if last point of transect is higher than mid point
4	true if the transect is extended in the direction of the first point
5	true if the transect is extended in the direction of the first point
6	true if extension of either line reaches a boundary.

Line Classification:

ridge	Both bounding points lower than middle	#FF##F
valley	Both bounding points higher than middle	#TT##F
slope break	One bounding point higher, the other	#TF### or
	lower than the middle;	#FT### or
	One end of the line extended, one not	###FT# or
		###TF #
flat	Both ends of the line extended	###TTT

Table 1. Transect classification rules (from Bennett and Armstrong, 1989).

Next, topographic significance is used to define "the difference in elevation between the central point and a line bounded by two points on opposite sides of the window" (Bennett and Armstrong, 1989:62). The consideration of topographic significance in the transect classification serves two purposes. First, by using an elevation threshold below which a point is considered topographically insignificant the algorithm can account for noise in the digital elevation model that could cause erroneous classification of cells. Thus, the possibility of misclassification is reduced and classification is restricted to the points for which there is strongest evidence for their hydrologic function. Second, the topographic significance threshold can be used as a tool to control the degree of detail represented in feature networks.

The algorithm also can be applied to areas of low relief unlike many other feature extraction algorithms based in local operators (Band, 1986). Since in areas of low relief there may not be sufficient information for cell classification in a 3 by 3 elevation window, the algorithm allows for "flexible windowing". When the elevation of the center cell is not significantly different from either ends of a transect running through it, the window is recursively extended in the direction of uncertainty (Figure 1). The fact that a transect is extended beyond the 3 by 3 elevation window is recorded in elements 4 and 5 of the transect boolean arrays.

Using the information derived from the form of the four transects, the center cell of the elevation window is classified into one of the following six hydrologic categories: 1) divide point, 2) drainage point, 3) pit, 4) pass, 5) slope break, or 6) plain. After all cells have been placed in a hydrologic category, an initial approximation of the basin's morphology is derived by establishing preliminary topological relationships between classified cells. This is achieved by linking adjacent cells that perform similar hydrologic functions.

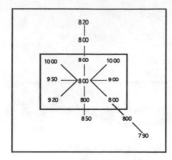

Figure 1. An example of transect extension in low relief areas (from Bennett and Armstrong, 1989).

PARALLEL PROCESSING

Parallel processing is a computing paradigm in which multiple computing resources are used to increase the amount of work performed per unit of time to solve a particular problem (Desrochers, 1987). Unlike sequential programming, however, an efficient implementation of a parallel algorithm is related to the architecture in which it is developed. To tailor an algorithm to the characteristics of the target architecture is a complex and tedious task (Polychronopoulos, 1988). Because of this complexity as well as the inadequacies of the software tools that are available for developing and debugging parallel programs (Mohiuddin, 1989), parallel processing has not been widely diffused. Despite these difficulties, parallel processing environments are now being developed by many computer manufacturers and will become increasingly widespread. Therefore, computationally intensive algorithms must adapt to the requirements of parallel processing.

Granularity of Parallelism
The decomposition of a problem into relatively independent tasks can be achieved in many ways. The grain size into which a problem is divided however, is fundamentally important. The granularity of an algorithm can be classified into three categories (Polychronopoulos, 1988):

- Fine grained algorithms have tasks equivalent to basic code blocks.
- Medium grained algorithms have individual tasks larger than a basic block but smaller than a subroutine.
- Coarse grained algorithms have tasks equivalent to one or more subroutines.

When deciding the most appropriate granularity for a problem two things should be considered in addition to what is the most natural decomposition strategy (Carriero and Gelernter, 1990: 14). First, the granularity of an algorithm is related to the granularity of the target architecture. "Fine grained" parallel computers with many single bit processors, for example, are not normally well-suited for coarse or medium grained parallelism. Second, it has been reported (Cok, 1991) that fine grained decomposition achieves a more efficient distribution of the workload than medium or coarse decomposition, but it usually requires a much greater communication overhead.

Parallel Programming Paradigms
Carriero and Gelernter (1991) introduced the term parallel paradigm to describe a distinct way of thinking about parallelism in an implementation independent context. There are three main parallel paradigms: event, geometric, and algorithmic parallelism. Two of these paradigms (event and geometric) are used to decompose the selected terrain feature extraction algorithm.

656

Event parallelism (also called the task or processor farm approach) involves the identification of a main process (master task or controller) and a set of specialized worker processes. The main task distributes data to worker processes and collects their results. The master task also schedules the work plan. Whenever the master task determines that a worker process is idle, it sends a new set of data. While event parallelism is inherently a load balancing paradigm, it incurs communication overhead as the controller must continuously send and receive data from the worker processes.

Geometric parallelism (also called data or result parallelism) involves the decomposition of the problem space into subregions within which local operations are performed (Healey and Desa, 1989). This is a popular paradigm as it "largely avoids the difficulty of finding a way to decompose a problem into parallel pieces" (Cok 1991). One very important issue that determines the efficient implementation of geometric parallelism in a particular problem is the interprocessor communication required. It is essential that the decomposition of the problem space maximizes processing within a region and minimizes interregional exchange of information.

Process and Data Dependencies
When designing a parallel algorithm, the process and data dependencies of the problem domain must by studied to identify interactions and interdependencies between data entities. Problems that have complex process and data dependencies are difficult to decompose, which may lead to poor performance or even incorrect results if the correct sequence of execution is not guaranteed.

Data dependencies involve the interaction and interdependencies of data items on the instruction or loop level (Desrochers, 1987; Armstrong and Densham, 1992). Data dependencies can occur: a) during data input, b) when nested indexing is used to address array elements, and c) when elements of the same array are needed by different processes. Data dependencies can be generalized on the process or procedure level (Williams, 1990; Evans and Williams, 1980). Williams (1990) identified several different types of relationships that can exist between two processes (P1 and P2) in shared and private memory systems. Since the parallel architecture that has been used in this research is a distributed memory parallel computer we will briefly discuss the potential relationships between processes in a private memory system. The main types of relationships between processes are:

• **Prerequisite**: process P1 must fetch what it requires before P2 stores its results. P1 must not require results produced by P2, as there is no guarantee that P1 will finish before P2.
• **Conservative**: process P1 must send its results to a process P3 before P3 receives the results of P2. This suggests that both processes modify the same data structures and the results of process P2 are required for subsequent processing. This type of relationship allows the unconstrained parallel execution of the two processes until the time that the two processes have to return their results.
• **Commutative**: process P1 may execute before or after P2, but not at the same time. This occurs mainly when both processes modify the same memory locations during their execution.
• **Contemporary**: processes P1 and P2 are completely unrelated and may execute at the same time, without any constraints.
• **Consecutive**: process P1 must send all or part of its results to process P2. This implies that the results of P1 are used by P2 so the execution of the two processes is inherently sequential.

PROBLEM DECOMPOSITION

Before deciding on a decomposition strategy that would fit the characteristics of the selected algorithm, two things should be considered: the computer architecture and the process and data dependencies of the algorithm. Our programming environment (a MIMD machine consisting of a 486 PC serially linked to 4 fully interconnected T800 Transputers) dictates that the problem domain should be decomposed into relatively large tasks (coarse grain parallelism) which require limited interprocessor communication. The data and process dependencies of the algorithm, on the other hand, complicate the decomposition process.

Dependency Analysis for the Feature Extraction Algorithm

In areas of low relief the determination of a transect's shape may require recursive extension. Since the length of the extension depends on local relief, it cannot be predefined. This causes inconveniences in distributed memory systems where each process can only access data from its own local memory. Also note that the relationship between different transect classifications is contemporary since each is independent and can execute in parallel. The information produced from the classification of the four transects (a bit-mapped boolean array) is used to classify the center cell of the elevation window. The type of relationship between these two processes (transect classification and cell classification) is consecutive, as the results of the classification of all four transects must be known prior to the classification of the center cell. The algorithm then connects adjacent cells that perform similar hydrologic functions. Within a DEM window the relationship between this connection step and the cell classification step is consecutive. However, when examined at a broader scale the two steps can be accomplished in parallel as the inter-window relationships are contemporary.

Because the algorithm exhibits different kinds of dependency relationships, several ways of organizing it for parallel processing can be devised. One way of decomposing the algorithm is into the three steps described earlier. This approach, however, creates relatively small tasks which require a large amount of communication (sending data and receiving results). Therefore, an alternative solution was developed to reduce overhead and improve performance. Since the first two steps of the algorithm, transect and cell classification, are closely related, and the results of the first are required by the second, they were combined together to form a single *cell classification* task. The third step of the algorithm, the *feature topology construction* task, is independent and executes after the cell classification has been completed. To coordinate the execution of these two worker tasks, a master task was also designed to distribute data to the worker tasks, receive their results and balance their workloads.

Problem Domain Decomposition

A simple way to decompose two-dimensional geographic data sets is geometric parallelism (Healy, 1989). In this approach the data set is divided in N equal parts, where N is the number of worker processors. Communication is required only for processing the boundary areas of each spatial subset. This approach works well for feature topology construction since simple data dependencies exist between the boundaries of the subsets. To process boundaries, boundary data between two adjacent subsets are made available to both of them and they are treated differently than interior cells. Furthermore, since each subset requires the same amount of processing, if data are equally distributed among worker tasks, the workload of the worker processors is well-balanced and the computing resources are used efficiently.

The application of geometric parallelism in the first part of the algorithm (cell classification) presented serious problems. In flat areas, the transect may have to be extended to determine its shape. This requires that the elevation values in the direction of the extension are available to the cell classification task. Since the length of a transect's extension cannot be known *a priori*, the size of the elevation window that may be required for the classification of a single cell cannot be predefined, and therefore the master task must send the whole data set to each worker processor. Though this may be feasible for small data sets, it degrades the algorithm's performance for large data sets. Consequently, whenever a transect must be extended, the worker task encountering the problem sends all needed information to the master task, which has access to the entire data set. The master task extends the transect and returns its shape to the worker that requested it. Transect extension also dictates that if the geometric paradigm is used, performance problems may occur from uneven workload balancing. If a purely geometric decomposition were used, a worker processor receiving data representing a flat area would be required to perform many more transect extensions than other worker processors and would continue to process data after other workers have completed their tasks. To deal with this problem, the cell classification task was implemented using event parallelism. This paradigm inherently balances processing loads and therefore, if one worker is required to do more processing than others, it will execute fewer times.

PERFORMANCE ANALYSIS

Before attempting to evaluate our algorithm, several implementation issues closely related to the algorithm's performance must be addressed. First, the size of the individual workload in the cell classification task must be determined. Since our parallel architecture is used efficiently when the work within each processor is maximized and interprocessor communication is minimized, this might lead us to try to increase the individual workload size as much as possible. Such action, however, causes the same workload balancing problems as the application of the geometric parallelism paradigm. The ideal workload size should be a function not only of the data set size, but also of local relief. To provide a simple demonstration of the importance of workload size we ran the application with 3 x 3 and 3 x 31 workload sizes. Using the latter workload size we achieved a 28% performance improvement which was due to the increase of the unit computational step and the consequent decrease of the communication load.

The performance of the parallel implementation of the algorithm was also improved by reducing the length of the records sent to the worker tasks. Time was saved not only by assembling smaller records but also by reducing communication overhead. To demonstrate this we executed the cell classification algorithm for the same data set using the old data structures of the sequential version and the new reduced ones. The results showed that the execution was 14.20% faster with the new data structures.

To evaluate other aspects of the performance of our parallel algorithm two measures were used: speedup and efficiency. **Speedup** (S) can be defined as the ratio of sequential run time (Tseq) to parallel run time (Tpar):

$$S = Tseq / Tpar \qquad (1)$$

Efficiency (E) can be defined as the ratio of the speedup (S) to the number of processors (N) running in parallel:

$$E = S / N \qquad (2)$$

We measured our algorithm's execution times when running it with one (sequential version), two and three worker processors (Table 2), then computed efficiency and speedup (Table 3). Figure 2 shows the graph of the execution time of the cell classification task and the feature topology construction task against the number of worker processors used. It is apparent from the timing results that by increasing the number of worker processors, a considerable speedup of the algorithm is achieved. In addition, Figure 2 indicates that by increasing the number of the worker processors, the relative performance improvement of the algorithm (efficiency) diminishes. Two factors explain this:

a) By increasing the number of the processors involved, more interprocessor communication is required.
b) The data set that has been used is quite small (27 X 31 points) and therefore the timing results are considerably influenced by communication overhead. If the program was executed with a larger data set the efficiency of the algorithm would not diminish as quickly.

Num. of Processors	Cell Classification	Feature Topology
1	9765	7390
2	5300	4700
3	3859	3526

Table 2. Timing measurements of the developed algorithm running the application with 1, 2 and 3 worker processors. Units are low priority Transputer clock tics (1 tic=64msec).

	Cell Classification		Feature Topology Construction	
Number of workers	Speedup	Efficiency	Speedup	Efficiency
2	1.84	0.92	1.57	0.79
3	2.53	0.84	2.10	0.70

Table 3. Measurements of the speedup and efficiency of the parallel version of the algorithm over its sequential counterpart, using 2 and 3 worker processors.

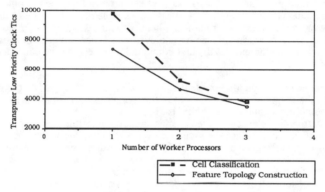

Figure 2. Graph of the algorithm's performance using 1, 2 and 3 worker processors

CONCLUSIONS

In this paper we briefly described the characteristics and principles of parallel processing that were applied to a terrain feature extraction algorithm implemented in a parallel environment. This application demonstrated the natural way that this particular problem can be decomposed into relatively independent tasks which are efficiently implemented in a parallel environment. The results showed the performance benefits of the parallel implementation over the traditional sequential approach. Also noted was the dependence of parallel algorithms on computer architectures, as a parallel algorithm has to be compatible with the processor granularity and the interprocessor communication topology of the target architecture. This hinders the wide application of parallel processing in many scientific fields such as geography where effort should be focused on analyzing the spatial interactions and dependencies of physical systems and processes to enable parallelism in an implementation-independent way.

ACKNOWLEDGMENTS

We wish to acknowledge comments and conversations with David Bennett and Paul Densham. Partial support for this research was provided by the National Science Foundation (SES-90-24278).

REFERENCES.

Armstrong M.P. and Densham P.J., 1992. Domain Decomposition for Parallel Processing of Spatial Problems. *Computers, Environment and Urban Systems* (forthcoming).

Band L.E., 1989. A Terrain-based Watershed Information System. *Hydrological Processes*, Vol. 3:151-162.

Bennett, D.A. and Armstrong, M.P. 1989. An Inductive Bit-Mapped Classification Scheme for Terrain Feature Extraction. *Proceedings GIS/LIS '89*, Orlando, FL.

Bennett, D.A. and Armstrong, M.P. 1992. A Knowledge-Based Approach to Topographic Feature Extraction.(unpublished manuscript)

Brawer S., 1989. *Introduction to parallel Programming.* Academic Press Inc., San Diego, CA.

Carriero N. and Gelernter D., 1990. *How to write parallel programs. A first course.* The MIT Press, Cambridge, MA.

Cok Ronald S., 1991. *Parallel Programs for the Transputer.* Prentice Hall Inc., New York.

Desrochers G.R., 1987. *Principles of Parallel and Multiprocessing.* Intertext Publications, Inc., McGraw-Hill Book Company, New York.

Evans D.J. and Williams S., 1980. Analysis and Detection of Parallel Processable Code. *Computer Journal* 23:66-72.

Healy R.G. and Desa G.B., 1989. Transputer based Parallel Processing for GIS Analysis: Problems and Potentialities. *Proceedings of the Ninth International Symposium on Computer-Assisted Cartography, AUTO-CARTO 9.* Bethesda, MD: American Congress on Surveying and Mapping: 90-99.

Holland J.H., Holyoak K.J., Nisbett R.E. and Thagard P.R., 1989. *Induction: Processes of Inference, Learning, and Discovery.* MIT Press, Cambridge, MA.

Hopkins S. and Healey R.G., 1990. A Parallel Implementation of Franklin's Uniform Grid Technique for Line Intersection Detection on a Large Transputer Array. *Fourth International Symposium on Spatial Data Handling,* Zurich, Switzerland: 95-104.

Jarvis R.S., 1977. Drainage Network Analysis. *Progress in Physical Geography,* 1:271-295.

Jenson S.K., 1985. Automated Derivation of Hydrologic Basin Characteristics from Digital Elevation Model Data. *Proceedings of Auto-Carto 7*:301-310.

MacDonald R., 1989. Parallel Processing-Powerful Tool for the Earth Sciences. *U.S. Geological Survey Yearbook.*

Mark D.M., 1983. Automated Detection of Drainage Networks from Digital Elevation Models. *Proceedings of Auto-Carto 6*, Ottawa, Canada:288-298.

Marks D., Dozier J. and Frew J., 1984. Automated Basin Delineation from Digital Elevation Data. *Geoprocessing* 2:299-311.

Mohiuddin K.M., 1989. Opportunities and Constraints of Parallel Computing. In *Opportunities and Constaints of Parallel Computing,* Sanz G.L.C. (ed), Springer-Verlag New York Inc.

Morris D.G. and Heerdegen R.G., 1988. Automatically Derived Catchment Boundaries and Channel Networks and their Hydrological Applications. *Geomorphology* 1, Elsevier Science Publishers B.V., Amsterdam:131-141

O'Callaghan J.F. and Mark D.M., 1984. The Extraction of Drainage Networks from Digital Elevation Data. *Computer Vision, Graphics, and Image Processing* 28:323-344.

Peucker T.K. and Douglas D.H., 1975. Detection of Surface-Specific Points by Local Parallel Processing of Discrete Terrain Elevation Data. *Computer Graphics and Image Processing* 4:375-387.

Polychronopoulos C.D., 1988. *Parallel Programming and Compilers.* Kluwer Academic, Boston.

Williams S.A., 1990. *Programming Models for Parallel Systems.* John Wiley & Sons, Chichester, England.

3L Ltd, 1989. *Parallel Pascal User Guide.* 3L Ltd.

THE VENTURA COUNTY PRIMARY CONTROL NETWORK:
A CASE STUDY IN A MULTI-DISCIPLINARY APPROACH
TO GPS NETWORK DESIGN AND IMPLEMENTATION

Jay Satalich, Assistant Land Surveyor
California Department of Transportation - District 7
Engineering Services Branch - GPS Unit
120 South Spring Street
Los Angeles, California 90012
(213) 724-0864

ABSTRACT

The California Department of Transportation (Caltrans), in cooperation with the County of Ventura and members of the geophysical community, has completed work on a geodetic control network to be used in transportation infrastructure projects, crustal motion studies, and GIS/LIS. This paper will demonstrate that through proper coordination, planning, and execution, a geodetic control network can be developed to respond to the needs of a variety of users. It will deal with such issues as the network design strategy, station accessibility, and standards for monumentation. Emphasis will be placed on how different user requirements were synthesized into a coherent whole. The paper will also discuss how this particular project, based upon the North American Datum of 1983 (NAD 83), is utilizing two new systems, the California High Precision Geodetic Network (HPGN) and the North American Vertical Datum of 1988 (NAVD 88).

INTRODUCTION

Caltrans District 7 in Los Angeles has been using Global Positioning System (GPS) survey techniques in support of transportation projects since 1988. GPS has proven to be an accurate, rapid, and economical method of providing coordinate control for a wide variety of route surveying tasks at District 7 (Launen 1991) and is regularly used.

Initially, these types of projects utilized NAD 83 horizontal control in the National Geodetic Reference System (NGRS) and in general, projects were widely separated. In 1990, California voters approved a number of bonds to upgrade and expand the state's transportation infrastructure. District 7, which consists of Los Angeles and Ventura Counties, has one of the most extensive and complex transportation systems in the country with over 2000 miles of highway corridor. By the spring of 1991, it had become apparent to the staff at District 7 that the existing method of control densification would be inadequate for the large volume of work which would be generated, thus a proposal was made to establish a district-wide primary control network. This would create a nested hierarchy of control surveys, minimize the effects of error propagation, and establish common control stations for both in-house and consultant projects. Also, there were large areas in the region where the reliability of the control was unknown. When coming into an area of unproven control, the prudent geodetic surveyor adopts a conservative approach and uses more control stations than necessary to complete a survey. This adds valuable time to all phases of a project by increasing the overall number of stations, and therefore, the total number of observations.

At about the same time, Caltrans at the state level was collaborating with the National Geodetic Survey (NGS) to establish a state-wide high accuracy reference network which would eliminate existing distortions and establish an accurate, continuous, and consistent datum for California (D'Onofrio and Nelson 1991). The HPGN arrived at a convenient time as it would provide the opportunity for District 7 to establish its own primary control network and densify the HPGN at the same time.

When the district-wide primary control network was first proposed, the decision was made to use individual counties as the fundamental building blocks for basic control surveys. This would keep the number of participants at a manageable level and afford flexibility in network design and project planning. The Ventura County Network would be the first half of a two part project to establish a District 7 Primary Control Network. Early in 1992, members of Caltrans, the County of Ventura, and the geophysical community held a series of meetings to define user requirements and discuss possible scenarios for establishing a primary control network in the county. At the project's conception, it was clear that the Ventura County Net must meet certain criteria.

GOALS

First, since the network would be based upon the California HPGN and would be a densification project of the state-wide network, the intent was to conduct the survey according to Federal Geodetic Control Subcommittee (FGCS) standards and specifications for GPS relative positioning and the project would eventually be submitted to NGS for "Blue Booking". First-order FGCS standards would be the minimum allowable accuracy, but it was felt that the dual-frequency P-code receivers which would be used for the project had the capability to achieve accuracies closer to the order B level. At the completion of the project, a record of survey would be filed with the County of Ventura.

Second, the network would use as much existing monumentation as possible with a spacing of 13-21 km between control stations. A station spacing such as this would enable users with single frequency GPS equipment to utilize the network as well. The ideal monument would be one which is safe, permanent, stable, and relatively free from overhead obstructions. The goal would be to have a geodetic quality mark with 2-wheel drive public access. Special consideration would be given to monuments which were considered to have historical significance and impeccable pedigree; perpetuating some of the older monuments would provide a link between the new network and the older surveys in the county. When a suitable monument could not be recovered, a new monument would be set which would consist of either an unsleeved NGS class A stainless steel type rod mark or a brass disk affixed to either bedrock or a massive reinforced concrete structure. No sleeve would be required for the rod marks because of the low expansive character of the soils and the lack of frost action in the project area.

Third, based on advice from NGS officials, as many monuments as practical would be tied to NAVD 88. This would be achieved by either occupying suitable benchmarks as primary control stations, or by establishing short level ties to selected primary control monuments from NAVD 88 benchmarks. Fortunately, numerous NAVD 88 level lines parallel the major transportation routes in the county, giving good vertical coverage of control stations throughout the network. First-order, class II differential leveling would be required for completion of the vertical portion of the project.

Although the primary goal of the project was to provide horizontal geodetic control, there were practical reasons to devote effort to the vertical aspects of the project. Through the use of established techniques, (Zilkoski 1990a, Zilkoski 1990b, Zilkoski and Hothem 1989, Milbert 1992) GPS derived orthometric heights could be established and compared to the leveling data. This would verify the accuracy of the NGS Geoid90 model in the region. If the model proved to be valid, it would give users of GPS in the region the option to exploit the use of GPS derived orthometric heights in a variety of surveying applications. If the model proved to be distorted, further gravity and/or leveling work could be targeted, giving geodesists the opportunity to further refine the geoid model in the area. Also, there are areas within Ventura County which are subject to subsidence, and GPS has proven to be a viable alternative to leveling in estimating subsidence movement (Strange 1989). The inclusion of leveling data and geoid modeling into the Ventura County Network would make the survey a fully integrated, 3-dimensional geodetic project.

Lastly and perhaps most importantly, the network would be designed to accommodate the greatest possible number of GPS users. There were distinct advantages to this approach.

District 7 needed to have a uniform and reliable control network to support the large amount of transportation related work which was going to occur in the region. Also, Caltrans had taken tentative steps to institute a state-wide GIS/LIS, therefore the coordinates generated from the HPGN and its densified counterparts would serve as the framework to spatially reference the future GIS/LIS. The County of Ventura eventually planned to implement a GIS/LIS of their own, so cooperating with Caltrans on the Ventura Control Net would establish their spatial reference network and eliminate the potential problem of surveyors using multiple datums in the same region.

As the project progressed, other public agencies became interested in the Ventura County Net for their own uses. The U.S. Geological Survey (USGS) was conducting subsidence studies in the Oxnard area, and requested information regarding the vertical portion of the project. The U.S. Department of Agriculture Forest Service (USFS) also inquired about the network to control holdings in the Los Padres National Forest. Finally, the U.S. Naval Civil Engineering Laboratory at Port Hueneme had an interest to control both the port district and offshore environs for navigation and engineering projects.

The future network would also try to include stations which would be suitable for use by the geophysical community. This meant that the monumentation would have to be of a stable, permanent type which is strategically located to straddle designated faults in the region. This would be achieved by utilizing a combination of existing and new monumentation. Scientists use geodesy to determine relative movement over time to monitor crustal motion. Ventura County is a region which is geologically active; using GPS techniques, movement on the order of 7 ± 2 mm per year has been recorded in the Ventura Basin (Donellan, Hager, and King 1992).

In this era of fiscal austerity, thought also had to be given to assure the longevity of the network. It is anticipated that maintenance of the control stations would primarily be the responsibility of Caltrans, but it is hoped that as others use the network, they would contribute to the long-term maintenance as well. It is inevitable that control stations will be destroyed, and these stations will have to be replaced and resurveyed.

Most of the control stations in the network lie astride long-standing NGS and County of Ventura level circuits, making vertical reobservation of these control marks very likely. The major factor in maintaining the long-term horizontal integrity of control networks in areas such as California is to develop and integrate a crustal motion model for the networks. REDEAM (REgional Deformation of the EArth Models) was developed for the original NAD 83 readjustment and applied to the older geodetic observations to update the coordinates of the control stations (Snay, Cline, and Timmerman 1987). These models had the ability to account for both secular and episodic movement. NGS has developed a new horizontal deformation model for California which is known officially as TDP-H91-CA, and known colloquially as the Time Dependent Positioning (TDP) model. TDP uses Very Long Baseline Interferometry (VLBI) and classical survey observations to model horizontal deformation. This model, which is still in the research stage, is a refinement of the earlier REDEAM models, and has the ability to determine crustal motion at the subcentimeter level (Snay 1992).

Application of this type of modeling can be an attractive alternative to constant reobservation of a network to obtain updated coordinates. If deformation modeling is deemed inappropriate, the crustal motion model will, at the very least, allow the geodesist to project ahead and determine when the distortion in a region has reached an unacceptable level where reobservations are necessary. One can see that deformation models such as TDP and REDEAM have the potential not only to be used as a tool for the geophysicist and geodesist, but also for the administrator as well; the administrator may now have a tool which can be used to forecast future work loads, and budget them accordingly.

Further refinement of horizontal deformation models would be accomplished by constantly integrating GPS data into the models and maintaining a database of these redundant observations. To this end, the District 7 GPS Unit has provided data to the Southern California Earthquake Center (SCEC) for its archives. For example, Caltrans has provided SCEC with GPS observations for a project which spanned the San Andreas Rift Zone near Palmdale, and for control stations in the Mojave Desert which were affected by the 1992 Landers Earthquake. Needless to say, episodic movement such as the Landers quake requires reobservation of the affected area before a valid model can be developed. Because of the Ventura County project, the staff at District 7 and members of the geophysical community have developed a close working relationship and regularly communicate with each other on topics of mutual interest.

Table 1 shows horizontal deformation velocities comparing measured SCEC GPS observations versus the TDP model at selected control stations which were common to the Ventura County Control Net (Snay 1992). These GPS observations were undertaken over the past three years by a team of scientists from NASA and MIT (Donellan, Hager, and King 1992). It is evident from Table 1 that the modeled results are comparable to the SCEC GPS measurements. Please note that the TDP model is still only a research tool, and is shown here to demonstrate the capabilities and potential applications of horizontal deformation modeling.

TABLE 1. Horizontal Deformation Velocities for Modeled Stations in the Ventura County Primary Control Network

NORTH VELOCITY (mm/yr)

STATION	SCEC OBSERVED	TDP MODELED	Δ SCEC-TDP	*SCEC S. DEV
SANTA PAULA NCMN	20.6	21.9	-1.3	0.3
SANTA CLARA	24.5	23.2	1.3	1.1
SOLIMAR	22.5	32.4	-9.9	0.4

EAST VELOCITY (mm/yr)

STATION	SCEC OBSERVED	TDP MODELED	Δ SCEC-TDP	*SCEC S. DEV
SANTA PAULA NCMN	-25.5	-27.9	2.4	0.4
SANTA CLARA	-22.2	-28.7	6.5	1.8
SOLIMAR	-24.1	-27.4	3.3	0.6

Assuming no motion at CIGNET station MOJAVE
* Denotes 1σ formal uncertainty

Modeled on June 23, 1992 by Richard A. Snay

PROJECT EXECUTION

After the initial meetings in early 1992, the District 7 GPS Unit began reconnaissance and monument recovery for the project. In all, twenty-nine control stations would be a part of the project (see Figure 1). The survey would utilize eight HPGN stations as control, thirteen existing monuments, and eight new monuments. Of the existing monuments, six were NGRS vertical control stations, four were NGRS horizontal control stations, and one mark was both an NGRS horizontal/vertical control station. Two non-NGRS horizontal control stations were also included. The average age of an existing monument in the control network was thirty years. The oldest horizontal geodetic mark in the county dating from the late

666

1800's, station SANTA CLARA, was also included. Of the new monuments, six were of the NGS class A rod type and two were brass disks. The majority of the new monuments were set in the westerly and north-westerly parts of the county where there is rugged terrain and a lack of suitable monumentation. All stations included in the network were accessible using 2-wheel drive vehicles.

When the recovery and reconnaissance for the network were completed, NAVD 88 benchmarks were recovered to support the vertical phase of the project. To augment the original seven existing NAVD 88 benchmarks, it was decided to establish vertical leveling ties to ten additional stations of the network. The distribution of vertical control points would give a good indication of the validity of the Geoid90 model in the more developed parts of the county, which is where the bulk of the production related GPS work would occur.

FIGURE 1. Stations in the Ventura County Primary Control Network

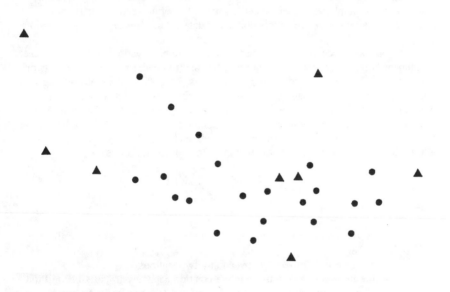

▲ **Denotes HPGN control station**
● **Denotes control station established by this survey**

The field observations were conducted in the early part of June using six 16 channel dual-frequency P-code receivers. The field work conformed to first-order FGCS specifications for static GPS relative positioning. Initially, the data was processed and quality control checked in the field. Reductions at this phase of the project were done to check for blunders and statistical outliers so that the progress of the field survey would not be slowed. Later, more rigorous processing techniques were applied to the data. Tribrachs were checked and adjusted on a daily basis as a regular part of the quality control routine.

When this manuscript was submitted to the editor (mid-August 1992), the differential leveling was in the process of being completed, precise ephemerides had become available for reprocessing, and the project deliverables were being compiled so that they may be submitted to Caltrans Headquarters in Sacramento for review and submission to NGS for inclusion in the NGRS.

PRELIMINARY RESULTS

The network design scheme consisted of 83 independent baseline vectors and 8 repeat baseline vectors. The network used a series of interlocking central polygons and triangles to prove both the internal consistency of the network and the published HPGN control values. FGCS defines (Eq. 1) that each observing session generates:

$$i = r - 1 \qquad (1)$$

where,

i = the number of independent, nontrivial baseline vectors per session
r = the number of receivers collecting data simultaneously during a session.

All stations were occupied on at least two separate observation sessions to detect instrument height blunders and to build redundancy into the survey. The network conformed to FGCS standards and specifications for GPS relative positioning and contained adequate redundancy for a survey of this type. FGCS defines that the maximum allowable linear error in a least-squares adjustment of the baseline vectors of a network will determine the relative positioning accuracy standard of a horizontal control survey using global positioning system methods (Eq. 2). The maximum allowable linear error for a first-order horizontal control survey is:

$$s = \pm \sqrt{[e^2 + (0.1pd)^2]} \qquad (2)$$

where,

s = maximum allowable error in centimeters at the 95 percent confidence level
d = distance in kilometers between any two stations
p = the minimum geometric relative position accuracy in parts-per-million (ppm) at the 95 percent confidence level (first-order = 10 ppm)
e = base error in centimeters (first-order = 1 cm).

The weakest station-to-station relationship in the minimally constrained adjustment had a line length of 5771.365 meters and had a maximum of 1.7 centimeters uncertainty in its adjusted position. The variance of unit weight for the minimally constrained network adjustment was 1.000. By applying Equation 2 developed above to the worst-case relationship (Eq. 3):

$$s = \pm \sqrt{[1^2 + (0.1 \times 10 \times 5.7)^2]} \qquad (3)$$
$$\underline{s = + 5.8 \text{ cm maximum allowable error.}}$$

FIGURE 2. Ventura County Network - FGCS First-Order and Achieved Closures for the Minimally Constrained Adjustment

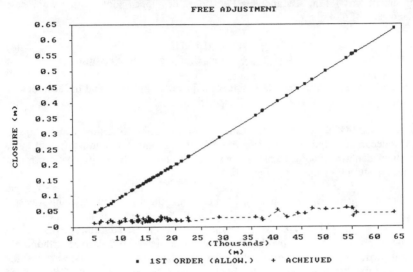

FIGURE 3. Ventura County Network - FGCS First-Order and Achieved Closures for the Constrained Adjustment

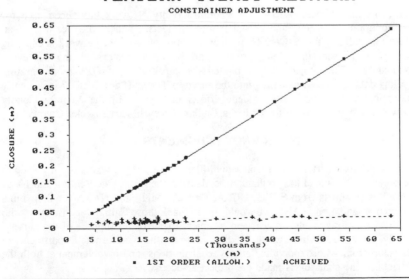

According to FGCS standards, the results of the minimally constrained least squares adjustment indicates that the precision of this survey is within first-order specifications.

The weakest station-to-station relationship in the constrained adjustment had a line length of 5771.367 meters and had a maximum of 1.7 centimeters uncertainty in its adjusted position. The variance of unit weight for the constrained network adjustment was 1.175. By applying Equation 2 developed above to the worst-case relationship (Eq. 4):

$$s = \pm \sqrt{[1^2 + (0.1 \times 10 \times 5.7)^2]} \tag{4}$$
$$\underline{s = +5.8 \text{ cm maximum allowable error.}}$$

The results of the constrained least squares adjustment indicates that the accuracy of this survey is within first-order FGCS specifications.

First-order FGCS closures became routine for all station-to-station relationships in the project (see Figures 2 & 3). The results of both the minimally and fully constrained network adjustments demonstrate that there is little systematic error in either the survey or the constraints.

When this manuscript was submitted to the editor, the differential leveling was still under completion. Preliminary analysis of the GPS derived orthometric heights for the seven existing NAVD 88 benchmarks show that the Geoid90 model fits well in a limited portion of the network. A full analysis of the GPS derived orthometric heights will occur when the leveling data becomes available. Because of the incomplete status of vertical portion of the project, the results of the GPS derived orthometric heights are being withheld until the fall conference.

CONCLUSION

Since its inception, subsets of the Ventura County Network have been used for transportation, cadastral, and crustal motion projects. Others have expressed an interest in utilizing the network for projects of their own and the network is enjoying use by GPS users in public, private, and scientific sectors of the surveying community. The project fully exploited the capabilities of GPS by integrating leveling data and geoid modeling into the survey. The staff at District 7 is currently examining other cooperative geodetic efforts in support of the Ventura County Network, and is preparing to conduct a similar survey in Los Angeles County.

ACKNOWLEDGMENTS

The Caltrans District 7 GPS Unit would like to thank those who were involved in the project. We would like to thank officials from the County of Ventura and NGS, as well as scientists from SCEC, UCLA, Caltech, MIT, and NASA for their input and recommendations. Also, we would like to thank those Caltrans' consultants who provided encouragement, advice, and technical expertise in support of the survey. Finally, we would like to thank the staff at the California Department of Transportation, whose individual efforts, contributions, and involvement at both the district and the state levels made the project a success.

DISCLAIMER

The opinions expressed in this paper are those of the author and do not necessarily reflect the views or policies of the California Department of Transportation.

REFERENCES

Donellan, A., Hager, B.H., and King, R.W. 1992. "Geodetic Determination of Rapid Shortening Across the Ventura Basin, Southern California." *Journal of Geophysical Research.* (Submitted March 6, 1992).

D'Onofrio, J.D. and Nelson, R.L. 1991. "High Precision Geodetic Network for California." *Presented at the ASCE Specialty Conference: GPS '91, Transportation Applications of GPS Positioning Strategy.*

Federal Geodetic Control Committee. 1989. *Geometric Geodetic Accuracy Standards and Specifications for Using GPS Relative Positioning Techniques.* National Geodetic Information Center, NOAA, Rockville MD 20852.

Launen, K.J. 1991. "GPS - Rapid Solutions for Transportation Management." *Presented at the ASCE Specialty Conference: GPS '91, Transportation Applications of GPS Positioning Strategy.* Sacramento, California, September 18-21.

Milbert, D.G. 1992. "GPS and Geoid90 - The New Level Rod." *GPS World.* January-February. pp. 38-43.

Snay, R.A., 1992. National Geodetic Survey, Rockville, MD. Personal Communication.

Snay, R.A., Cline, M.W., and Timmerman, E.L. 1987. *Project REDEAM: Models for Historical Horizontal Deformation.* National Geodetic Information Center, NOAA, Rockville, MD 20852. pp. 4 & 20.

Strange, W.E. 1989. "GPS Determination of Groundwater Withdrawl Subsidence." *Journal of Surveying Engineering.* Vol. 115, No. 2, 198-206.

Zilkoski, D.B. 1990a. "Establishing Vertical Control Using GPS Satellite Surveys." *Proceedings of the 19th International Federation of Surveying Congress (FIG).* Commission 5. pp. 281-294.

Zilkoski, D.B. 1990b. "Minimum Steps Required When Estimating GPS-Derived Orthometric Heights." *Proceedings of the GIS/LIS '90 Fall Convention,* Anaheim, California, November 7-10.

Zilkoski, D.B. and Hothem, L.D. 1989. "GPS Satellite Surveys and Vertical Control." *Journal of Surveying Engineering.* Vol. 115, No. 2, pp. 262-282.

A "BOTTOM-UP" APPROACH TO GIS WATERSHED ANALYSIS

Jeffrey A. Schloss, Freshwater Biology Group and Cooperative Extension Water Resources
and
Fay A. Rubin, Complex Systems Research Center, Institute for the Study of Earth Oceans and Space

University of New Hampshire, Durham, NH 03824

ABSTRACT

Typical use of GIS for watershed management has relied on a more or less "Top Down" approach (ie: slope, soils and buffer areas around waterfront and wetlands) for developing management decisions. This method produces "developable areas" in terms of Best Management Practices or local zoning laws but does not take the particularities of the lake into consideration. An alternative method is a "Bottom Up" approach that uses the GIS to consider lake morphometry and water quality data, as well as biological and habitat parameters. Such a method was used on the multi-basin Squam Lakes combining extensive volunteer monitor data (NH Lakes Lay Monitoring Program) with the comprehensive information base entered into the New Hampshire GRANIT GIS system. "Bottom Up" analysis of the Squam Lakes indicates that the many basins (bays and embayments) act as separate "lakes" throughout most of the year. Mean depth of the basins, as well as the ratio of basin surface area to area of the sub-watershed, determines each basin's water quality to a large extent. These relationships, past and present water quality, and in-lake wildlife habitat, are used to calculate a set of weighting factors to define critical areas within the lakes. Applications are discussed for future planning, conservation and management decisions.

BACKGROUND

The Squam Lakes watershed in central New Hampshire was the subject of a study to develop a model watershed management plan. The study was coordinated by the NH Office of State Planning (OSP), assisted by a multi-agency technical advisory group and under the review of a local citizen advisory committee. A comprehensive information base was entered into New Hampshire GRANIT (Geographically Referenced Analysis and Information Transfer), the state's GIS. Data layers analyzed included geology, soils, hydrology, elevation, land use, vegetative cover, and wildlife habitat. Also included was ten years of water quality data collected throughout the lakes by volunteer monitors of the Squam Lakes Association in collaboration with the Freshwater Biology Group at the University of New Hampshire as part of the New Hampshire Lakes Lay Monitoring Program (see Ellett and Mayio 1990).

The resulting plan (Scott et al 1991) emphasized the need for balancing further development of the watershed with the natural capability and limitations of the land. The watershed of the two lakes covers 42,418 acres of which 7,847 is water area. A land capability analysis, performed by removing wetlands and shoreline buffer areas, steep slopes, poor soils as well as already developed and protected lands and then overlaying the current zoning requirements for each of six towns bordering the lake, disclosed that over a third of the watershed (13,476 acres) could still be developed under current regulations. Total build-out of the watershed would greatly affect the lake water quality which is currently in the excellent category but showing subtle changes suggesting a slow decline (Schloss et al 1990). To assist communities in preventing this total build-out scenario the plan included a GIS analysis of land containing productive natural resources (groundwater availability, agricultural soils and productive forest soils) and suggests protective measures to be incorporated into local plans and reviews. The plan further recommended that analysis be conducted to locate critical areas of the lake that need protection.

This present study was undertaken to develop such a method for locating critical lake areas and developing additional GIS products to assist the local communities in their watershed management decisions.

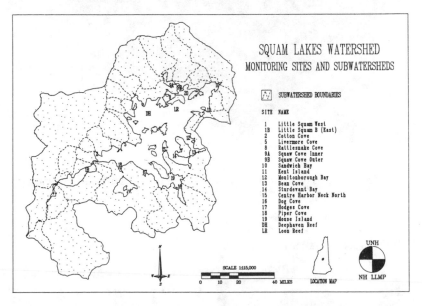

Figure 1. Map of the Squam Lakes Watershed with monitoring sites and subwatersheds.

BOTTOM UP APPROACH

The suite of watershed analyses conducted in the OSP study involve a more or less "Top Down" approach which, except for shoreline and buffer consid-erations, employed land based constraints and did not incorporate any par-ticularities of the lake directly into analysis. For a lake system like the Squam Lakes where the shoreline is quite irregular with many bays and coves, (Figure 1) it was decided to investigate possible critical lake areas using a "Bottom-Up" approach. Lake bathymetry was input into the GRANIT system and the generated three dimensional computer perspective reveals a very complicated system of sub-basins throughout the lakes separated by sills (Figure 2). Many of these sills rise high enough to effectively separate the bottom waters of these basins when the lake is thermally stratified during the summer and winter (Figure 3). Thus instead of acting as a single reaction vessel or "bathtub", the separate basins of the lake may show different water quality, dependent on basin characteristics and the characteristics of the abutting sub-watersheds. Combining the bathymetric data with temperature profiles taken by the lake monitors throughout the decade allowed for the delineation of 18 lake basins of which 17 had been consistently monitored (see Figure 1).

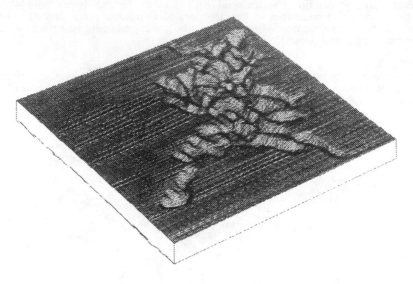

Figure 2. Three dimensional computer perspective of the Squam Lakes bathymetry.

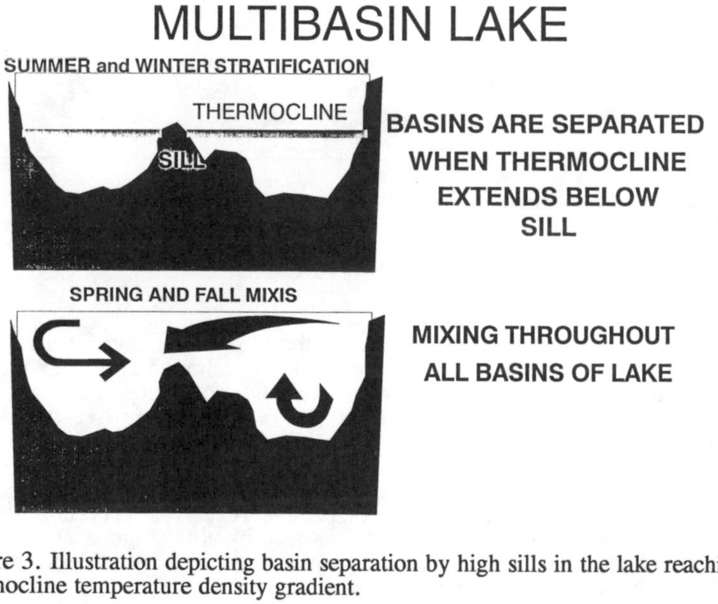

MULTIBASIN LAKE

SUMMER and WINTER STRATIFICATION

THERMOCLINE

SILL

BASINS ARE SEPARATED WHEN THERMOCLINE EXTENDS BELOW SILL

SPRING AND FALL MIXIS

MIXING THROUGHOUT ALL BASINS OF LAKE

Figure 3. Illustration depicting basin separation by high sills in the lake reaching the thermocline temperature density gradient.

Areal and Morphometric Determinants

An intensive analysis of basin water quality utilizing the many information layers entered into GRANIT for the Squam Lakes watershed study (listed above) is nearing completion (Schloss and Rubin in preparation). For wider applicability preliminary effort was devoted to areal and morphometric parameters as those data are more readily available for most lakes. By utilizing Secchi Disk water transparency, seasonal average chlorophyll a and the frequency of relative "bloom events" (greater than twice the typical average chlorophyll levels; generally > 6 ppb chlorophyll a) it was found that the mean depth (LAKE VOLUME to LAKE SURFACE AREA) of the majority of lake basins was an indicator of how well the basin was assimilating nutrient inputs from the abutting sub-watersheds (Figure 4). A secondary factor found to be indicative of water quality was the size of the sub-drainage area relative to the basin area; Those basins with very large abutting sub-watersheds relative to the basin area displayed poorer water quality (higher productivity, lower water clarity). Combining these findings the GIS can now produce a water quality assessment indicating lake areas with current water quality concerns, and sub-watershed locations where future development would have the greatest effect (Figure 5). For the Squam Lakes, the most critical basins were those with less than 2 meters (6.6 feet) mean depth and those with a subwatershed area to basin area ratio greater than 10. Least critical were those basins with mean depth greater than 8 meters (26.5 feet) or subwatershed to basin ratios less than 2.

SECCHI DEPTH vs MEAN DEPTH
Squam Lakes Watershed Analysis

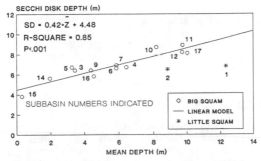

CHLOROPHYLL vs MEAN DEPTH
SQUAM LAKES WATERSHED ANALYSIS

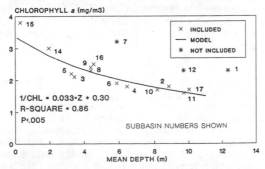

Figure 4. Relationship between mean depth of basin and water quality parameters secchi disk transparency (top) and chlorophyll (algae) concentration (bottom).

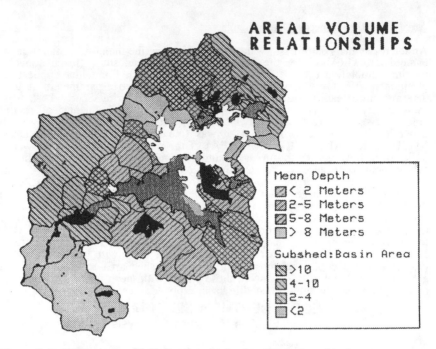

Figure 5. In-lake water quality assessment and sub-watershed areal/volume relationships. For the lake basins, the darker the basin color, the more productive the basin. For the sub-watersheds, the line direction denotes the mean depth or area ratio constraint while the darkness of the line denotes the severity.

In-lake Wildlife Assessment

Along with water quality, in-lake wildlife resources of the lakes are a major concern. The Squam Lakes offer some of the state's best cold water and warm water recreational fishery. In addition, the Common Loon, found throughout the lake, is a very charismatic species throughout the state and is given high priority for protection. Locational and extent information on GRANIT of warm water fish nesting areas, cold water fish basins, spawning reefs and brooks, and loon nesting areas provided by NH Fish and Game and the Loon Preservation Society allows for the development of an In-Lake Wildlife Assessment.(Figure 6). The map displays those sub-watersheds that drain into lake areas with various extent of critical wildlife habitat.

Extent of loon habitat was determined by examining the amount of shoreline bordering each basin which offers suitable loon nesting areas. The extent ranges from 3,500 to 14,800 linear feet for basins that contained any habitat (14 out of 18 basins) for a total of almost 114,000 linear feet around the lake. High extent basins (Extent 3 in Figure 6) were those with greater than 10,000 linear feet of habitat. Low extent basins were those with less than 5000 linear feet of habitat. Warm water fish extent was determined in a similar fashion using the total area of bass nesting sites which ranged from 1.3 to 200 acres for a total of about 400 acres at 9 of 18 basins. Greater than 60 acres was considered high extent and less than 20 acres was considered low extent. For the coldwater fishery, extent of habitat was determined by adding together the number of salmon and whitefish "holes" (deep sites known to contain populations), the number of smelt brooks and the extent of trout and whitefish reefs. Basins containing all three habitat and those with high incidence of any habitat were considered high extent basins.

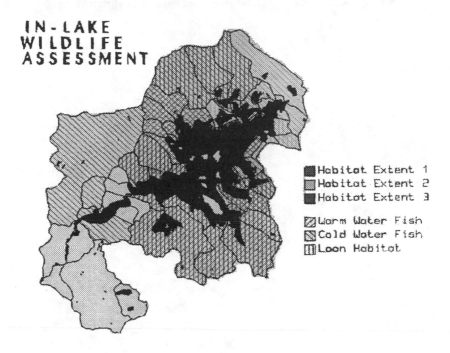

IN-LAKE WILDLIFE ASSESSMENT

Habitat Extent 1
Habitat Extent 2
Habitat Extent 3

Warm Water Fish
Cold Water Fish
Loon Habitat

Figure 6. In-lake wildlife assessment. For explanations of extent see text. Direction of line indicates the specific wildlife component while the intensity of the line indicates extent.

This information can be useful in determining where in the sub-watershed care must be taken to protect these wildlife resources. Conservation commissions and planning boards can then take this into account when reviewing development projects and specifically address potential impact to the various wildlife components of each basin. Knowledge of the type of wildlife may also allow for the use of seasonal management plans where clearing and construction are limited during nesting, spawning, migration or other critical times.

CRITICAL BASIN DETERMINATION

By combining the current water quality, basin assimilation potential (Areal-Volume Assessment) and wildlife assessments, the critical areas of the lakes (on a relative scale) can be delineated (Figure 7). In the Squam Lakes situation, equal weight was given to each of the three assessments but this can be modified dependent on the specific concerns of the community. This information can then be integrated with the "Top Down" land based concerns to facilitate protection or management strategies sensitive to the lake. In addition, this approach can be used to locate impotant areas for protective easements or land purchases.

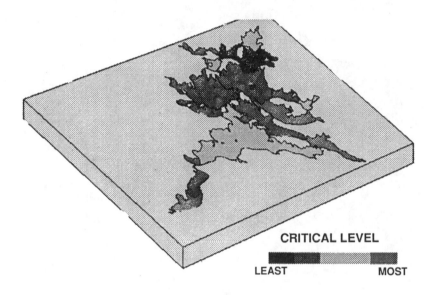

Figure 7. Critical basin areas in the Squam Lakes as determined by giving current water quality, basin assimilation characteristics and in-lake wildlife resources equal importance. The smaller bays and coves tend to be most critical with the deeper water basins being less critical.

CONCLUSION

The "Bottom-Up" approach to GIS watershed analysis disclosed the correlation between water quality in the various basins of the lake with specific basin and subwatershed characteristics. While the breadth of water quality and other lake data for this particular lake well exceeds information available for most lakes, this study has shown that readily available data on lake bathymetry and subwatershed areas should be incorporated into studies of multi-basin lakes and reservoirs. Dependent upon the local priorities and concerns other in-lake data should be collected and assessments should be created to insure critical lake areas are delineated. With this information available to the local decision makers, protection and management of the important watershed and lake resources can be facilitated. The appraoch can also delineate in lake and abutting sub-watershed "hot spots" that require protection or further investigation.

ACKNOWLEDGEMENTS

The authors wish to thank Dave Scott and Jim McLaughlin of the Office of State Planning for use of the GRANIT facilities and data. Molly Boutwell assisted in the map preparation. This study would not have been possible without the efforts of the volunteer monitors of the Squam Lakes Association in compiling the extensive water quality data-base.

REFERENCES

Ellett, K. and A. Mayio. 1990. VOLUNTEER WATER MONITORING: A Guide for State Managers. United States Environmental Protection Agency. Office of Water. Washington DC. EPA 440/4-90-010

Schloss, J.A., A.L. Baker and J.F. Haney. 1990. The Squam Lakes: Annual Report of the New Hampshire Lakes Lay Monitoring Program. Freshwater Biology Group, University of New Hampshire, Durham.

Schloss, J.A. and F.A. Rubin. in preparation. Areal and Morphometric Determinants of Primary Productivity in a Multi-basin Lake.

Scott, D., D. McLaughlin, V. Parmele, F. Latawiec Dupee, S. Becker and J. Rollins. 1991. Squam Lakes Watershed Plan. Office of State Planning, Concord, NH.

INTEGRATION OF A COUNTY AND CITY GIS
OVER FIBER-OPTIC LAN

David E. Schmidt
Winnebago County Planning Department
415 Jackson Street
Oshkosh, Wisconsin 54903-2808
(414) 236-4837

John E. Lee
Genasys II, Inc.
2629 Redwing Road, Third Floor
Fort Collins, Colorado 80526
(303) 226-3283

ABSTRACT

The Winnebago County Geographic Information System (WINGS) project was conceived and designed to be an integrated multiparticipant GIS involving a Wisconsin county and its constituent cities. The concepts of data and technology sharing were considered to be both desirable and paramount. Following selection and implementation of the GIS hardware and software by WINGS participants, a necessary task was the preliminary integration of the separate GISs installed by Winnebago County and the City of Oshkosh.

This paper describes the design and implementation of a cost-effective, state-of-the-art, near-media speed, fiber-optic based, local area network used to connect the County and City GISs. Benefits of performance, cost, data sharing, technology transfer, and future connectivity options are discussed. Integration options for inclusion of future, remote WINGS participants are also presented.

INTRODUCTION

The Winnebago County Geographic Information System (WINGS) project was conceived and designed to be an integrated multiparticipant GIS involving a Wisconsin county and its constituent cities. Winnebago County is located in east central Wisconsin and occupies an area of 578 square miles. Oshkosh, the county seat, is the largest center of population in the county with approximately 56,000 residents. The WINGS project was undertaken with the objective of developing and implementing an accurate and up-to-date common GIS for the County and several area municipalities. The concepts of data and technology sharing were considered to be both desirable and paramount. One project goal was to develop a common GIS database which could be utilized by all the participants.

Potential participants in the WINGS project included Winnebago County, the City of Oshkosh, the City of Neenah, the City of Menasha, the Town of Menasha, the Village of Winneconne, the City of Omro, East Central Wisconsin Regional Planning Commission, and Wisconsin Bell. A phased implementation approach was conceived with one major problem – how to integrate the individual GISs to achieve the common database goal. While the participants could easily share the digital database using tape transfer, a physical connectivity between systems was preferred. Connectivity would minimize tape back-ups and increase the synchronaity of the database, which was constantly undergoing changes.

Phase One was the integration of the separate GISs initially installed in the courthouse and the Oshkosh City Hall. It had been decided that the County would be responsible for the maintenance and repository of the WINGS database. These participants were physically separated into two buildings several blocks (600 feet) apart. Other participants, who were to acquire systems in Phase Two, also posed integration problems due to their physical locations, often up to 10-20 miles from the county seat.

THE HARDWARE AND SOFTWARE

After issuance of an RFP, evaluation of the vendor responses, and evaluation of the technical benchmark results of the finalists, the WINGS participants chose the GenaMap GIS from Genasys II, Inc. The GenaMap software is a full-functionality, vector-based GIS that stores data in a topologically structured edge/node format and allows for easy and rapid data capture, edit, display, and analysis. The GenaMap GIS operates in the UNIX operating system environment.

The UNIX-based computing environment offered several advantages to the WINGS participants, including expandability, versatility, modularity, and connectivity. UNIX-based workstations from Hewlett-Packard (Model 9000 Series 700) were chosen as the computer hardware platform for the GenaMap GIS. The HP workstations offered several advantages in terms of price, performance, and connectivity solutions. The HP 700 workstations are based on HP's Precision Architecture Reduced Instruction Set Computing chips and operate with HP's implementation of the UNIX O/S, HP-UX Version 8.

In addition to the line of HP 700 workstations available to the WINGS participants, X-terminals (HP 700/RX) and low-end 386-based PCs (running Interactive 386ix UNIX) were considered for expanded capacity and remote sites.

THE CONNECTIVITY PLAN

It was envisioned that all proposed communications between workstations/ participants would be based on a standard IEEE 802.3 Ethernet Local Area Network (LAN) with transport protocols, or on 9600 baud modem telecommunications. This environment is based on industry standards and was designed to be robust, high speed, and effective. The objective was to tie all eventual workstation locations together into a single, integrated network so that resources and data sharing were facilitated. It would also take advantage of existing 9600 Baud communication (phone) lines.

Any location running a workstation would be connected via Ethernet. Ethernet operates at 10 Mb/sec and is extremely reliable. With the TCP/IP and NFS protocol software layers, 100 percent data reliability is maintained. A ThinLAN coaxial cabling system would be used for the physical connections. Network servers would be located initially in the County facility, and then later in the City of Oshkosh and the City of Neenah facilities. Each Network Server would service the local LAN for that governmental unit. The other towns and cities would use the County server as their network/communications server. Within each server location, all workstations, X-terminals, and PCs to be used for GIS or related activities would also be connected via the LAN.

Other workstations/PCs at remote locations could be connected via phone lines and simple dial-in modems. Some communications package, such as Reflections 7 Plus, would be required. This configuration would allow interactive file transfer between the remote sites and the County; however, interactive graphics would not be supported.

This proposed LAN environment, while more expensive than shipping tapes back and forth, was deemed to be the best long-term solution for the WINGS participants. One power of a LAN is its flexibility and extensibility. The initial workstation would require only a LAN terminator. The network could then be quickly expanded as stations are added by simply running LAN cable and hooking up other machines within the County building. The TCP/IP and NFS protocol utilities handle all communication, file sharing, and file transfer. In this way, only one version of the geographic database need be maintained on the network. The model used in the County facility could be directly translated to the City of Oshkosh and City of Neenah. In this way, totally self-contained networks could be established locally.

Connecting these participants would require a different approach. To do this, Genasys originally proposed HP Remote Bridges, CODEX 3500 Data Service/Channel Service (DSU/CSU) digital modems, AST-5 transceiver units, and AUI drop cables for interconnecting the individual Ethernet LANs into a single enterprise-wide LAN. The HP Remote Bridge can connect one local and one remote LAN to form a single, integrated communications network across geographically dispersed sites. A standard V.35 interface to an external DSU/CSU, such as the proposed CODEX 3500, provides access to common telephone services.

The HP Remote Bridge (HP 28674A) is a two-port, transparent learning bridge that filters each packet from the local and remote segments which can operate at full media speed (14,880 64-byte packets per second). Filtering improves throughput of the wide area network by forwarding only the necessary traffic that must go to/from the remote sites, thus reducing network bottlenecks. Unlike LAN repeaters, the HP Remote Bridge does not propagate corrupt packets between LANs, thus maintaining end-to-end integrity. The HP Remote Bridge incorporates the latest version of the IEEE spanning tree algorithm for support of redundant/alternative links, thereby ensuring continued data transmission between LANs in the event of a primary link failure. The HP Remote Bridge includes an RS-232C console port for monitoring and control functions within the bridge. This interface allows network administrators to check bridge status, spanning tree configuration, network traffic statistics, collisions, and perform basic diagnostics.

HP recommends the CODEX 3500 DSU/CSU digital modems for direct connection to telephone services such as the DATAPHONE Digital Service (DDS). The CODEX 3500 combines all data rates in a single unit and supports 2.4Kb, 4.8Kb, 9.6Kb, 19.2Kb, and 56Kbs operation speeds. However, for near-media speed with the GIS application, the CODEX 3500 digital modem connection would require WINGS participants to contract with their telephone service provider to select a dedicated digital line at the 56Kbs data rate.

Using this configuration, workstations would be able to perform file transfers and processor communications across the network. Benefits to WINGS participants would include: files that reside on any workstation and/or server would be available to any other workstation and/or server in the same manner as local disks via the NFS Yellow Pages facilities; users would be able to log into any physical workstation node and have their customized user environment and files downloaded onto that workstation as if they were working at their personal system; users would be able to remotely execute applications programs, system commands, and utilities on any system as if they were working locally; programs executing on any workstation or server would be able to communicate with programs on any other workstation or server on the network; users would have access to use the 'lp' and/or optional HP spooling systems for all GIS peripheral devices across the network; and the system administrator would be able to prepare a new version of the operating system for any node on the network from any other node, and to install the operating system remotely on the node's local disk.

The benefits of the LAN bridge configuration were clearly the ability for the County and the City GISs to interoperate at near-media speed for a low communications cost. Total costs for the LAN bridge equipment components were less than $10,000 retail. However, the stumbling block came when the County investigated the costs to provide a dedicated 56Kbs phone line between the County and City facilities. To be provided by the local phone company, these costs approached $1,000/month and were deemed too excessive by the WINGS participants. Faced with use of standard phone line speeds (9.6 Kbs) and the commensurate drop in communications speeds, power capabilities, and flexibility, Genasys was asked to investigate alternative approaches.

THE CONNECTIVITY SOLUTION

Following an additional site visit and further inquiries, it was learned that situated between the County and City facility was a County/City law enforcement building. Utilizing an underground tunnel, PCs were apparently connected between this building and both the County and City buildings. Thus, there was a direct underground connection that could be used. While the distance through the tunnel (<1200 feet) did not prohibit the use of ThinLAN cabling with repeaters, it was found that AT&T 62.5/125 micron 12 strand fiber optic cable could be acquired, pulled, and fitted with connectors, couplers, and splitters for less than $7,000. This physical connection media offered several advantages in terms of shielding, security, and media bandwidth.

The move from a telephone-based to a fiber-optic-based physical media necessitated a different approach for the LAN connectivity. It was desirable, however, that the new approach provide functionality and benefits similar to the remote bridge solution. HP's recommendation was to use the 10:10 LAN Bridge (HP 28673A) to provide the interconnectivity between the LANs. Like the remote bridge, the 10:10 LAN bridge is a two-port learning bridge that provides address filtering for traffic isolation, uses the spanning tree protocol for alternative/redundant paths, operates at media-speed, maintains end-to-end data integrity, and has an RS-232 console port for network monitoring. In addition, HP Ethertwist Transceivers (HP 28683A) are used on both ends to connect the fiber-optic cable to the AUI port of the bridge. The fiber-optic transceivers connect to the cable via two ST connectors (Tx and Rx), support both 62.5/125 and 50/125 micrometer fiber-optic cable, and support cable distances of up to 2 km. Use of the 10:10 LAN bridge thus provided functionality identical to the remote bridge for an identical price – less than $10,000 retail.

Immediately prior to installation, Genasys discovered that it could substitute an HP EtherTwist hub (HP 28688B) for one of the 10:10 LAN bridges. The EtherTwist hub is a Type 10BASE-T multiport repeater that works in conjunction

683

with the LAN bridge to support growing LAN topologies. The EtherTwist hub provides 12 twisted-pair (10BASE-T) ports, with the BNC and AUI ports all simultaneously active; thus, after installation, there would be 12 unused twisted pair ports. The EtherTwist hub also provides autosegmentation capabilities to identify and disconnect any segment disturbing the network (fault isolation) and provides a network management RS-232 console port. Use of this device also resulted in a cost reduction of approximately $2,200.

Further cost savings were realized prior to implementation with a newer 10:10 LAN Bridge from HP (HP 28681A). This bridge is a local bridge that does not offer the network management console port or use of the spanning tree algorithm, but does filter network traffic. Operating at 90 percent of media speed (13,373 packets/sec), this hub is slightly slower, but this "near-media speed" was not perceived to significantly affect performance. The obvious advantage was a $2,500 further reduction in cost.

The final configuration was implemented as diagramed below.

Winnebago County   City of Oshkosh
415 Jackson Street   215 Church Avenue

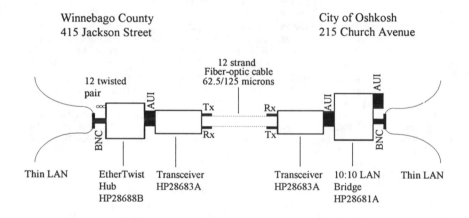

Figure 1. WINGS Fiber-Optic LAN Implementation

FUTURE SOLUTIONS

Integration of additional WINGS participants involves two strategies. For those participants large enough to have their own LAN (Cities of Menasha and Neenah), the phone-based remote bridge connection originally conceived is still viable. As before, the cost of the dedicated high-speed phone lines will be the major impediment to this strategy. However, the City of Menasha is already connected to Winnebago County via T1 speed telecommunications, which should have sufficient unused bandwidth to accommodate the GIS LAN.

Remaining participants with only standalone workstations (low-end HPs or PC-based UNIX) are expected to implement standard non-digital 9600 baud modem communications. Coupled with some form of communications/terminal emulation package, this connection would at least allow participants to transfer files remotely, without the need for tape transfer.

CONCLUSION

The final connectivity solution as implemented was a cost-effective, state-of-the-art, near-media speed, fiber-optic based, local area network used to connect the County and City GISs. Benefits were realized in terms of performance and cost, and the objectives of data sharing and technology transfer were met. This implementation also provided for future connectivity options for the inclusion and integration of additional, remote WINGS GIS participants.

DATA QUALITY AND CHOROPLETH MAPS: AN EXPERIMENT WITH THE USE OF COLOR

Diane M. Schweizer and Michael F. Goodchild
University of California Santa Barbara
Santa Barbara,CA 93106-4060
diane@geog.ucsb.edu
good@geog.ucsb.edu

ABSTRACT

Thematic attribute uncertainty is an inherent feature of maps produced from geographic databases since all spatial data are of limited accuracy. To minimize the incorrect interpretation of GIS output, accuracy information needs to be readily available. The purpose of this research is to determine if the connotative implicatons of gray are effective in displaying data uncertainty. The Hue Saturation Value (HSV) color model is tested for its ability to simultaneously map data attributes and corresponding data uncertainty. Saturation is used to symbolize data attributes. Value is used to symbolize uncertainty rankings as the value scale corresponds to the gray scale. An experiment designed using this color theory was administered to 101 students. Results of the experimentation indicate that value, when correlated to grayness, is not an effective means of displaying data uncertainty. The authors suggest the inability of value to represent data uncertainty may be attributable to the relationship between grayness and the value scale defined for this experiment. As a result, the authors further suggest maximum grayness occurs towards the center of the gray scale rather than at an endpoint.

DATA UNCERTAINTY

All spatial data are of limited accuracy as spatial databases are abstractions (Goodchild and Gopal 1989) of actual spatial, locational (thematic) and temporal variation. It follows that thematic attribute uncertainty is an inherent feature of maps produced from geographic databases. Uncertainty arises according to the particular data collection and statistical methods employed in producing a map. Within a GIS, the accuracy of spatial data is changed through such manipulations as scale change and data generalization and such operations as buffering and overlay, among others.

While the degree of uncertainty can be minimized through careful attention to data collection and data manipulation practices, data uncertainty cannot be eliminated. The effective and proper use of GIS output is a function of decision-makers' awareness of the error component of GIS output and the decision-makers' ability to interpret such information. To minimize incorrect interpretation of GIS output, accuracy information needs to be readily available. Possibilities for

686

the communication of data quality information include the use of metadata and attribute tables. An alternative method is to communicate data quality information visually. Visual techniques include the use of color as a means of displaying data quality information. (Buttenfield 1991, Goodchild 1991, MacEachran 1991, McGranaghan 1991).

According to Robinson (1967), when perceiving color on a map, map readers do not clearly distinguish between psychological reactions to color and the color representation as described by the legend. It follows that an effective method of displaying data uncertainty is to use an idea that triggers the notion of uncertainty in the map users' mind. The concept of uncertainty invokes images of fog and haze, as "it is not clear to me, it is foggy". As such, fog is a metaphor for uncertainty. The color which best represents fog is gray. The purpose of this research is to determine if the connotative implications of gray are effective in displaying data uncertainty.

COLOR

Color, or visible light, can be measured according to three physical properties: wavelength, intensity and purity (Thorell and Smith 1989). Wavelength represents the segment of the visible spectrum being reflected, intensity is the strength of the light and purity is the existence of one or more wavelengths. Humans perceive color according to psychological rather than physical properties. (Hilbert 1987, Travis 1991). Three psychological dimensions of color are hue, the perception of wavelength; saturation, the perception of purity; and brightness, the perception of intensity (Evans 1974, Thorell and Smith 1989). A one-to-one correlation between the physical properties of color to the psychological properties of color does not exist (Thorell and Smith 1989). For example, changing the wavelength of visible light may or may not result in a change in hue. The lack of a one to one relationship between the physical and perceptual properties of light makes it difficult to alter color according to color perception on a computer screen. The RGB (Red-Green-Blue) color model is often used to represent color on computer screens, however, it is not perceptually intuitive.

The Hue-Saturation-Value (HSV) color model is a transformation of RGB space and has been developed according to the three psychological dimensions of color. The HSV color model is represented as a hexacone. Hue is measured as an angle between 0 and 360 degrees about the vertical or value axis. Saturation is measured from 0 at the center of the hexacone to 1 along the perimeter of the hexacone. At a saturation of 1, a color is fully saturated and at a saturation of 0, a color is said to be desaturated and the gray scale exists along the value axis. Value ranges from 0 to 1. A value of 0 is black and a value of 1 is white. The HSV model is considered to be a perceptually intuitive color model (Travis 1991, Thorell and Smith 1989) as humans are able to perceive differences along the three individual components.

Evans (1974) identifies grayness as darkness and suggests grayness increases along the gray scale with black as the limit for increasing grayness. The goal of this paper is to determine if value, as represented by the HSV color model, is an effective tool for displaying data quality information. The structure of the HSV model appears to lend itself to the representation of data quality because of the ease of manipulation along the value axis. A cross-section of the HSV model, with the following parameters: hue equal to 265 (blue), saturation ranging from 0 to 1 and value ranging from 0.3 to 1 (see Figure 1) is examined in this research.

Cross-section of HSV Hexacone

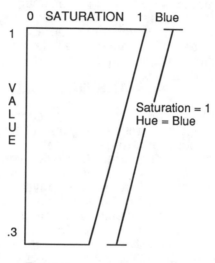

Figure 1

BIVARIATE CHOROPLETH MAPS

The problem cartographers are being faced with when displaying data values and data quality is a bivariate mapping problem. Two distinct sets of variables, as well as their relationship to one another need to be represented in a perceivable way (Carstensen, 1984). (For the purpose of this research, data values shall be referred to as data quantity to prevent confusion between data values and the value component of the HSV model).

Guidelines for color use suggest saturation (Cuff 1972) and value (Cuff 1972, Antes and Chang 1990) are effective in displaying quantitative differences. Attribute data quality and data quantity can both be defined in quantitative terms. Thus, saturation and value are employed to map these two variables.

In this research, value, as described by the HSV color model, is being tested for its ability to map data certainty and saturation is used to map data quantity. If the use of value is effective in displaying data certainty, the shading schemes for bivariate maps displaying data quality and data quantity can be considered to belong to a separate group from other bivariate choropleth shading techniques. If however, the use of value is ineffective in communicating data uncertainty, data quality and data quantity choropleth maps can benefit from bivariate choropleth shading research.

EXPERIMENT

An experiment was designed to determine if users can effectively interpret data quality when encoded as value and data quantity information when encoded as saturation and displayed simultaneously on a bivariate choropleth map. One hundred and one students verbally screened for color blindness participated in the experiment. Subjects were categorized as experienced or inexperienced map users. Experienced map users were represented by graduate geography students as well as undergraduate students who completed at least one advanced cartography course. Inexperienced map users were represented by undergraduate students from a variety of disciplines taken from introductory level geography classes.

The same bivariate choropleth map of the United States was used for the entire experiment. Participants were informed that all data was simulated and decisions were to be based on the color scheme of the map. To avoid any biases associated with the United States, the quantity variable was identified as an imaginary "tribble" population density. Customary statistical measures of accuracy are described in terms of standard deviation and standard error. For the purpose of this experiment, data quality was presented as a percentage of data quantity to eliminate heteroskedasticity and described as how well the mapped data conforms to actual data.

Although hue, saturation and value are based on the three psychological dimensions of color, increments along the individual axes are not visually equidistant. As such, saturation varied from 0 to 1 and value varied from 0.3 to 1 according to Stevens' equal value gray scale as suggested by Kimerling (1985). Hue was held constant at 265 degrees (blue).

To give the appearance of continuous shading, 15 saturation and 15 value levels were used . The use of 15 value levels and 15 saturation levels requires 225 distinct colors to be displayed on the screen simultaneously. The maximum number of allowable color on the particular hardware device used for this experiment is 256. 225 colors were selected to allow additional colors to be used for other purposes such as text. Continuous mapping rather than discrete mapping was chosen for this experiment as the authors feel continuous mapping provides a more complete understanding of the data being displayed. The concept of choropleth maps without class intervals was introduced by Tobler (1973) as a means of enhancing map readability. Additional research provided by Muller (1979) suggests map readers are able to perceive geographic patterns on continuously shaded maps and can consistently and logically interpret the information being displayed.

The experiment consisted of having students respond to a series of 31 questions administered from IBM PS/2 computers. For each question, subjects were asked to make a selection of either a higher quality value or a higher quantity value from a selection of two states. More specifically, along the bottom of the screen either the phrase "Choose the more certain tribble population density" or "Choose the higher tribble population density" was displayed. The two possible states to chose from were highlighted in white to eliminate requiring prior knowledge of state names. Selections were made using a mouse. The experiment consists of two parts.

The purpose of Part I was to determine if there is an intuitive association between value and data quality. Additionally, results of this portion of the experiment will determine if users were differentiating between saturation and value and if they were consistently associating saturation with one variable and value with the other.

In Part I of the experiment, legend boxes did not appear. Subjects were presented with a series of 16 questions. For each question, subjects were asked to select the state with either higher data certainty or higher population density from a choice of two states. Before the maps appeared, subjects were required to read an information screen explaining the nature of the experiment. The information screen familiarized the subjects with the concept of mapping data quality as it is a relatively unfamiliar concept. Of 90 subjects participating, 40 were experienced map users and 50 were inexperienced.

Part II of the experiment was designed to determine if subjects associated the amount of value with data uncertainty despite the symbolism presented by the legend. This portion of the experiment consisted of a control legend, Legend A and a test legend, Legend B. See Figure 2 for a description of each legend. In Legend A, saturation represented population density and value represented the level of data uncertainty. In Legend B, the axes are reversed. That is, value represented population density and saturation represented data uncertainty. Forty-nine subjects, 20 experienced and 29 inexperienced, were administered control legend A against 52 subjects, 20 experienced and 32 inexperienced, administered test legend B.

Subjects were presented with a series of 15 questions. The legend was constant throughout this part of the experiment. As in Part I, subjects were asked to select the state of either higher data quality or higher quantity from two possible choices.

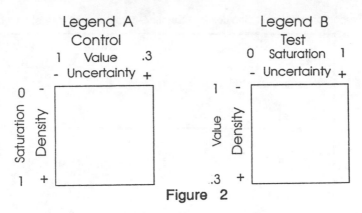

Figure 2

RESULTS

Part I : No Legend

The mean of percent correctly scored for experienced users is 54% with a standard deviation of 21%. For inexperienced users, the mean is 48% with a standard deviation of 20%. The similarity between mean scores and standard deviations for the two populations raises the question of whether or not choices were made randomly. If the selections made were random, a probability of .5 can be assigned to obtaining a correct choice and a probability of .5 can be assigned to obtaining an incorrect choice. Random selection by users would show a proportion of correct choices for each of the 16 questions clustered about .5.

Plots of the proportion of correct choices (see Figures 3 a,b) do not show a distribution clustered about .5 indicating the choices made were not random. The pattern of choices for the two populations resemble each other indicating certain criteria existed when making selections. Questions 10, 11 and 14 were examined to determine if users were attempting to differentiate between saturation and value. For each of these questions, one state had a value of 1 and the other state had a saturation of 0 providing a clear distinction between saturation and value. In other words, one choice appeared blue, with no indication of gray and the other choice appeared gray with no indication of blue. If users were making a consistent association between saturation and value and the two variables, either consistently high or consistently low scores are expected for each of these questions. However, this is not the case, questions 10, 11 and 14 received scores of 68%, 52% and 50%, respectively for experienced users and 56%, 32% and 43%, respectively for inexperienced users.

Proportion of Correct Choices By Question--Experienced Users, Part I

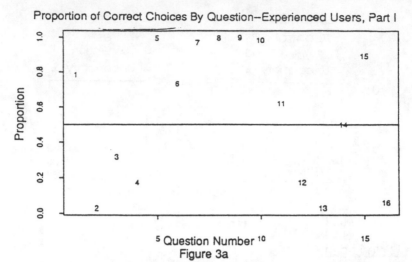

Figure 3a

Proportion of Correct Choices By Question--Inexperienced Users, Part I

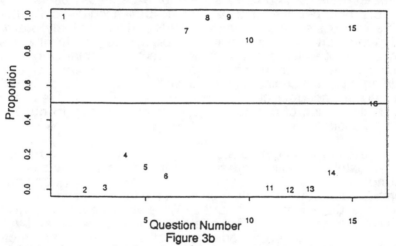

Figure 3b

To determine methods of selection made by both populations, individual questions were examined. The pattern of response between the two populations is evident from the proportions of correct choices of questions 1, 2, 4, 7-10, 12, 13 and 15. Although a legend did not appear in Part I, results were scored according to the legend scheme assigned to the control legend. Questions were categorized as high scoring, low scoring or undetermined. The criteria for the categorization are as follows: if 33% or less of the population answered the question correctly, the question is low scoring; if 67% or more of the population answered the question correctly, it is high scoring; if more than 33% and less than 64% of the population answered the question correctly, it is undetermined.

Of the questions which had similar response patterns between the two populations, questions 8 and 9 are categorized as high scoring and questions 2 and 13 are categorized as low scoring. The remaining questions, 1, 4, 7, 10, 12 and 15 are undetermined. Examination of the high and low scoring questions suggests users may have been making some selections based on the common cartographic assumption, "dark is more". A general agreement exists within the cartographic community that regions of greater data magnitude should be represented by darker symbols (Cuff 1974, Keates 1973, Robinson et al. 1984, Robinson 1967)

Based on the assumptions that less value is darker than more value (Evans 1974), more saturation is darker than less saturation and value is darker than saturation, all responses could be categorized as one of the following: a) both selections had equal amounts of saturation with differing amounts of value in which case the state with the lower value was termed darker; b) both selections had equal amounts of value with differing amounts of saturation in which case the state with the higher saturation was termed darker; c) the relative amounts of saturation and value were symmetrical about the upper left to lower right diagonal in which case the state with the lower value was termed darker.

The correct choice for question 9 (high scoring) could also be categorized as "darker". The other high scoring question, question 8, did not follow the pattern of "dark is more". Question 8 asked for the more certain data. Although the incorrect selection is the "darker" state, the correct selection is a bright and vivid blue, with no gray content and appears to be the obvious choice. The questions which received low scores had "lighter" colors representing more.

The next step in this analysis is to determine why the two populations did not follow a similar pattern for the remaining questions, questions 3, 5, 6, 11, 14 and 16. The particular colors chosen to represent the two states for questions 3, 5, and 6 were located close to each other in the control legend thus making it difficult to perceive a color difference. For questions 11 and 14, one choice was gray with no blue and the other blue with no gray and neither state appeared "darker". Thus, these two questions may have been answered arbitrarily. The choices for question 16 both contained value components and saturation components. As is evident from the previous selections, value has not been associated with data quality and thus this selection may have been made arbitrarily. Figure 4 shows where the color symbols would appear on the control legend for each question along with the correct choice. The two possible choices for each question are the endpoints of each line. The correct choice is denoted by a circle and the users' choice is identified with an 'X'. For the undetermined questions, the users' choice is omitted.

As data quality information is not customarily presented on choropleth maps, question type was compared to score to determine if users had more difficulty answering those questions asking for data certainty. Both populations scored slightly lower on those questions asking for data certainty. However, the differences between the two question

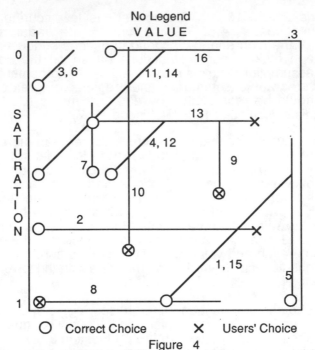

No Legend

VALUE

O Correct Choice   X Users' Choice

Figure 4

types are not great enough to suggest users are unable to interpret data quality information.

Part II: With a Legend Box

Legend A: Control Legend

A two-sample t-test based on a 95% confidence level indicates there is not a significant difference between the two populations for Legend A. The mean score is 72% with a standard deviation of 23% for experienced users and the mean score is 67% with a standard deviation of 19% for inexperienced users.

As in part I, a similar pattern of responses exists between the two populations. The responses for questions 1-7, and 9-13 identify this pattern. This group of questions can be divided into high scoring questions, questions 2, 3, 6, 9, 10 and 12 and undetermined questions, questions 1, 4, 5, 7, 11 and 13. See Figure 5 for the location of the color symbol on the legend for the two choices for each question.

Of the high-scoring questions, two of the correct selections (2 and 12) can also be categorized as "darker". For the remaining high scoring questions, 3, 6, 9 and 10, the correct choice is "lighter". This indicates that map readers were paying more attention to the legend box than to individual biases. However, since the overall scores indicate some misinterpretation of the legend box, individual biases may have influenced some of the decisions more than the legend.

Legend B: Test Legend

As for Legend A, a two-sample t-test based on the 95% confidence level indicates there is not a significant difference between the two populations. The mean score is 79% with a standard deviation of 19% for experienced users and the mean score is 70% with a standard deviation of 24% for inexperienced users.

The similar pattern of responses between the two populations is not as evident for test Legend B as for the absence of a legend and the presence of control Legend A. The larger difference between the mean scores for Legend B than for Legend A is an indication of the weaker response pattern.

Further examination of the results suggests users were making selections based on the "dark is more" assumption. Based on the test legend, 67% of the correct answers (questions 1, 3, 4, 5, 6, 7, 8, 9, 10, 15) could also be categorized as "darker". Questions 1, 3-10, and 12-14 are high scoring questions and questions 2, 11 and 15 are undetermined questions. Except for questions 12, 13, and 14, all of the high scoring questions have "darker" colors representing more. Of the undetermined questions, questions 2 and 11 have "lighter" colors representing more and question 15 has a "darker" color representing more. See Figure 6 for the location of the color symbol on the legend for the two choices for each question.

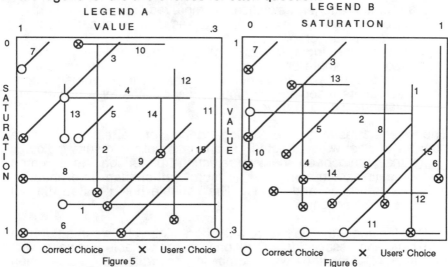

Figure 5

Figure 6

COMPARISON OF LEGEND A AND LEGEND B

An important factor in this work is the comparison of the two legends. If users score high on the control legend and low on the test legend then the assertion can be made that value is effective in displaying data quality. Previous analysis of the individual legend schemes does not support this hypothesis. For both legends, neither consistent high scores nor consistent low scores were recorded for individual questions providing a clear distinction between saturation and value. This indicates that the legends were not easy to interpret and that users were not able to differentiate between saturation and value when making selections.

As there is no difference between the experienced and inexperienced users for the two legends, a two-sample t-test was taken to determine if there is a difference between the scores for Legend A and Legend B. The computed test statistic indicates there is not a significant difference between the two legend types based on a 95% confidence interval.

DISCUSSION

The inconsistency among users' selections in the absence of a legend is not unreasonable. Maps do not customarily display quality information and the introduction of such a variable can be confusing to map readers. Verbal communication with participants indicated the mapping of data quality is an unfamiliar concept and selections were made arbitrarily or according to individual biases. This suggests education of users may improve the effectiveness of displaying data quality.

The addition of a legend significantly improved the results with minor differences according to which legend was used. Thus, there is not an intuitive association between value and uncertainty. The increase in mean scores indicates data uncertainty can be effectively portrayed on bivariate choropleth maps. However, data uncertainty should be treated as would any other variable. Improvements upon current bivariate mapping techniques may provide for a higher level of understanding among users.

The lower scores for the control legend than for the test legend were contrary to what was expected. Examination of the results indicates users may have been making selections based on the common cartographic assumption "dark is more" rather than differentiating between saturation and value. The variable being tested did not affect users' selections.

For the test legend, users were not differentiating between saturation and value and associating saturation with one variable and value with the other. Rather, they were combining saturation and value when making selections for both variables.

Researchers of color vision have determined that the human perception of color is psychological and physiological, not physical. (Travis 1991, Hilbert 1987). Although humans can distinguish between variations in hue, saturation and value, people do not equate color with amounts of these three variables. Thus, individuals perceive color according to individual ideas and concepts. In order for value to successfully represent data quality, map readers would need to become familiar with this association through repeated use. The learned association between value and data uncertainty would not be successful or practical for cartography. By requiring value to be representative of data certainty, an unnecessary restriction is placed on the cartographer.

CONCLUSION

Although the use of value as described by the HSV model cannot be accepted as an effective measure of mapping data quality, the following conclusions can be drawn from this experimentation.

1. There is not an intuitive association between data quality and value as displayed by the HSV color model.
2. There is not a significant difference in scores according to the particular legend displayed or the level of experience of map users.
3. Although data uncertainty is an unfamiliar concept, there is not a significant difference in how users responded to data quality questions as compared to data quantity questions.

These results indicate users are able to interpret data quality information. As there is not a significant difference of scores according to the particular legend scheme nor the particular variable, bivariate choropleth maps which display data quality and data quantity should not be treated differently than bivariate choropleth maps in general. Thus, maps displaying data quantity and data quality should benefit from bivariate mapping research. Additionally, as the scores increased significantly between the absence and the presence of a legend, the particular legend employed is an important factor to the degree of understanding of map interpretation.

A reason for the lack of an association between data quality and value is the relationship between value and gray. As suggested by Evans (1974), grayness increases along the value scale with black as the limit. From the results of this study, the authors suggest an alternate interpretation of grayness. That is, maximum grayness is achieved toward the center of the gray scale with grayness decreasing from this point towards white and from this point towards black. The point of maximum grayness is not yet defined. In this regard, the representation used in this experiment associates data quality with darkness rather than grayness. An interesting addition to this experiment would be to see how map users' interpretations of data quality information change when mapped according to the definition of maximum grayness occuring towards the center of the gray scale as well as when mapped according to previously defined bivariate mapping techniques and different color models. Results of such experimentation would determine if the outcome from this experiment could be attributed to the particular definition of grayness, the color model or the shading technique.

REFERENCES

Antes, James R. and K. Chang, 1990, "An empirical analysis of the design principles for quantitative and qualitative area symbols", *Cartography and Geographic Information Systems,* Vol. 17, No. 4 pp. 271-277

Burger, Peter and Duncan Gillies, 1989, <u>Interactive Computer Graphics</u>, Addison Wesley Publishing Company, Inc.

Buttenfield, Barbara P., 1991, "Visualizing Cartographic Metadata", NCGIA Technical Paper 91-26.

Carstensen, Laurence W., Jr., 1984, "Perceptions of Variable Similarity on Bivariate Choroplethic Maps", *The Cartographic Journal,* Vol. 21, pp. 23-29

Cuff, David J., 1972, "Value versus chroma in color schemes on quantitative maps", *The Canadian Cartographer,* Vol. 9, No. 2, December 1972, pp. 143-140

Cuff, David J., 1974, "Impending conflict in color guidelines for maps of statistical surfaces", *The Canadian Cartographer,* Vol. 11, No. 1, June 1974, pp. 54-58

Evans, Ralph M., 1974, <u>Perception of Color</u>, John Wiley & Sons, Inc., New York, NY.

Goodchild, Michael, F. and Sucharita Gopal, Editors, 1989, <u>Accuracy of Spatial Databases</u>, Taylor and Francis

Goodchild, Michael, F., 1991, "Position Paper for I7 Specialist Meeting",NCGIA Technical Paper 91-26.

Hilbert, David R., 1987, <u>Color and Color Perception</u>, Center for the Study of Language and Information, Lecture Notes Number 9

Keates, J.S., 1973, <u>Cartographic Design and Production</u>, Longman Press, New York.

Kimerling, A. Jon, 1985, "The comparison of equal-value gray scales", *The American Cartographer,* Vol. 12, No. 2, 1985, pp. 132-142.

McGranaghan, Matthew, 1991, "A view on visualizing spatial data quality", NCGIA Technical Paper 91-26.

MacEachran, Alan M., 1991, "Visualization of Data Uncertainty: Representational Issues", NCGIA Technical Paper 91-26.

Muller, Jean-Claude, 1979, "Perception of Continuously Shaded Maps", *Annals of the Association of American Geographers*, Vol. 69, No. 2, pp. 240-249.

Robinson, Arthur H., 1967, "Psychological Aspects of Color in Cartography", *International Yearbook of Cartography*, 7, pp. 50-61.

Robinson, Arthur H., Randall D. Sale, Joel L. Morrison and Philip C. Muehrcke, 1984, Elements of Cartography, fifth edition. John Wiley & Sons, New York.

Thorell, L.G. and W.J. Smith, 1989, Using Computer Color Effectively, Prentice Hall, Englewood Cliffs, New Jersey

Tobler, Waldo R., 1973, "Choropleth Maps Without Class Intervals?", *Geographical Analysis*, Vol. 5, pp. 262-265.

Travis, David, 1991, Effective Color Displays, Academic Press Inc., San Diego, CA

TESTING A SEGMENTATION PROCEDURE FOR DEFINING REGIONS IN A DIGITAL IMAGE CORRESPONDING TO NATURAL VEGETATION STANDS

Joseph Shandley and Janet Franklin
Department of Geography
San Diego State University
San Diego, CA, 92182
(619) 594 5491

Tom White
Cleveland National Forest
10845 Rancho Bernardo Drive
Rancho Bernardo, CA 92127-2107

Jennifer Rechel
US Forest Service Fire Lab
4955 Canyon Crest Drive
Riverside, CA 92502-6099

ABSTRACT

Remotely sensed satellite imagery can provide information on vegetation cover and other aspects of the biophysical environment that are useful in a GIS for ecological and natural resource modelling. In this study, an image segmentation procedure developed by Woodcock and Harward (in press) for forest vegetation was tested on Landsat Thematic Mapper imagery for an area of southern California chaparral vegetation in the Pine Creek watershed, Cleveland National Forest. The approach had not been tested in an area of Mediterranean-type scrub vegetation. The resulting stand maps and a per-pixel classification were compared to reference maps produced by air photo interpretation for several test areas. The segmented classification showed an improvement in map accuracy over the per-pixel classification through logical spatial generalization, although map accuracies were still generally low.

INTRODUCTION

One of the goals of the US Forest Service is to develop a technique for mapping ecological types using satellite imagery. Remotely sensed imagery, when combined with other geographically referenced data (i.e. soils, topography), has proven to be a useful tool for vegetation mapping and analysis in Southern California chaparral, scrub, and forest communities (Davis et al., 1986; and Yool et al., 1985). The key to a useful ecological type map will be the accurate delineation of vegetation stands. Since vegetation stands are, by definition, contiguous areas sharing the same attributes (i.e. vegetation composition and physiognomy, cover, successional stage, and soil substrate), an ecological type map based on accurately delineated stands should be more useful than a map based on conventional per-pixel classification of satellite imagery.

In the past, land cover or vegetation type maps derived from satellite imagery have been produced using statistical classification techniques, where the minimum mapping unit (MMU) is determined by the spatial resolution of the sensor. However, the pixels in the image are often smaller than the desired MMU. For USFS management, the MMU is between 2 and 4 hectares (Strahler, 1981), which corresponds to 22 to 44 pixels for Thematic Mapper data. Even if the per-pixel methods produce accurate maps, they are not useful for forest management or research since a larger MMU is required (for example, see Chou et al., 1990). To overcome this problem, images have often been processed on a per-pixel basis and then generalized by some type of smoothing or filtering method to produce a final map.

In this study an image segmentation procedure developed by Woodcock and Harward (in press) for forest vegetation was tested on Landsat Thematic Mapper imagery for an area of southern California chaparral vegetation in the Pine Creek watershed, Cleveland

National Forest. Their region growing algorithm is designed to define multipixel regions in an image corresponding to natural vegetation stands. It differs from other segmentation approaches in that 1) it allows conservative region growth, 2) it does not rely on boundary definition, and 3) it does not constrain the objects to have homogeneous properties (i.e. multispectral reflectance), but allows for some heterogeneity in the formation of regions, and 4) the minimum and maximum size of the regions can be specified. If stands could be automatically delineated from satellite imagery, management decisions regarding fire management, grazing practices, wildlife habitat conservation, watershed/riparian monitoring, and other goals could be implemented more efficiently and effectively.

STUDY AREA

The Pine Creek watershed of the Cleveland National Forest encompasses approximately 85,000 acres with an elevation ranging from 1,500 to over 6,400 feet. The topography is characterized by steep slopes and rugged terrain, although some gentle undulating terrain exists in smaller areas. Precipitation at the lower elevations (500-2500') is in the range of 12 to 20 inches/year while the higher elevations (>4500') average 25 inches/year with occasional snowfall.

The vegetation consists mainly of chaparral types (dominated by Ceanothus, Chamise, Mountain Mahogany, Manzanita, and Scrub Oak) with some soft chaparral components such as California Buckwheat and Coastal Sage Scrub. The higher elevations are characterized by conifer and broadleaf woodland types. There are meadows dominated by herbaceous vegetation at the higher elevations, and some areas of "type conversion" (where chaparral has been cleared for fire control and replaced by annual grasses or sparse herbaceous and subshrub vegetation), especially in Pine Valley. The general classification system included: 1) Conifer woodland 2) Riparian woodland 3) Wet meadow 4) Dry meadow 5) Mixed chaparral 6) Chamise chaparral 7) Scrub oak chaparral 8) Sparse/Barren/Disturbed 9) Water 11) Coastal sage scrub 12) Coast live oak woodland and 13) Wright's buckwheat/Great basin sage.

METHODS

A Landsat Thematic Mapper (TM) image, acquired in September (1988), was used to test the segmentation algorithm. The image was rectified and georeferenced to the UTM coordinate system at 24.68 m resolution. The image was then masked to produce a sub-image which depicted only the study area.

A classification procedure was applied to TM bands 4, 3, and 2 of the sub-image using an unsupervised approach. The same 3-band image was also segmented, classified, and labelled for comparison to the per-pixel classification. The accuracy of the resulting thematic maps was then assessed by statistical comparison of each map to a reference map. Each of these steps will be described in more detail below.

Per-pixel Processing

The 3-band TM image was classified using unsupervised clustering and a minimum distance classifier within the ERDAS image processing software. Each of the 100 clusters generated was assigned a vegetation class label based on interpretation of aerial photos (1980 1:24,000 CIR and 1990 1:62,500 positive transparencies) and knowledge of the study area. Clusters were then grouped into vegetation types.

Segmentation Procedure

The image segmentation software was written to interface with the Image Processing Workbench (IPW) developed at the University of California, Santa Barbara (Frew, 1990). The algorithm was designed to conservatively define regions in an image by controlling a number of parameters including the rate of pixel merging, the degree of merging, and minimum and maximum region size constraints.

The 3-band TM image was converted from an ERDAS image file to one compatible with IPW, where the segmentation algorithm could be applied. A number of

701

segmentation trials were run to test for the optimal parameter settings for vegetation stand delineation in our study area. Preliminary tests indicated that the initial global threshold (which limits adjacent region merging based on interregion spectral distance) could vary considerably (i.e. between 2-8 units in multispectral space) without significantly affecting the outcome of the segmentation. Therefore, in order to avoid over-merging errors at this stage, the value was set to 4 units for all further trials. The merge coefficient (which calculates a new pass tolerance for each pass and controls the rate of merging) was found to be most effective at 0.1 (Woodcock, pers. comm.). The optimal minimum region size was based on user needs (i.e. minimum management units) and the complexity of the landscape. The segmentation tests were run with the minimum requirement of 30 pixels per region, which corresponded to 6.7 acres. The maximum region size criteria was 100 pixels, with no merging allowed between regions containing 100 or more pixels.

Classification of the segmented image followed a two-step process. The output from segmenting the TM image was a 'region_map', which contained the region identification number (region_id) for every pixel in the image (Woodcock, 1991). Regions were assigned to vegetation types by overlay of the region map on the per-pixel classified image described above. The most frequently occuring land cover class in each polygon determined the vegetation type label for that stand. A plurality option was also tested in which a specified minimum percentage of the pixels in a region must fall within one class if a class label was to be assigned. If none of the vegetation classes were able to meet this percentage, then the region was given a unique class label and later manually edited (assigned a vegetation class label). The plurality lower bound for this study was 30%, which resulted in a large number of regions which had to be labelled manually.

By calculating a vector of means, a variance-covariance matrix, and the number of pixels for each region, it would be possible to perform the classification by clustering the regions (Woodcock, 1991). However, the aggregate spectral signature of a region is not always representative of a vegetation stand. For example, the spectral mean of a sparse scrub oak chaparral stand could be similar to the mean of a dense mixed chaparral stand and would be erroneously grouped in the same cluster. Therefore, the overlay of the per-pixel classified image was the chosen method for classification in order to avoid this type of error (following the suggestion of Woodcock, pers. comm.).

The second step of the segmentation classification involved label editing of the individual polygons (or regions) using a program (region_tool) developed by Harward (see Woodcock, 1991). This software allowed the regions to be displayed on a background color composite image with the ability to manually edit the region labels with the aid of aerial photos. This option was only used when regions were not dominated by one particular vegetation class and the plurality rule was employed.

Reference Map Processing

In order to statistically assess the accuracy of the segmented stand map and compare it to the per-pixel class map, a reference map had to be generated. Since one of the USFS objectives is to utilize the vegetation maps for management purposes, one of us (T. White), a Forest Service ecologist familiar with the study area photointerpreted three subareas in the watershed. The goal was to delineate the boundaries of vegetation stands in three elevation zones representative of the range of vegetation types in the watershed. However, the preliminary results of only the middle elevation subarea will be discussed in this paper.

The polygons were delineated and labelled on 1:24,000 CIR (July, 1990) aerial photos which had been enlarged from 1:48,000 positive transparencies. The photointerpretation was restricted to the center portion of the photos to reduce the effects of geometric distortion. The photointerpreter was instructed to meet the MMU of the segmented image and allow for heterogenous stands to exist (heterogeneous in cover, or tone and texture, but not in vegetation label). The reference maps were then digitized.

The reference map presented in this paper was spatially registered to the 3-band TM image with 11 ground control points. The data were transformed using a first-order transformation matrix with a total root mean square (RMS) error of 0.76 pixels. The classified segmented and per-pixel images were each masked to correspond to the same area as the reference map.

702

The accuracy assessment involved comparing the class areas and classification accuracies of both the segmented and per-pixel images to the photointerpreted subarea. The ERDAS (SUMMARY) program was used to overlay each classified image on the reference map and compute cross-tabulation statistics. The number of points were arranged in error matrices where measures of commission error (user's accuracy), ommission error (producer's accuracy), and overall map accuracy were determined. The class area distributions for each of the three images are depicted in histograms.

RESULTS AND DISCUSSION

As shown by the classification error matrices, overall the segmented classified image produced higher accuracies than the per-pixel image in almost every class (Tables 1 & 2). Individual class accuracies for the per-pixel image were low, ranging from 20.5% for riparian to 80.4% for wet meadow (user's accuracy). The producer's accuracies varied greatly as well, from 5.7% for riparian to 100% for wet meadow. The overall map accuracy for the per-pixel classified image was 49.6% (Table 1). The segmented classified image improved the individual class accuracies for every category except wet meadow. The individual user's class accuracies ranged from 32% for scrub oak chaparral to 90.5% for water. Producer's accuracies varied from 21.4% for riparian to 100% for wet meadow and water. The overall map accuracy for the segmented classified image was 60.7% (Table 2).

Since Class 2 was a thin intermittant riparian corridor, it was difficult to distinguish it from the adjacent chaparral because of its high spectral variance and limited spatial extent. Therefore, the riparian area was underestimated in both classified images. Confusion between the three chaparral classes and coast live oak woodland accounted for lower class accuracies, between 32-74%. Most of the error between mixed and scrub oak chaparral resulted from the disagreement between the per-pixel and reference map over the dominant vegetation type in many of the polygons. It was not uncommon to find polygons on the north-facing slopes in the reference map labelled mixed chaparral which had a mixture of both classes. The most common error was that areas of chamise and scrub oak chaparral and oak woodland in the per-pixel and segmented classifications were mapped as mixed chaparral in the reference map. It is possible that the area of mixed chaparral had been overestimated in the reference map.

One advantage the segmentation procedure had over the per-pixel approach was the ability to better discriminate between spectrally similar classes. For example, dry meadow in the per-pixel image was commonly mistaken for mixed or chamise chaparral, sparse, or oak woodland. Similarly, sparse/barren was confused with dry meadow, mixed and chamise chaparral, and oak woodland. The segmentation procedure greatly reduced the commission error in both of these examples.

Another factor which contributed to the error was the difference in the scale of the mapping units between each of the three images. The segmented and per-pixel images were mapped at a finer scale than the reference image which was more simplified, resulting in a number of misclassifications. Because the segmented image was closer in scale to the reference image, as discussed above, accuracies tended to be higher although many stands were still misclassified. For example, many stands classified in the segmented map corresponded spatially to the delineated stands on the reference map (Figure 1), but were given a different label. In this sense, most of the inaccuracies of the segmented image could be attributed to identifying and labelling, not necessarily to stand delineation.

The area of each class in each image is shown in Figure 2. The area of conifer woodland (Class 1) was about the same in the segmented and reference images. The per-pixel image, however, contained no Class 1 because all woodland areas were initially classified as coast live oak in this elevation zone. Wet and dry meadow, and mixed and chamise chaparral were all similar in terms of total area. Scrub oak chaparral (Class 7) was overestimated (2X) by both classified images in comparison to the reference map. This was due to north-facing slopes being classified as scrub oak chaparral that were photointerpreted as mixed chaparral. It was not unexpected that sparse/barren (Class 8) was overesimated for the per-pixel classified image since many of the dry meadow

(Class 4) points were in the same cluster. Riparian (Class 2) covered a larger area in the reference map than in the two classified images. This was because many of the pixels were classified differently due to the patchiness of this zone. As a result, the segmentation algorithm created regions where the upland classes dominated and integrated riparian pixels into the adjacent vegetation type, usually mixed or chamise chaparral.

CONCLUSION

The objective of this study was to test an image segmentation algorithm in mediterranean-type scrub vegetation. Although classification accuracies were relatively low compared to accepted thematic mapping standards, the procedure did improve the map accuracies over the per-pixel classification method. A more accurate per-pixel map for labeling the segmented image would improve individual class and overall accuracies. Also, a more objective yet efficient method of evaluating the segmentation accuracy would be useful and needs to be addressed. Our continuing work includes testing the segmentation algorithm with different input data (transformed spectral channels and topographic aspect derived from digital terrain data) and evaluating the map accuracy for all three photointerpreted test areas within the study area.

REFERENCES

Chou, Y., R.A. Minnich, L.A. Salazar, J.D. Powe, and R.J. Dezzani, 1990. Spatial autocorrelation of wildfire distribution in Idyllwild Quadrangle, San Jacinto Mountain, CA, *Photogrammetric Engineering and Remote Sensing*, **56**(11):1507-1513.

Davis, F.W., S. Goetz, and J. Franklin, 1986. The use of digital satellite and elevation data in chaparral ecosystems research, *Proc. of the Chaparral Ecosystems Research Conference*, May 16-17, Santa Barbara, CA.

Frew, J.E., 1990. The Image Processing Workbench, PhD Dissertation, University of California, Santa Barbara, 303 p.

Strahler, A.H., 1981. Stratification of natural vegetation for forest and rangeland inventory using Landsat digital imagery and collateral data, *Int. Journal of Remote Sensing*, **2**(1):15-41.

Woodcock, C.E., 1991. GIS operations using Image Formats in IPW, Paper presented at *GRASS '91 Users Conference*, Berkeley, CA, March 6.

Woodcock, C.E. and Harward, V.J., in press. Nested-hierarchical scene models and image segmentation, *Int. Journal of Remote Sensing*.

Yool, S.R., J.L. Star, J.E. Estes, D.B. Botkin, D.W. Eckhardt, and F.W. Davis, 1986. Performance analysis of image processing algorithms for classification of natural vegetation in the mountains of Southern California, *Int. Journal of Remote Sensing*, **7**(5):683-702.

Reference Data

Per-Pixel Classified Data	Background	Conifer Woodland	Riparian	Wet Meadow	Dry Meadow	Mixed Chaparral	Chamise Chaparral	Scrub Oak Chaparral	Sparse /Bare	Water	Defunct	Coastal Sage Scrub	Coast Live Oak Woodland	Wright's Buckwheat	Classified Totals	User's Accuracy
Background	0	0	0	0	0	0	0	0	0	0	0	0	0	0	0	NA
Conifer Woodland	0	0	0	0	0	0	0	0	0	0	0	0	0	0	0	NA
Riparian	0	0	119	0	22	97	287	7	0	0	0	9	40	0	581	20.5
Wet Meadow	0	2	0	1181	0	4	1	0	81	0	0	5	194	0	1469	80.4
Dry Meadow	0	14	3	0	1966	195	158	12	115	0	0	28	903	0	3394	57.9
Mixed Chaparral	0	1017	668	0	1206	34927	6546	3320	728	0	0	289	4759	0	53460	65.3
Chamise Chaparral	0	62	420	0	420	4581	6107	216	744	0	0	572	1822	0	14944	40.9
Scrub Oak Chaparral	0	158	337	0	86	12714	191	5835	83	0	0	2	975	0	20381	28.6
Sparse/Bare	0	18	41	0	333	2203	526	49	1826	0	0	73	597	0	5666	32.2
Water	0	0	0	0	0	23	0	1	0	81	0	0	0	0	105	77.1
Defunct	0	0	0	0	0	0	0	0	0	0	0	0	0	0	0	NA
Coastal Sage Scrub	0	0	110	0	0	129	1009	9	0	0	0	357	11	0	1625	22.0
Coast Live Oak Woodland	0	86	394	0	113	3973	506	1179	75	0	0	5	2321	0	8652	26.8
Wright's Buckwheat	0	0	0	0	0	0	0	0	0	0	0	0	0	0	NA	NA
Reference Totals	0	1357	2092	1181	4146	58846	15331	10629	3652	81	0	1340	11622	0	110277	45.17
Producer's Accuracy	NA	0.0	5.7	100.0	47.4	59.4	39.8	54.9	50.0	100.0	NA	26.6	20.0	NA	54730 / 54.4	49.6

Percent Accurate: 49.6

Table 1. Error matrix for Per-pixel classified image vs. middle elevation reference map showing cross tabulation of all pixels. Note that no areas were labelled conifer woodland in the per-pixel classification for the middle elevation zone.

705

Reference Data

Segmented Classified Data	Background	Conifer Woodland	Riparian	Wet Meadow	Dry Meadow	Mixed Chaparral	Chamise Chaparral	Scrub Oak Chaparral	Sparse /Bare	Water	Defunct	Coastal Sage Scrub	Coast Live Oak Woodland	Wright's Buck- wheat	Classified Totals	User's Accuracy
Background	0	0	0	0	0	0	0	0	0	0	0	0	0	0	0	NA
Conifer Woodland	0	1093	0	0	265	0	0	0	0	0	0	0	0	0	1358	80.5
Riparian	0	3	448	0	0	102	261	53	0	0	0	0	21	0	888	50.5
Wet Meadow	0	0	0	1061	3	0	0	0	0	0	0	0	307	0	1373	77.3
Dry Meadow	0	0	0	0	2446	141	0	58	0	0	0	0	808	0	3453	70.8
Mixed Chaparral	0	0	373	0	1125	37905	3442	3095	539	0	0	255	4550	0	51284	73.9
Chamise Chaparral	0	0	547	0	125	3808	10480	154	480	0	0	229	1205	0	17028	61.5
Scrub Oak Chaparral	0	143	528	0	1	12126	131	6425	18	0	0	3	704	0	20079	32.0
Sparse/Bare	0	0	6	0	27	958	0	0	2265	0	0	0	305	0	3561	63.6
Water	0	0	0	0	0	0	0	0	0	143	0	0	15	0	158	90.5
Defunct	0	0	0	0	0	0	0	0	0	0	0	0	0	0		NA
Coastal Sage Scrub	0	0	32	0	0	277	589	3	0	0	0	853	16	0	1770	48.2
Coast Live Oak Woodlan	0	0	158	0	215	3526	428	841	348	0	0	0	3809	0	9325	40.8
Wright's Buckwheat	0	0	0	0	0	0	0	0	0	0	0	0	0	0		NA
Reference Totals		1239	2092	1061	4207	58843	15331	10631	3650	143		1340	11740	0	110277	62.6961
															66928	
Producer's Accuracy	NA	88.2	21.4	100.0	58.1	64.4	68.4	60.4	62.1	100.0	NA	63.7	32.4	NA	65.4	60.7

Percent Accurate: 60.7

Table 2. Error matrix for segmented classified image vs. middle elevation reference map showing cross tabulation of all pixels. Note that regions labelled conifer woodland in the segmented image were labelled during the manual editing process.

a)

b)

c)

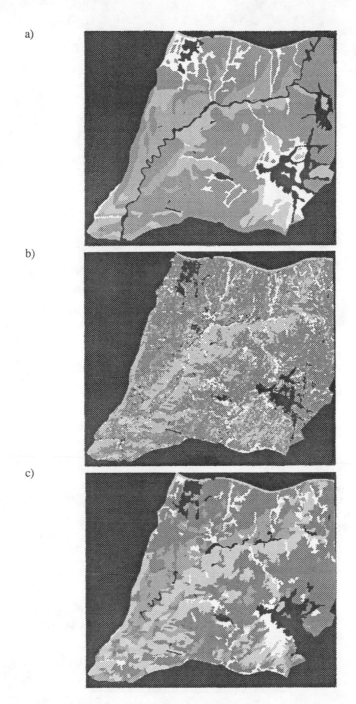

Figure 1. a) The reference map, b) the per-pixel classified image and c) the segmented classified image for the middle elevation subarea. The size of the area is approximately 41.64 square miles.

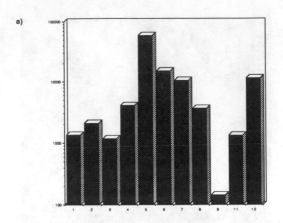

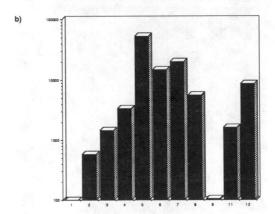

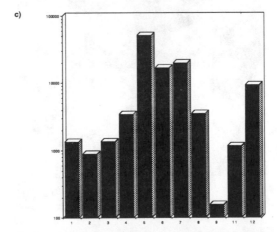

Figure 2. Histograms showing the area of each vegetation class in the a) reference map, b) the per-pixel classified image and c) the segmented classified image. See text for vegetation class names.

AM/FM/GIS PIONEER PROJECT CHRONICLE: CAMRIS

Joseph J. Stefanini
Commonwealth Gas Company
157 Cordaville Road
Southborough, MA 01772

ABSTRACT

For most organizations, implementation of an AM/FM/GIS project is an undertaking that entails sailing into the uncharted waters of data compilation and conversion, graphic computer system selection, integration of other data sets, applications development, end user training, and data maintenance. While many consultants and vendors can help provide guidance for this journey, COM/Gas, as an early pioneer in this field with its CAMRIS (COM/Gas Automated Mapping Record Information System) project, can provide the 'voice of experience' in navigating these potentially stormy seas.

Nearing the completion of data compilation and conversion of the 56 cities and towns in its operating area, COM/Gas is also well on the way towards the implementation of one of the largest gas distribution AM/FM/GIS data models. Already underway, the first major data application, map maintenance, will preserve the large data conversion investment. This application, and the many others designed to follow, will utilize the best available computer technologies, with a core operating system in a distributed environment crossing VAX and UNIX based platforms. In addition, COM/Gas has successfully worked with its state utility regulatory agency for project support and has also received outside interest in some of the CAMRIS map products.

In this paper, COM/Gas will share its rich vein of experience in an attempt to show where the favorable routes of AM/FM/GIS implementation lie, while avoiding the hazards that can await the unwary.

COMPANY OVERVIEW

Commonwealth Gas Company was incorporated in 1851 as the Worcester Gas Light Company. Through acquisitions of smaller companies, COM/Gas has grown to become Massachusetts's second largest natural gas distribution company providing service to over 215,000 residential, commercial, and industrial customers within 56 cities and towns. This service area covers 1,177 square miles in the central and eastern portions of the state with a resident population of more than one million. COM/Gas presently employs nearly 800 people and utilizes and maintains approximately 2,600 miles of mains and 1,840 miles of service piping.

709

CAMRIS PROJECT: CONVERSION

Background

The individual companies that now make up COM/Gas each had developed their own map and record sets within seven different office locations. Consequently, company-wide, no standard mapping format existed, and few record sets referred to a common map system. The creators and users of the map information were finding it increasingly difficult to use and maintain the data. Consolidating and standardizing these disparate record sets became paramount to management following the consolidation of general office operations at the Company's new Southborough, Massachusetts headquarters in the mid-1970's.

Thus, in 1981 COM/Gas conducted an internal review of its existing mapping and records system. The internal review led to a formal external Audit in 1983 which detailed not only the shortcomings of the existing maps and records but also presented alternatives for consideration. The alternatives included an improved manual system, a computer-aided drafting system, and an integrated automated mapping and facilities management ("AM/FM") system.

The Audit concluded that any one of these alternatives to convert to an upgraded system would require a similar level of effort as any of the other alternatives. However, the adoption of an AM/FM system would provide the most advantageous end results: a more manageable, cost-effective, and accurate maps and records system achieved through computer-based consolidation and standardization.

This computerized-mapping alternative was a strategic match with COM/Gas' vision for greater use of advanced technologies to reduce costs while achieving higher levels in system reliability; thus, COM/Gas decided to follow this recommendation. From these humble origins, CAMRIS (COM/Gas Automated Mapping and Record Information System) was born.

Planning/Testing

To reduce risk, and to maintain manageability of the project, COM/Gas decided to use a phased approach for the testing of the various steps towards full AM/FM system implementation. In this manner a small, dedicated internal staff would be able to concentrate on and effectively oversee the individual project activities. This tied-in with the Company's early decision to use outside consultants and vendors wherever possible to take advantage of the specialized knowledge and equipment that would be needed to implement the project.

As the base for all facility data overlays and other mapping applications, the various phases of landbase conversion were tested first. COM/Gas initiated aerial photography, map digitization, and QA/QC inspection to obtain digital

710

maps for two representative towns. Through this effort, COM/Gas received valuable knowledge in developing mapping and QA/QC specifications and procedures. More importantly, COM/Gas had the opportunity to build and maintain working relationships with specialized mapping vendors. The importance of these relationships cannot be overstated - each AM/FM/GIS implementation project is unique, and the technologies involved mandate that clients and vendors work together towards common goals to make these projects work.

In short order, COM/Gas proceeded with testing overall computer-mapping functionality with a prototype based on a one-quarter square mile facility area, followed by a pilot conversion test on a full town to fine-tune the conversion specifications and methodologies and to provide an actual data model for end-user and management review. The pilot was conducted on the town of Natick, a town covering sixteen square miles with representative mapping records dating from the post-civil war era to the introduction of advanced plastic piping systems in the 1970's and early 1980's.

The pilot was completed in early 1987 with the delivery of a computer master tape and over 200 hard copy map plots that made nearly 8,000 non-standard and cumbersome manual records obsolete. With positive end-user feedback from review of the pilot computerized maps, the Company decided to proceed with full-scale implementation of CAMRIS.

Full Scale Conversion

Following an extensive RFP process that began with invitations to fifteen qualified vendors, seven bids were received for the conversion of the landbase of all 56 cities and towns in the COM/Gas service area and all existing manual COM/Gas maps and records to the CAMRIS digital format. In 1988, the successful bidder (Baymont, Inc.) began this work with completion scheduled for mid-1993. The work includes: Aerial Photography, Land Data Entry, Facility Records Copying and Facility Data Entry.

A second vendor (Intelligent Infrastructure Technologies, Inc.), has been retained since 1986 to perform any necessary technical consulting support work for both the conversion and system development efforts. This vendor is also responsible for providing the independent QA/QC inspection of all converted landbase and facility data produced under the total conversion contract.

The consulting work also includes support for selection of an operating system and producing the applications necessary to retrieve, update and edit the converted maps for the ultimate end-users.

In 1987, COM/Gas pro-actively went before the Massachusetts Department of Public Utilities (the 'DPU') to bring an awareness to that regulatory agency about this AM/FM/GIS technology and the benefits that would accrue to the ratepayers and general public at-large from the CAMRIS automated mapping system. As a result of this process, the DPU gave its support to the CAMRIS project, and has remained steadfast in that support to date.

OPERATING SYSTEM

The Data Model

At the time full-scale data conversion began for CAMRIS, none of the available computer systems were fully capable of being employed as a core operating system for COM/Gas' mapping requirements. As a result, the Company committed itself to developing a strict empirical data model that could later be used with any future operating system with minimum re-work of the converted data. The data model was thus designed to allow facility and landbase information to be converted in vector-properties form with automatic data verification capabilities. The end result was a logical data model capable of nearly-effortless migration to any intelligent data management system capable of meeting COM/Gas' operational needs.

This data model allowed the creation of the first end-user functionality for CAMRIS, the View/Plot query tool. From graphic workstations, and with a minimum of training, COM/Gas employees can view, and plot, any part of the CAMRIS map data at the push of a button, from a single building address or street intersection to an entire town. With View/Plot, plots can even be launched from non-graphic ASCII terminals. Many disparate devices are supported, including Intergraph workstations, X-terminals, IBM 6000 workstations, Hewlett-Packard plotters, and laser printers.

FRAMME

After data conversion was substantially underway, COM/Gas became aware of the efforts to bring the FRAMME (Intergraph Facility Rulebased Application Model Management Environment) application development tool to the workstation level. COM/Gas subsequently selected FRAMME as the core of its intelligent data management system for CAMRIS because the rulebased model could provide solutions to its most fundamental needs: centralized geo-based facility data, rulebase controlled data maintenance with full integrity, and a user-defined distributed database scheme. In addition, the extensive programming tools inherent in this product provide a suitable environment for integration of stand-alone application needs without compromising the integrity of the data.

When migration of all of the landbase and facility conversion data to the rulebased environment has been completed, COM/Gas will have over ten gigabytes of graphic and non-graphic data available in the system covering the 1,177 square miles of its service territory.

Application Development

To preserve the integrity of the converted map data, all map and record updates will be accomplished through an application called Add/Delete developed on the rulebased environment tool. This application will allow end-users to quickly add, delete, or modify landbase and facility information as parts of a digital facility network reflecting the day-to-day field operations of Commonwealth Gas such as new pipe installations, pipe retirements, and other maintenance activities.

Other applications are currently in the phased-integration stage, including an interface to the gas network simulation software model (Stoner GASS) in use at COM/Gas. In addition, linkages to AutoCAD for Engineering support have been developed that allow CAMRIS and general engineering data to be freely and quickly available on either platform. Note that in general, however, unlike the Add/Delete tool, most of these other applications are limited to 'read-only' access to the central CAMRIS database to protect the integrity of the converted map data.

Future Applications

Applications are being considered for other interested departments within the company which will be delivered based on need and cost benefit analyses. One such application is for Conservation Load Management. In today's economic and political climate it is increasingly important for utilities to find a balance between conservation programs designed to reduce individual customer energy usage and increasing the total customer base at minimum cost. The most efficient way to do this is to gain new customers in areas where load conservation has freed up existing pipeline capacity; in these areas, new customers can be added at minimal expense to build new plant.

Using the CAMRIS maps, such potential customers can be quickly identified and their locations quickly exported to a sales database. In the same vein, existing commercial and industrial customers presently not using gas for all of their energy needs can be targeted by comparing current gas usage to the actual capacity of the existing plant servicing those customers. The Company's sales and marketing teams can then concentrate their efforts on those potential customers that have the best revenue/cost-to-serve ratio. The information on the CAMRIS maps makes this identification and analysis feasible on a wide-scale basis.

Another application being considered is to combine the CAMRIS geographic mapping system with the Company's SCADA technology, which is currently being upgraded. Whereas most SCADA systems deal only with a 'skeletonized' representation of the pipe network, the tie-in to CAMRIS information will give the SCADA operating personnel access to all of the piping network detail information down to the service pipe level, thereby enhancing the overall accuracy of the SCADA model.

In the future, the use of CAMRIS data will not be limited to office personnel. It is envisioned that field crews will have immediate access to this information through use of PC-like data stations in their truck, or by portable laptop or notepad computers. It is becoming increasingly more likely that within a few short years, field personnel will even be able to immediately data-enter maintenance activity information directly onto palm-sized computers for later direct uploading to the CAMRIS database.

Host-Client Environment

Presently, the CAMRIS data is stored in Oracle and Intergraph IGDS formats and is distributed over the COM/Gas computer network. Those parts of the network used by CAMRIS include a DEC VAX cluster (VAX 9210, 8650, and 6240), Intergraph servers (InterServ 6505 and 305), Intergraph workstations (InterPro 6505's, 225's, and 220's), and an IBM RISC System/6000 with X-Terminals. These systems are connected through Ethernet, and include the operating systems DEC VAX/VMS, Intergraph CLIX UNIX and IBM AIX UNIX, as well as access through an SNA gateway to the corporate IBM mainframe.

Through years of dedicated commitment, well-planned designs, and technical wisdom, CAMRIS now utilizes a fully-integrated Host-Client environment that effectively eliminates multi-hardware and multi-operating system concerns.

CAMRIS PRODUCTS: USED AND USEFUL

During the conversion process COM/Gas obtained several mapping products that are being used extensively in day-to-day operations inside and outside of the company. These products include 1"=800' aerial photography, precision 1"=50' landbase maps, and 1"=50' scale landbase/facility maps of the entire gas distribution network of mains, services and fittings. Each of these products cover all 56 cities and towns in the Company's service territory.

The aerial photography has been used within the company to perform preliminary layouts for pipeline projects. In the future, the aerial photographs may be scanned into the system in raster form to be used as background overlays to the CAMRIS maps, thereby providing a single-source map record on-screen for analysis of these piping projects. Outside of COM/Gas, the engineering/photogrammetric firm that originally captured the photography has been using the photographs as the basis for client projects such as easement layouts and identifying areas for potential development.

The landbase maps are used extensively within the Company. The planners in the Distribution department of the Company's operations group use the land maps as the base for overlaying facility information in areas of new activities, eliminating the significant manual effort used to create similar maps in the past. These maps are also being used when considering the extension of gas lines to proposed developments or existing areas that currently do not have gas service available. The maps help the planners layout various pipe routing options while simultaneously identifying new sales leads for the Sales and Marketing group along the route options proposed.

COM/Gas' Engineering staff also make extensive use of the map products as the foundation plans for the layout of all pipeline design and construction projects. With in-house developed translation routines and symbol libraries, CAMRIS map data is easily imported from the central CAMRIS database for use on the Engineering personal computers running AutoCAD. Incorporating the highly accurate CAMRIS land information with PC-based Computer Aided Design and Drafting (CADD) technology greatly reduces expensive and time consuming on-site survey work. The linkage works both ways, in that the detail regulator and take station drawings created by the engineers on the PC's can be accessed through the CAMRIS network for later instant reference by all users of the system at a touch of a button.

Enhancing the Company's use of the CAMRIS mapping data is the rapid search, retrieval, and plotting of landbase and facility information through the View/Plot tool described earlier. At the touch of a button, any authorized user can view, print, or plot any location from a single building to an entire town.

COM/Gas' landbase mapping products have also been noticed by outside companies and municipal government agencies as having great use for them. As such, these products are being made available through a third party (James W. Sewall Company) who provides the support and distribution of this digital data.

LESSONS LEARNED

In developing CAMRIS, COM/Gas has gained many valuable insights on successfully implementing an AM/FM/GIS system. With these insights, the following advice is freely given:

First, know exactly what your goals are. Then work toward them through thorough planning, testing, and documentation.

Second, use a phased approach while implementing an AM/FM/GIS system. This will limit the risk in achieving each goal, while maintaining manageability of the project.

Third, realize the uniqueness that is inherent in each project and that consistent and effective communications between the client and the vendors are necessary to achieve the stated goals. Without effective communication, what the client wants may not be what the vendor delivers.

Fourth, gain the support of executive management from project inception and work diligently to maintain the support throughout the project. Since AM/FM/GIS project implementations usually span long periods of time and require significant financial investment, on-going executive backing is needed to keep the project moving.

Finally, since these types of projects need maintenance once in the hands of the users, they need to be developed using the best and most widely-recognized technologies, without being totally dependent on one source. This will allow for the future growth of the project in customization and other applications as they become needed, while minimizing the risk of being tied to an orphaned hardware and/or software platform.

SUMMARY

The successful implementation of an AM/FM/GIS project does not come from luck alone. The journey can be fraught with frustration and hidden pitfalls. Extensive planning, testing, and documentation in a phased approach is needed, not just for data conversion, but for data maintenance and application development as well.

We at COM/Gas wish you the best on your AM/FM/GIS endeavors.

Joseph J. Stefanini
Director of Engineering
Commonwealth Gas Company

As Director of Engineering at COM/Gas, Joe Stefanini is responsible for managing the engineering support to the company's various mechanical, civil and architectural projects, including CADD and PC applications. One of the major engineering support functions under his direction is the implementation of CAMRIS, the company's state-of-the-art Automated Geographic Information System.

Joe received his formal education on Geographic Information Systems from the University of Wisconsin and has over 12 years experience with AM/FM/GIS technology. He holds a Bachelor of Science degree in Management from Lesley College in addition to an Architectural Engineering Degree from Wentworth Institute.

BEYOND THE TRADITIONAL VEGETATION MAP TOWARDS A BIODIVERSITY DATABASE

David M. Stoms, Frank W. Davis, Peter A. Stine

Department of Geography
University of California, Santa Barbara
Santa Barbara, CA 93106-4060
Internet: stoms@rocky.geog.ucsb.edu

and Mark Borchert

Los Padres National Forest
6144 Calle Real
Goleta, CA 93117

ABSTRACT

Relational GIS database management capabilities offer an exciting, but essentially untapped, alternative to the traditional paper vegetation map. These capabilities are illustrated in the context of a new digital vegetation database being compiled for a statewide biodiversity assessment in California. Thematic Mapper imagery is used to segment a region into ecological landscapes of uniform geology and physiology that support recurring complexes of soils and vegetation. Whereas a traditional small-scale vegetation map would assign a single class type to each map unit, we are encoding detailed floristic and structural information, including wetland types and regional endemic species contained within map units. We demonstrate how this database can be manipulated in various ways. First, the distribution of a single plant species is extracted. Next, we produce a custom classification of habitat types. We also present a numerical analysis to illustrate the variability in species composition across regional environmental gradients within a single community type (i.e., chamise chaparral). Thus, this relational vegetation database supports many more user views than the traditional thematic vegetation map.

INTRODUCTION

Land use/land cover is among the most common thematic layers in GIS databases. There are nearly as many types of vegetation maps, however, as there are applications. Some vegetation maps represent floristic variation, others physiognomic or structural type, and some portray a combination of the two (Kuchler and Zonneveld, 1988). Recent examples of vegetation maps include timber inventory (Franklin et al., 1986) emphasizing commercial conifer species and the size and density of their stands; wildlife habitat using broadly defined physiognomic types at the subformation level (Stoms et al., in press) or using more detailed classes (Johnson et al., 1991); and military route planning (Garvey, 1987) in which cover density is most important. Biodiversity databases, supporting multidisciplinary research and land use planning, must be designed to accommodate diverse (and possibly unforeseen) user views (Davis et al., 1990).

DeMers (1991) contrasted the traditional vegetation map produced under the communication paradigm with GIS vegetation databases operating under the analytical paradigm. The traditional area-class vegetation map was created to communicate the distribution of predetermined thematic classes. Conversion of such a paper map into a digital GIS map layer increases the utility of the map for comparison with other environmental factors or for quantitative inventory of amount and size-distribution of vegetation patches. The map, however, only represents one cartographer's view of the original data, with the classification designed for a single purpose. GIS users now recognize the flexibility of computer technology and have identified new applications that could be better served by providing users as much of the original data as feasible. Given the large investment required to compile a GIS data layer, accommodating many user views into the same database can increase its efficiency and effectiveness.

In this paper, we describe a vegetation database being compiled as part of a conservation assessment of the entire state of California (i.e., the California Gap Analysis Project: Davis and Stoms, in press) and more broadly as part of a statewide GIS to support biodiversity research at a biogeographic scale. This biodiversity database, and particularly the vegetation data, must support many user views. The vegetation layer will be used initially both to assess the representativeness of existing nature reserves relative to the diversity of natural communities of the state and to model the distribution of wildlife species based on their habitat preferences. Examples of several uses of this multivariate vegetation database are presented for a region covering the southern coast of California for which mapping has been completed. Besides maintaining flexibility in providing various user views of the database, this database design also allows numerical taxonomical methods of classification and ordination that produce more explicit and repeatable results than having a cartographer assign unique classes to map units.

CONSTRUCTION AND CONTENTS OF THE DATABASE

The vegetation map had to meet several standards: 1) the map had to be detailed enough to represent a landscape scale view of the distribution of species and plant communities and yet, for practical reasons, be generalized so that the entire state could be mapped within time and budget constraints; 2) the vegetation classification system had to be compatible with the habitat classification system used in modeling wildlife species' distributions; and, 3) it had to be flexible enough to accommodate different user views of the biodiversity database. In response to these objectives, we selected a minimum mapping unit (MMU) size of 100 hectares (250 acres) and a nominal mapping scale of 1:100,000 for upland vegetation. Major wetland areas are mapped using a 40 hectare (100 acre) MMU, but smaller wetlands are encoded as attributes of larger upland map units.

The vegetation map of California is being produced by visual photointerpretation of 1990 digital Thematic Mapper satellite imagery (details are given in Davis and Stoms, in press). Images are resampled to the Albers equal-area projection with a 100 meter resolution (i.e., 1 hectare pixels). Map unit boundaries are then digitized directly on the computer screen over a display of the TM data to enclose landscapes that are spectrally homogeneous in terms of color and texture. For the coastal area presented here, TM bands 3, 4, and 5 are displayed in the blue, red, and green color planes, respectively.

A variety of data sources relating to vegetation are used to attach thematic attributes to map units, based on the idea that converging lines of evidence will increase the accuracy of and confidence in the map. These sources range from 1990 high altitude aerial photography, small scale vegetation maps such as CALVEG (Parker and Matyas, 1981), and large scale vegetation maps of local areas. The Vegetation Type Map (VTM) Survey maps developed from field surveys by the United States Forest Service in the 1930's (described in Critchfield, 1971) have been especially useful for labeling map units that have not been converted to urban or agricultural uses or have not been recently burned as evidenced from the satellite images. Based on these sources, large and heterogeneous map units are split or redrafted if necessary. Rather than attempting to generalize a map unit to a single type, we assign both primary and secondary cover types as attributes. Up to three dominant overstory plant species are recorded for both the primary and secondary types. Small wetlands and species of special interest, i.e., regional endemics, are also encoded as attributes of upland landscape mosaics. The proportion of each map unit in the primary and secondary types, and the canopy density are also recorded, usually on the basis of aerial photography. Although canopy height can be an important habitat element for wildlife, we assume that at this scale of mapping, vegetation is a constantly shifting mosaic of seral stages. In other words, seral stages are below the resolution of the database. See Figure 1 for an example of the ARC/INFO attribute coding form.

Example of Vegetation Attributes Coding Form

Polygon-ID	907
Primary Series	32001 Chamise
Primary Species 2	32036 Chaparral Whitethorn
Primary Species 3	32024 Scrub Oak
Primary Series Percent	40-49 %
Primary Canopy Cover	40-59 %
Secondary Series	32305 White Sage
Secondary Species 2	None
Secondary Species 3	None
Secondary Series Percent	30-39 %
Secondary Canopy Cover	40-59 %
WHR Wetland Types	2 Valley Foothill Riparian
Wetland Series	41004 Coast Live Oak
Special Interest Series	42007 Bigcone Douglas Fir
Aerial Photo #	90-115-2832
Labeler's Initials	PAS
Field Check Status	0: not visited
Disturbance	1: undisturbed
Comments	Dense oak riparian

Figure 1. Example of the coding form used to edit map unit attributes. Data encoded as numerical codes are expanded here to their corresponding text values.

Three indices of landscape pattern--dominance, contagion, and fractal dimension (O'Neill et al., 1988)--will also be encoded for each map unit. The indices are derived from an unsupervised classification of the TM imagery and stratified by map unit boundaries. Initial map units with extreme index values were checked for potential editing, such as splitting heterogeneous units. Once the ecoregion-wide vegetation map is completed, final indices will be recorded as additional database attributes. Thus, these landscape indices are useful both for error detection and revision and for providing complementary information on the spatial patterning of vegetation stands within landscape-scale map units.

Mapping to date has been completed for the south coast of California and is underway for the Sierra Nevada, the Mojave and Sonoran deserts, and the central coast. Illustrations of different user views in the remainder of the paper are based on the vegetation database for the completed region, from the San Bernardino Mountains to the Mexican border.

APPLICATIONS OF THE VEGETATION DATABASE

Biogeographical Distribution of a Plant Species

A simple user view of the database would be to ask for the distribution of a single species. Bigcone Douglas-fir (*Pseudotsuga macrocarpa*) is of conservation interest because it is restricted to mid-elevations of montane environments of southern California. It generally occurs in small patches in protected canyon sites or as a minor canopy species, and as such is not represented in conventional small-scale vegetation maps. In the vegetation database, however, the presence of this species is encoded as an attribute, so map units where it is present or is dominant can be identified (Figures 1 and 2). Although the precise location of small patches of Bigcone Douglas-fir may not be determined at this map scale, such a representation is suitable for delineating its biogeographical range or focusing conservation field inventories.

Wildlife Habitat Modeling

The translation of plant communities into wildlife habitat types is a common user view of a biodiversity database. The California Wildlife-Habitat Relationships (WHR) database developed for all wildlife species in the state describes suitability of 48 habitat types (Mayer and Laudenslayer, 1988). WHR uses a coarser classification than the CALVEG series schema, but there is not a simple many-to-one mapping of CALVEG into WHR (Mayer and Laudenslayer, 1988). By retaining floristic detail in the database, we are better able to assign a primary, secondary, and riparian habitat type to each map unit. By relating the WHR database of habitat suitabilities to the GIS database of habitat types, and limiting the distribution by wildlife species' range boundaries, we can predict the species' potential distributions. Figure 3 depicts potential habitat of the Orange-throated Whiptail, a reptile restricted to California's south coast.

Within-class Variation of a Vegetation Class

Another limitation of traditional vegetation maps is the lack of information on variability of species composition within a class. Even at the series level of classification, map units can have quite variable mixes of dominant species. In a vegetation database, on the other hand, data are provided either to measure the variability or to classify the map unit to satisfy a user's specific need.

Chamise chaparral (*Adenostoma fasciculatum*), for example, is the most widespread chaparral type in California (Parker and Matyas, 1981). However, there is considerable floristic and ecological variation among stands dominated by this species. To analyze the variation within the chamise chaparral series, the composition of map units where this evergreen shrub occurs as one of the primary species was extracted from the

721

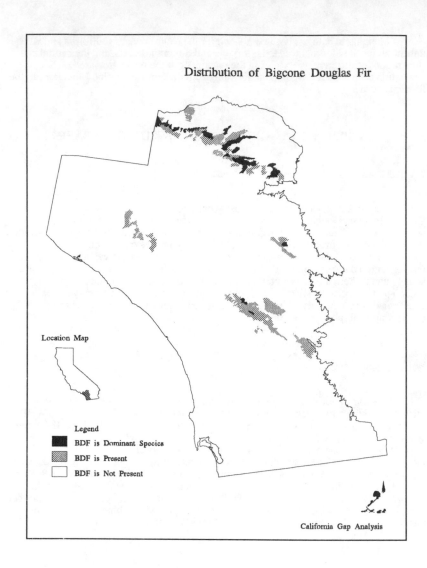

Figure 2. Map of distribution of a single species, Bigcone Douglas Fir (BDF)
(*Pseudotsuga macrocarpa*).

database. These presence data were then classified by Two Way Indicator Species
Analysis (TWINSPAN: Hill, 1980). This classification method is usually applied to field
plot data to order a dataset of species by plots (or by map units in this case) so that similar
map units and species are grouped together. The samples are classified in a divisive
hierarchy, based on indicator species that are strongly grouped in the sorted table.

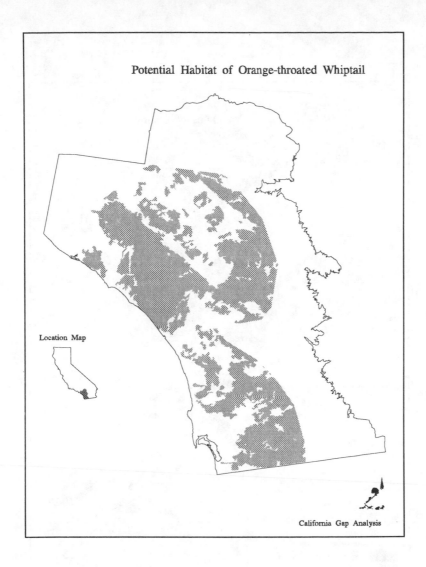

Figure 3. Map of potential habitat of the Orange-throated Whiptail as constrained by its known range limits.

The sample classification shown in Figure 4 distinguishes two lowland coastal forms of chamise chaparral from higher elevation, interior forms, generally separated by the 750 meter contour. The class containing *Ceanothus greggii* and *Adenostoma sparsifolium* along the eastern edge of the study area is associated with a drier and more continental climate.

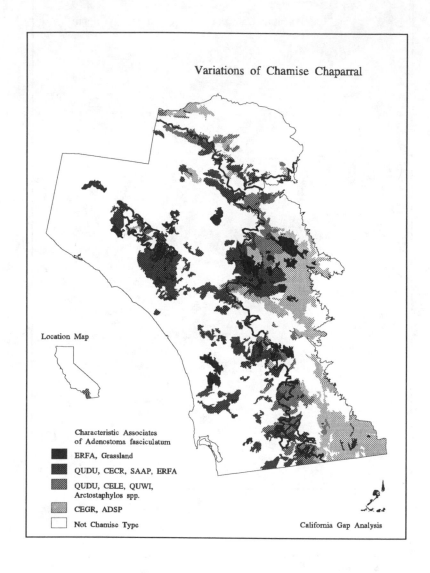

Figure 4. Map of a TWINSPAN classification of types of chamise chaparral. Associated species codes: ADSP = *Adenostoma sparsifolium*; CECR = *Ceanothus crassifolius*; CEGR = *Ceanothus greggii*; CELE = *Ceanothus leucodermis*; ERFA = *Eriogonum fasciculatum*; QUDU = *Quercus dumosa*; QUWI = *Quercus wislizenii* (shrub form); SAAP = *Salvia apiana*. The bold line through the center of the study area is the 750 meter contour.

DISCUSSION AND CONCLUSIONS

Localized versions of similar vegetation databases have been pioneered for much smaller sites. DeMers (1987) compiled a GIS of a 133 acre nature preserve in Ohio that contained data on plant composition, stem density, and average diameter for each tree species within 10 meter grid cells. He later generated a GIS database from field data on floristics and canopy structure for a 7 by 7 mile study area on Mount Desert Island, Maine (DeMers, 1991). To the best of our knowledge, our project represents the first such multivariate GIS vegetation database over the domain of an entire region.

Clearly, such an ambitious enterprise at regional scale will not be practical in all circumstances. Because detailed floristic information can not be extracted over large areas from satellite data, large scale vegetation maps must be available to supplement the subformation level such imagery does contain. In most cases, however, large-scale maps present several problems. Most importantly, existing maps do not exhaustively cover the entire study area, leaving gaps where other methods must be utilized (e.g., environmental modeling, extensive field sampling, or generalization of attributes). Classification schemas of maps covering different portions of the region are seldom developed for the same objective, forcing the "crosswalking" of various schemas. Similarly, these maps were typically compiled at different times with different methods, which again makes mosaicking them into a regional map difficult.

To date, the vegetation database of California has been completed for the southern coastal area, but is also in progress for the rest of the state as well. Our goal is to complete mapping of the entire state at the same level of spatial and taxonomic resolution by the end of 1993. Database performance will be assessed within each ecoregion through a combination of techniques, including a systematic sampling by transects and by measures of agreement with large scale vegetation maps of local areas. Once the vegetation database is completed, it should support many other user views, including natural resource inventories, correlation of plant communities with environmental factors, gap analysis, land cover change detection (monitoring), modeling distributions of individual species and vegetation types, and predicting effects of climate warming (Davis et al., 1990).

ACKNOWLEDGMENTS

Support for the California Gap Analysis Project was provided through grants from the National Fish and Wildlife Foundation and Southern California Edison Company. We thank Barbara Allen-Diaz, Janine Stenback, and the UCSB Map and Imagery Laboratory for providing data critical to the project. Mike Bueno, Violet Gray, Josh Graae, John Kealy, and Steve Sherrill assisted in compiling the vegetation database. Allan Hollander translated dominant canopy composition into habitat classes.

REFERENCES

Critchfield, W. B., 1971. *Profiles of California Vegetation*, U. S. Forest Service, Pacific Southwest Forest and Range Experiment Station, Research Paper PSW-76, Berkeley, 54 pp.

Davis, F. W., D. M. Stoms, J. E. Estes, J. Scepan, and J. M. Scott, 1990. An information systems approach to the preservation of biological diversity. *International Journal of Geographical Information Systems*, 4: 55-78.

Davis, F. W. and D. M. Stoms, 1992. Gap analysis of biodiversity in California, in *Proceedings of the Symposium on Biodiversity of Northwestern California*, in press.

DeMers, M. N., 1987. Sliced vegetation maps. *Proceedings of IGIS Symposium*, Association of American Geographers, Washington, D. C., vol III, pp. 487-498.

DeMers, M. N., 1991. Classification and purpose in automated vegetation maps. *The Geographical Review*, **81**: 267-280.

Franklin, J., T. L. Logan, C. E. Woodcock, and A. H. Strahler, 1986. Coniferous forest classification and inventory using Landsat and digital terrain data. *IEEE Transactions on Geoscience and Remote Sensing*, **GE-24**: 139-149.

Garvey, T. D., 1987. Evidential reasoning for geographic evaluation for helicopter route planning. *IEEE Transactions on Geoscience and Remote Sensing*, **GE-25**: 294-304.

Hill, M. O., 1980. *TWINSPAN and indicator species analysis*. Ecology and Systematics, Cornell University, Ithaca, New York.

Johnson, L. B., G. E. Host, J. K. Jordan, and L. L. Rogers, 1991. Use of GIS for landscape design in natural resource management: habitat assessment and management for the female black bear, in *Proceedings of GIS/LIS'91*, pp. 507-517.

Kuchler, A. W. and I. S. Zonneveld, editors, 1988. *Vegetation Mapping*. Kluwer Academic Publishers, Dordrecht, The Netherlands.

Mayer, K. E., and W. F. Laudenslayer, Jr., 1988. *A Guide to Wildlife Habitats of California*. California Department of Forestry and Fire Protection, Sacramento, CA.

O'Neill , R. J., J. R. Krummel, R. H. Gardner, G. Sugihara, B. Jackson, D. L. DeAngelis, B. T. Milne, M. G. Turner, B. Zygmunt, S. W. Christensen, V. H. Dale, and R. L. Graham, 1988. Indices of landscape pattern. *Landscape Ecology*, **1**: 153-162.

Parker, I., and W. J. Matyas, 1981. *CALVEG: a classification of Californian vegetation*, USDA, Forest Service, Regional Ecology Group, San Francisco.

Stoms, D. M., F. W. Davis, C. B. Cogan, M. O. Painho, B. W. Duncan, J. Scepan, and J. M. Scott, 1992. Geographical analysis of California condor sighting data. *Conservation Biology*, in press.

AN INITIAL INVESTIGATION OF INTEGRATING NEURAL NETWORKS WITH GIS FOR SPATIAL DECISION MAKING

Daniel Z. Sui
Department of Geography
The University of Georgia
Athens, GA 30602

ABSTRACT

This paper presents an initial investigation of integrating a standard back-propagation artificial neural network with GIS for development suitability analysis. It was found that the three-layer network developed for this study can be incorporated into the GIS modeling process, and achieved a Root Mean Square Error (RMSE) of 0.1 after 7,200 training cycles. The neural network-based GIS modeling approach is able to make a close approximation of experts' decisions without the explicit elicitation of experts' knowledge into production rules. The bottleneck problems of traditional cartographic modeling techniques, such as weight determination, selection of summation functions, and the inability to handle noisy and missing data have been resolved through the training process and hidden layers in an artificial neural network. **Key Words: Neural Networks, GIS Modeling, Developmental Suitability Analysis, Artificial Intelligence.**

INTRODUCTION

The applications of Artificial Intelligence (AI) technology, particularly expert systems, for spatial data processing have been very popular within the GIS community in recent years (Smith et al., 1987; Fisher et al., 1988; Kim et al., 1990; Smith and Yang, 1991; Peuquet, 1991). Although the integration of Artificial Intelligence/Expert Systems with GIS has achieved considerable success during the last several years, knowledge acquisition and representation remain a headache for most knowledge engineers. It was found that some human knowledge is inexpressible in the form of rules and sometimes may not be understandable even though it can be expressed (Hoffman, 1987). Furthermore, most human experts have difficulties in describing their knowledge explicitly and completely. Rigid rule-based expert systems do not always function as their designers expected, and many failures have been documented in the recent literature (Bell, 1985; Sherald, 1989). Therefore an automated learning strategy that can simulate the experts' learning and reasoning process would be highly desirable.

The development of neural computing technology, which is inspired by the studies of the human brain and nervous systems, seems to offer a new alternative to the rule-based expert system. Recently, several efforts have also been made by our colleagues in the remote sensing community to apply neural network modeling for remote sensing image processing (Benediktsson et al., 1990; Hepner et al., 1990; Ryan et al., 1991). These pioneering works have confirmed that the neural network-based image

727

processing approach usually achieves a higher classification accuracy than most traditional statistical methods. The potential application of neural networks in map design is also discussed in cartographic literature (Johnson and Basoglu, 1989). However, such a fascinating field has not been discussed in the GIS literature so far, not to mention actual applications.

The objective of this paper is to explore the possibility of incorporating neural computing technology into the GIS modeling process and to examine the implications of neural computing for the further development of GIS technology. The rest of this paper is organized into five sections. A brief introduction to the development suitability problem is given in part II. General principles of neural networks and neural computing are described part III. Part IV covers the methodology about the integration of neural networks with GIS modeling for developmental suitability evaluation. Results and discussions are contained in part IV, which is followed by some concluding remarks in Part VI.

The DEVELOPMENT SUITABILITY PROBLEM

Developmental suitability may be defined as the fitness of a given land parcel for a specific use based on the conditions of the site (Steiner, 1983). In essence, the developmental suitability evaluation involves two steps: the selection of criteria and the combination of the selected criteria to obtain an aggregate fitness measure for each individual evaluation unit (Hopkins, 1977). Once the evaluation criteria are determined, the landscape attributes based on these criteria will be assigned a score and a weight. The developmental suitability assessment then becomes how to apply an aggregation method to combine these evaluation criteria cartographically or algebraically to obtain the composite suitability index.

In order to facilitate the comparison of results of the neural network-based approach and those of the conventional cartographic modeling approach, the data sets for Del Norte, Texas, offered by Strategic Mapping Inc. (Laserna and Landis, 1989), are used for this case study. This data set includes the six criteria for development evaluation, land use, soil type, sewer line, and the proximity to major freeways, highways, and railroads. There are a total of 218 land parcels. All land parcels in the study area were first evaluated using the traditional cartographic modeling technique based on the same criteria, and the neural network-based GIS modeling approach was then applied for the evaluation. Details of GIS-based cartographic modeling techniques can be found in Berry (1987) and Tomlin (1990).

NEURAL NETWORKS AND NEURAL COMPUTING

Neural computing is a new kind of AI technology which can perform brainlike functions after being trained using inputs and desired outputs. An artificial neural network is composed of many highly interconnected processing elements arranged in a way analogous to the biological neurons (Fig. 1a). A processing element has many input paths and combines, through a summation function, the value of these input paths (Fig. 1b). The result is an internal activity level for the processing element. The combined input is then modified by a transfer function which passes information only if the combined activity level reaches a certain threshold. The output value of the transfer function is generally passed directly to the output path of the processing element. The output path of a processing element can be connected to input paths of other processing elements through connection weights. Since each connection has a

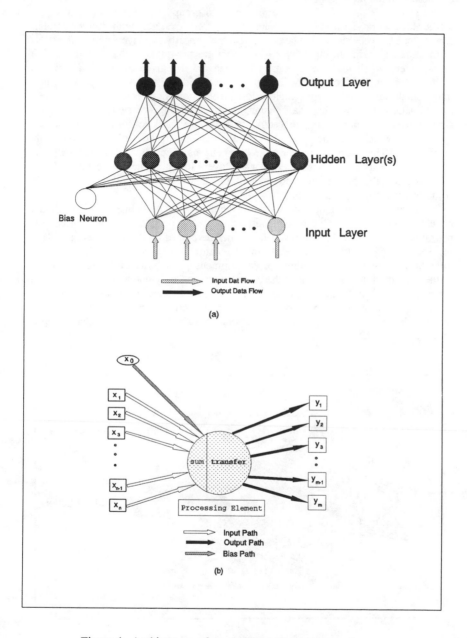

Figure 1. Architecture of an Artificial Neural Network:
(a). Macro-structure (b). Micro-structure

corresponding weight, the signal on the input lines to a processing element is modified by these weights prior to being summed.

Processing elements are usually organized into groups called layers or slabs. An artificial network consists of a sequence of layers or slabs with full or random connections between successive layers. There are typically two layers with connections to the outside world: an input layer where data are presented to the network, and an output layer, which holds the response of the network to a given input. Layers distinct from the input and output buffers are called hidden layers. Typically these intermediate layers of neurons capture low-level features, such as the presence of a simple pattern in single criteria or some weighted average of several inputs. The signals from neurons in the preceding layers were used to form more complex "percepts," which may be associated closely with the desired classification in the training set. There are also one or more bias neurons which always have an output of 1. They serve as the threshold units for the layers to which they are connected, and weights from the bias neurons to each of the neurons in the following layer are adjusted exactly like other weights.

At present, there are more than thirty different artificial neural network models that have been developed. One of the most famous, and perhaps the most successful, neural paradigms -- the back-propagation model developed by Rumelhart et al. (1986), is used in this study to illustrate how the neural computing technology can be integrated with the GIS modeling for developmental suitability evaluation. Because of the computational limitations of the PC environment, only one hidden layer with ten processing elements is included in the back-propagation neural network (Fig.2). The six elements in the input layer correspond to the six criteria imported from GIS, and the five elements in the output layer correspond to the ordinal evaluation scale adopted in urban planning.

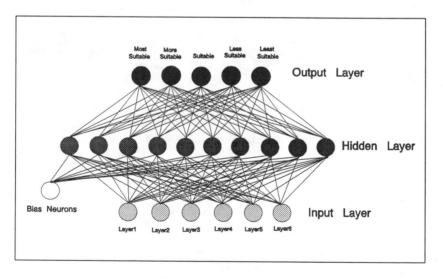

Figure 2. The Neural Network Architecture for Developmental Suitability Evaluation

The methodology of the neural network-based GIS modeling for developmental suitability analysis consists of three stages. These three stages are connected by a data format conversion program between GIS and Neural Network simulator.

Stage One: Data Preparation Basic GIS routine functions were used during this stage. Land use types and soil categories within a specific land parcel are extracted using the overlay and polygon selection capabilities of Atlas*GIS. The corridor and polygon selection capabilities of Atlas*GIS are used to identify parcels from the land parcel layer within a specified distance to main sewer trunk lines, freeways, highways, and railways. These six new data layers derived from overlay and buffer operations are then exported to ASCII format using the import/export facilities in Atlas*GIS. A Pascal program was written to convert the exported ASCII data file into the data format suitable for neural network modeling. The six GIS data layers now become the six processing elements in the input layer of the artificial neural network.

Stage Two: Neural Network Modeling The PC version NeuralExplorer 2.0, a comprehensive neural network modeling shell developed by NeuralWare (NeuralWare, 1990), was employed to implement the standard back-propagation neural network.

The training of the neural network is the most critical step in neural network modeling. The training file was prepared using the same criteria as used in the traditional cartographic modeling approach. Twenty-five examples are included in the training file. It is very important to have approximately the same number of training sites in each output category, because many neural networks are basically "lazy." A neural network usually attempts to find the easiest way to solve a particular problem (Klimasauskas, 1991). This kind of swamping effect is even more pronounced when boundaries between "good" and "bad" decisions are fuzzy.

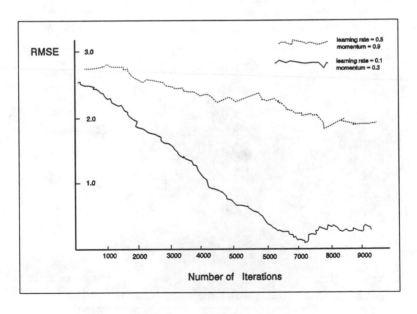

Figure 3. The Neural Network Converging Rates

The training of the neural network is very sensitive to learning rate ϵ and momentum term α. Interested readers may consult Rumelhart and Hinton (1986) for the definition of these two terms. Inappropriate starting values for ϵ and α often throw the network into oscillation or saturation situations. When saturation occurs, the outputs of the training and testing tend to be about the same, "suitable" ("0 0 1 0 0" in the output file) for all the land parcels. After a fairly long period of experiments to fix the oscillation and saturation problems, the learning rate and momentum were finally set to 0.1 and 0.3 respectively. After these changes in parameters were made, the convergence rate is much faster. As shown in Figure 3, at 7,200 cycles of iteration, the Root Mean Square Error (RMSE) is about 0.15. After 7,200 iterations, the RMSE begins to deviate from the converging trend again.

After the learning process through the training file, the entire study area was evaluated through the NeuralExplorer's recall process. The result is saved as spreadsheet format using the input/output (I/O) facilities in NeuralExplorer. The result is then converted back to the GIS format for output purposes.

Stage Three: Output Generation During this stage, the thematic mapping functions of Altas*GIS were used to prepare the final evaluation maps based on the evaluation results from the neural network modeling.

RESULTS AND DISCUSSIONS

The result from neural network modeling is shown in Figure 4. Compared to the result of the traditional map overlay-based cartographic modeling (Fig. 5), the distribution of less and least suitable land parcels is quite similar on the two maps. Since two transportation criteria (accessibility to freeways and highways) are given more weights in the cartographic modeling process than in the neural network modeling, the most and more suitable land parcels in Figure 5 are apparently leaning to the proximity to the major transportation routes. The most and more suitable lands on the neural network modeling result map are confined to the lower right area, as the

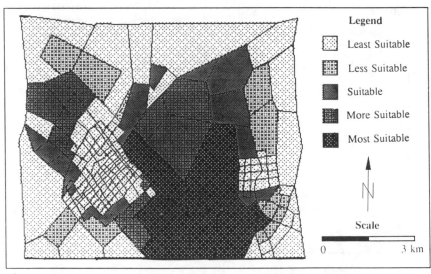

Figure 4. The Development Suitability Map: Neural Network Modeling

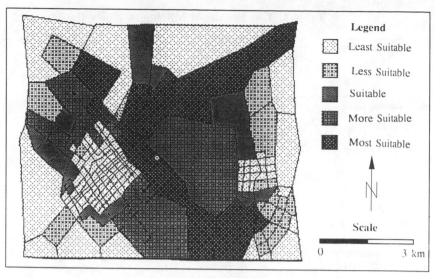

Figure 5. The Development Suitability Map: Cartographic Modeling

weights of the six evaluation criteria are more balanced and distributed after the learning process based on the training example. Instead of subjectively assigning weights to each criterion, the neural network modeling will automatically determining weights of each factor through back-propagation based on representative training sets. Therefore, in contrast to the cartographic modeling which is characterized by the subjective linear summation process based on user's experience, the neural network-based GIS modeling can be described as a more objective, nonlinear learning process based on examples.

Three kinds of experiments were conducted to test the fault tolerance ability of the neural network by deleting some data items or using a noisy value instead. As shown in Table 1, the results from neural network modeling are surprisingly consistent to the expected results. In contrast, cartographic modeling either gave strongly biased results (caused by noise data, as in experiment 2) or terminated because of data type error (as in experiments 1 and 3). This indicates that the neural network approach can perform more realistically on the diversified GIS data layers than the traditional approach.

However, the neural network-based GIS modeling is not without its own problems. Right now it is still unclear what kind of network architecture and training algorithm should be used for a particular application. Although there are some intuitive thoughts about applicability of various neural networks (Baily and Thompson, 1990; Nelson and Illingworth, 1991), the selection of a particular type of network and the configuration of the network architecture are determined by most neural network users in an ad hoc manner. Besides these issues, the troubling local minima problem, the possible network paralysis, and the unpredictable convergence rate may also potentially threaten the successful application of a neural network.

Although neural computing technology has created a series of new questions, it is nonetheless one of the most dynamic and exciting fields developed in this century. It is on its way toward maturity. The potential of neural computing as a new alternative for spatial decision making should be widely recognized and deserves more attention in the GIS community.

Table 1. Test Results of Missing or Noisy Data

Input for the Six Criteria	Evaluation Output	
	Neural Network	Cartographic Modeling
A. missing data		
5 3 1 1 1 _	1 0 0 0 0	Data Error
_ 3 0 _ 1 1̄	0 0 1 0 0	(No further analysis)
(_ is missing data)		
B. noisy data		
4 3 1 10 1 0	0 1 0 0 0	1 0 0 0 0
1 2 5 1 10 1	0 0 0 0 1	1 0 0 0 0
(10 and 5 are noisy data)		
C. Wrong data type		
1 1 1 W 1 0	0 0 0 0 1	Data Error
3 W 1 1 W 1	0 0 0 1 0	(No further analysis)
(W is wrong type of data)		

CONCLUDING REMARKS

This paper has demonstrated how neural network modeling technology can be incorporated into the GIS modeling process to guide decision makers in assessing the suitability of land resources. It was found that the neural network employed in this paper can perform the developmental suitability analysis at an expert's level when an appropriately configured neural network is supplied with sufficient representative training data. By repeatedly adjusting weights to a myriad of interconnections, the system can gradually learn the combination that leads to close approximation to expert decisions.

This initial investigation of integrating neural networks with GIS has also implied the tremendous potential of neural computing to assist spatial decision making in a GIS environment. Bottleneck problems of conventional cartographic modeling techniques such as weight determination, selection of summation functions, and inability to handle noisy and missing data have been resolved through the learning process and hidden layers in the artificial neural network. The intensive knowledge acquisition process in the development of rule-based expert systems has been replaced by the hidden search for connections between input and known output in neural network modeling.

Although this study also left a series of questions for future investigation, the results suggest that the neural network-based GIS modeling can be a powerful alternative approach toward automated spatial decision making.

ACKNOWLEDGEMENT

The permission given by Strategic Mapping Inc. for the use of their data is gratefully acknowledged. The author would like to thank Drs. Jay Aroson, C.P. Lo, Ronald McCllendon, Anthony Stam, and Jean-Claude Thill for their constructive comments on an earlier draft of this paper. The author also wishes to thank Elizabeth Hovinen for her extensive and professional editorial help.

REFERENCES

Bailey, D., and D. Thompson, 1990, How to develop neural network applications, *AI Expert*, 6:38-46.

Bell, M.Z., 1985, Why expert system fail? *Journal of Operations Research*, 6:613-19.

Benediktsson, J.A., P.H. Swain and O.K. Ersoy, 1990, Neural networks approaches versus statistical methods in classification of multisource remote sensing data, *IEEE Transactions on Geoscience and Remote Sensing*, 28:540-51.

Berry, J.K., 1987, Fundamental operations in computer-assisted map analysis, *International Journal of Geographical Information Systems*, 1:119-36.

Eberhart, R.C. and R.W. Dobbins (eds.), 1990, *Neural network PC tools: A practical guide*. San Diego, CA: Academic Press Inc.

Fisher, P.F., W.A. Mackaness, G. Peacegood, and G.G. Wilkinson, 1988, Artificial intelligence and expert system in geo-data processing, *Progress in Physical Geography*, 12:371-88.

Hepner, G.F., T. Logan, N. Ritter and N. Bryant, 1990, Artificial neural network classification using a minimal training set: Comparison to conventional supervised classification, *Photogrammetric Engineering and Remote Sensing*, 56:469-73.

Hopkins, L., 1977, Methods of generating land suitability maps, *Journal of American Institute of Planning*, 43:386-400.

Hoffman, R., 1987, The problem of extracting the knowledge of experts from the perspective of experimental psychology, *AI Magazine*, 8:53-67.

Johnson, D.S. and U. Basoglu, 1989, The use of artificial intelligence in the automated placement of cartographic names, *Arto-Carto 9*, 225-30.

Klimasauskas, C.C., 1991, Applying neural networks, *PC AI*, 5:31-34.

Kim, T.J., L.L. Wiggins and J.R. Wright, *Expert systems: Applications to urban planning*. New York: Springer-Verlag.

Laserna, R. and J. Landis, 1989, *Desktop mapping for planning and strategic decision-making*. San Jose, CA.: Strategic Mapping Inc.

Nelson, M.M. and W.T. Illingworth, 1991, *A Practical guide to neural nets*. Reading, MA.: Addison-Wesley Publishing Company.

NeuralWare Inc., 1991, *Applications in neural computing*. Pittsburgh, PA.: NeuralWare Inc.

Peuquet, D.J., 1991, An overview of the applications of artificial intelligence approaches for Geographic Information Systems, *Proceedings of the seventh annual conference on interactive information and processing systems for meteorology, oceanography and hydrology, new orleans*, Louisiana.

Rumelhart, D.E., G.E. Hinton, and R.J. Williams, 1986, Learning internal representations by error propagation, in Rumelhart, D., J. McClelland and PDP Research Group (eds.), *Parallel distributed processing: Exploration in the microstructure of cognition*, Vol. 1, 169-183, Cambridge, MA.: MIT Press.

Rumelhart, D.E. and G.E. Hinton, 1986, Learning representation by back-propagation errors, *Nature*, 323:533-36.

Ryan, T.W., P.J. Sementilli, P. Yuen and B.R. Hunt, 1991, Extractions of shoreline features by neural nets and image processing, *Photogrammetric Engineering and Remote Sensing*, 57:947-55.

Sherald, M., 1989, Neural networks versus expert system: Is there room for both? *AI Expert*, 5:10-15.

Smith, T.R., D.J. Peuquet, S. Menon and P. Agarwal, 1987, KBGIS- II: A knowledge-based geographic information system, *International Journal of Geographical Information System*, 1:149-72.

Smith, T.R. and J. Yiang, 1991, Knowledge-based approaches in GIS, in Maguire et al. (eds.), *Geographical information systems: Principles and applications*, Vol. 1, 413-25. London: Longman Press.

Steiner, F., 1983, Resource suitability: Methods for analysis. *Environmental Management*, 7:401-420.

Tomlin, C.D., 1990, *Geographic information systems and cartographic modeling*. Englewood Cliffs, N.J.: Prentice-Hall.

Wasserman, P.D., 1989, *Neural computing: Theory and practice*. New York: VNR Press.

A MODEL-BASED GIS APPROACH
FOR URBAN DEVELOPMENT SIMULATION

Daniel Z. Sui and C. P. Lo
Department of Geography
The University of Georgia
Athens, GA 30602

ABSTRACT

The 1980s witnessed a great proliferation and adoption of GIS technology in the urban and regional planning community. However, most GIS packages do not adequately support spatial decision making because of their lack of sophisticated analytical and modeling capabilities. This paper develops a model-based GIS approach for the urban development simulation of Hong Kong. A modified version of the Lowry-type urban model was constructed and integrated with a vector-based GIS to simulate the impact of the relocation of Kai Tak International Airport on the future development of Hong Kong. The results demonstrate that the integration of urban models with GIS not only facilitates the input of variables into the urban model and the automatic mapping of the modeling results, but also improves the analytical and modeling capabilities of the GIS. Policy implications of the modeling results are also addressed. **Key Words: Urban Modeling, GIS, Spatial Impact Analysis.**

INTRODUCTION

In recent years, GIS technology has become increasingly important to urban planners all across the world as witnessed by a flood of publications (Huxold, 1990; Scholten and Stillwell, 1990; Worral, 1990; Parrott and Stuz, 1991). While the benefits and advantages of using GIS in solving urban planning problems are very well documented in the literature, complaints and dissatisfactions about the performance of GIS in the urban planning process have also been voiced by both academics and urban planners (Openshaw, 1987; Harris, 1989; Goodchild, 1991). A general consensus within the GIS community is that GIS is more successful as an information management rather than as a spatial analysis and modeling tool. To alleviate this situation, the integration of GIS with sophisticated analytical and modeling techniques has been the focus of research among academics, urban planners, and software developers in recent years. Although some conceptual discussions are now available (Dangermond, 1987; Clarke, 1990; Brail, 1990), the integration of GIS with various urban modeling techniques is still at its infant stage and much research effort needs to be expended in that direction. As Harris and Batty (1992) observed, "...in general at present, linking GIS to system models of all kinds is an activity which has hardly

737

begun with few urban models taking on the vestiges of GIS and vice versa."

The primary objective of this paper is to explore the possibility of integrating GIS with urban modeling for simulating the effects of different urban development policies. The city under study is Hong Kong, which has planned to relocate its international airport from the congested urban area to the rural Chek Lap Kok area of North Lantau Island in the west by 1997 -- the year when sovereignty over Hong Kong reverts to China. Specifically, a model-based GIS approach is developed and applied to simulate the effects of the Kai Tak International Airport relocation on the future development of Hong Kong. By doing so, we also want to identify the problems and prospects of integrating urban models with GIS.

STUDY AREA

Hong Kong, a British Crown colony located on the southeast coast of China, is chosen as the study area (Fig. 1). With only 1076 square kilometers two-thirds of which is hilly area, where 5.76 million people live, Hong Kong's development has always been limited by the availability of developable land to meet the needs of industry, commerce, and housing as its economy took off. The Hong Kong government's development philosophy has been characterized by the so-called "laissez-faire" capitalism with minimal government intervention. It is therefore not surprising that urban planning has only played a supporting role of the "invisible hand" of markets in the past. Only since the 1980s, when Hong Kong assumed the role of a world financial center and its citizens became more affluent, has the government begun to promote urban planning in the development of Hong Kong, as evidenced by the

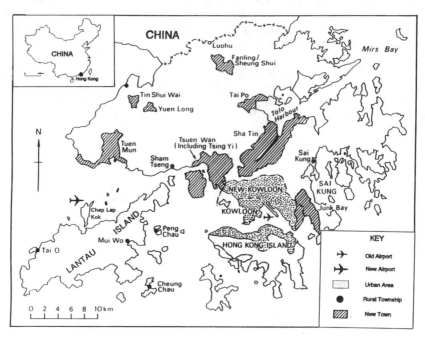

Figure 1. Location of the Study Area

738

elevation in 1990 of the Town Planning Office to a Planning Department under the Planning, Environment and Lands Branch of the Hong Kong Government (Lo, 1992).

One of the key decisions made by the Hong Kong government in recent years is to relocate the Kai Tak International Airport from its current urban location in Kowloon to Chek Lap Kok off the island of Lantau, a very remote, rural area located to the west of the main urban area (Fig. 1). Accompanying the relocation of the international airport, a number of highways, bridges, and tunnels to connect the island with the main metropolitan areas will have to be constructed as an integral part of the airport relocation project. This gigantic project will not only cost a huge amount of money (approximately U.S. $ 20 billion) but also will have tremendous impact on the future development Hong Kong. The model-based GIS approach developed in this paper is to simulate the impact of the airport relocation on the spatial structure of Hong Kong.

The URBAN MODEL

To evaluate the effects of alternative urban policies requires the development of an urban model in which the urban structures are synthesized into an abstract form (Foot, 1981). The model developed in this paper is a modified version of the Lowry-type model. The original Lowry model was developed to simulate the different development scenarios for the renewal of Pittsburgh (Lowry, 1964). The Lowry model was later modified and applied in both developed and developing countries for urban development simulations (Putman, 1983; Echenique, 1986). The Lowry model assumes that the residential location and service employment of a subregion can be described in terms of the spatial and functional interaction between the Non-Locally Dependent Employment (NLDE or basic employment) whose major function is to generate products and services for the people outside the region, and the Locally Dependent Employment (LDE or service employment) whose main function is to offer various services for the local residents.

The Lowry-type model has two sub-models: one for residential location and the other for service center location. The model can be conceptualized as an economic base mechanism coupled with spatial interaction modeling (Foot, 1981). The basic employment in the model is given exogenously (controlled by planners). The residential location of basic employment within each subregion is allocated by a spatial interaction model. The future population and service employment within each region are predicted through the economic base mechanism. The demands of service employment in one region by all other regions are determined through the service location sub-model. The service employment also requires accommodation, thus interacting with the residential model. The whole procedure is then iterated until the system converges to equilibrium. For more details of the Lowry model, see Wilson (1974), Batty (1976), and de la Barra (1989).

In considering the rapid development of the urban economy in Hong Kong, several modifications of the original Lowry model have to be made before it can be applied to Hong Kong. First, the categorization of basic versus service employment should be modified. Manufacturing employment was treated as the driving force of the economy and used as basic employment in most previously developed Lowry-type models. While the manufacturing sector still plays a very important role in Hong Kong's economy, some sectors in advanced services have also become increasingly important. In fact, Hong Kong's economy has experienced a transformation from the manufacturing-based economy to a more service-oriented economy (Lo, 1992). As

Hong Kong has become the third largest financial center in the world, 67 percent of Hong Kong's Gross Domestic Product (GDP) are contributed by the finance industry and other advanced business services. In view of the growing importance of the several advanced service sectors in Hong Kong, the basic employment in the Lowry model should also include Finance, Insurance, Real Estate (FIRE) employment, transportation and communication employment plus the traditional manufacturing employment. Second, the attraction index for residential location should also be redefined. The definitions in some of the previous urban modeling efforts (Cripps and Foot, 1970; Piasentin et al., 1978) are too simplistic and thus unrealistic to be applied in Hong Kong. With the support of GIS, we can develop a more appropriate attraction index through the combination of multiple factors. This can be achieved through a GIS cartographic modeling process based on buffer generations, overlay analysis, and database operations. The attraction index H_i for census district i is defined as follows:

$$H_i = I_{MR} * I_{TLQ} * I_{ACCESS} \qquad (1)$$

where I_{MR} is the monthly rent index; I_{TLQ} is the index of total living quarters; I_{ACCESS} is the railway (including subways) accessibility index for district i. I_{MR} and I_{TLQ} are derived by assigning an ordinal score to each category from the optimal classification. I_{ACCESS} is obtained by buffer analysis of the major transportation routes using GIS.

Mathematically, the spatial interaction model derived from Wilson's entropy maximization is employed (Wilson, 1967). The entropy-maximizing model as used in the residential allocation model is defined as

$$T_{ij} = A_i E_i H_j \exp(-\lambda c_{ij}) \qquad (2)$$

where $A_i = \{\sum H_j \exp(-\lambda c_{ij})\}^{-1}$; T_{ij} represents the number of people who work in zone i but live in residence zone j; E_i is employment at workplace zone i to be allocated to residence districts; H_j is the residence attraction index defined in equation (1). λ is a residential location factor to be determined during model calibration, and c_{ij} is the average traveling distance from district i to district j. The number of residents and service employment in each district are then determined through the economic base mechanism:

$$P_i = \alpha E^b_i (1-\alpha\beta)^{-1} \qquad (3)$$
$$E_i = E^b_i (1-\alpha\beta) \qquad (4)$$
$$E^s_i = E_i - E^b_i \qquad (5)$$
$$\alpha = P/E^b \qquad (6)$$
$$\beta = E^s/P \qquad (7)$$

where P_i is the population in district i; E^b_i is the basic employment in district i; E^s_i is the service employment in district i; E_i is the total employment in district i; P is the total population; E is the total employment; α is the employment participation rate; β is the service employment scale factor; and E^b is the total basic employment; E^s is the total service employment.

The service location model is defined as

$$S_{ij} = A_i E^s_i F_j \exp(-\mu c_{ij}) \qquad (8)$$

where $A_i = \{\sum F_j \exp(-\mu c_{ij})\}^{-1}$; S_{ij} represents the number of service employees in zone i demanded by the population of zone i; E^s_i is the service employment in zone i; F_j is the service center attraction factor for zone j, the population in each district is used as a surrogate for F_j; μ is service location parameter to be determined during model calibration; and c_{ij} is the average travelling distance from zone i to zone j.

METHODOLOGY

The methodology can be summarized in the following four general stages:

The Construction of the Database The database is composed of two parts: spatial database and attribute database. The spatial database includes the boundaries of census districts (also called secondary planning units), and major transportation routes (main roads and railways/subways) in Hong Kong. There are a total of 42 census districts. The spatial database was created through the manual digitization of the census district boundary map and the 1:25,000 topographic map using the Atlas*GIS input module. The attribute database related to employment, population, average monthly rent, total and living quarters was constructed using the built-in relational database in Atlas*GIS. Both the spatial and attribute data were for the year 1986, the most recent census data available to the authors at the time of research.

GIS Analysis and Modeling After the construction of the database, GIS analysis and modeling were performed based on the data available in the database. The process involves the following operations: (1) Database operations were performed to derive the NLDE and LDE as defined above from the major employment data. (2) Buffer zone operations were performed on railway routes. The railway accessibility is defined as whether the census tract is within the certain buffer zone (more accessible: less than 1 mile; accessible: 1-3 miles; less accessible: over 3 miles). The results of buffer operations are stored as a new attribute for each census district in the attribute file. (3) An overlay operation was performed based on the transportation accessibility, average monthly rents, and total number of living quarters in each census district to derive the attraction index as defined in equation (1). The result of the overlay analysis is stored as the attraction index for each census district for urban modeling.

Urban Modeling The attribute data and the results from GIS analysis were exported from Altas*GIS to an ASCII file for urban modeling. The urban model was implemented using Turbo Pascal 6.03 running on a 486 personal computer. The simulation period was from 1986, the base year, to 1997, when the new airport will have been completed and operational. The model was calibrated using the journey-to-work and the journey-to-shop data from the special transportation survey conducted by the Department of Transportation of Hong Kong in 1981. The Newton-Raphson method available in SAS 6.03 NLIN procedure was used for the automatic calibration of the non-linear entropy maximizing model. From initial estimates, the Newton-Raphson method searches iteratively for improved values until eventually the equation has been solved. For more technical details about Newton-Raphson method, see Batty (1976).

According to the Port and Airport Development Strategy (PADS) (Hong Kong Government, 1989), 25 percent of the forecasted territorial need for industrial land will be located in the surrounding area of the new airport and ports. A community of 150,000 people is planned at Tung Chung next to the airport. Additional population will be located at Tai Ho. Because of the various environmental constraints, the Tai Ho community can only support 110,000 people. The vacated area in the old airport area has been planned for various light industrial use. Based on these initial options, it is assumed that the basic employment in the surrounding areas of new airport and ports will double by 1997. Then what are the territory-wide effects of such a dramatic development on the spatial structure of Hong Kong? The model was then run to simulate the effect of the airport relocation on the demands of housing and services in each census district.

Output Generation The results from the urban modeling were converted back to

GIS for output purposes. A HP LaserJet III was used to produce maps of modeling results automatically. By comparing the results with the 1986 resident and employment patterns, some policy implications for the future development of Hong Kong are revealed.

RESULTS

As shown in Figure 2, there are three centers of residential growth by 1997: the surrounding area of the container port (Kwai Chung and Tsing Yi), the surrounding

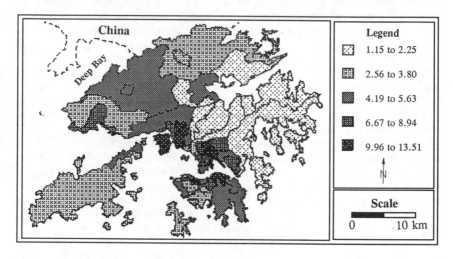

Figure 2. Percentage Growth of Residents by 1997

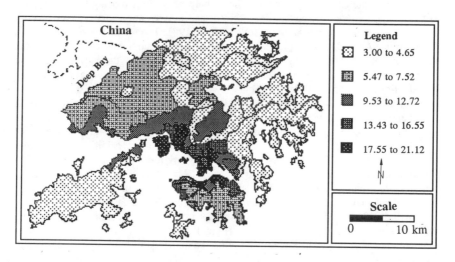

Figure 3. Percentage Growth of Service Demand by 1997

area of the old airport (Kowloon city and Wong Tai Sin; Hong Hom; Kwun Tong; Yau Tong), and the urban areas of the northern part of Hong Kong Island (North Point and Central District). There will be an increase of residents of approximately 10 percent by 1997. The growth of residents in the rest of Kowloon and Hong Kong Island will also be quite substantial, an average increase of 5 percent by 1997. In contrast, the growth of residents in the new territories and areas surrounding the new airport is not very significant (Fig. 2). Likewise, there will be tremendous demand for services in the main urban area, particularly in North Point/Quarry Bay, Central District, Hung Hom, Mong Kok, and Kwai Chung/Tsing Yi (Fig. 3). Compared to the more diversified distribution of service employment in 1986 (Fig. 4), the demand for services is apparently more concentrated in the urban core areas. The growth of demand for services in the New Territories is also quite substantial, which means that there will be more journey-to-service trips taking place between and within the urban areas. Obviously, this will increase the pressure on the transportation facilities.

In summary, the modeling results indicate an extremely strong pressure towards the residential development and demand for services in the urban areas (Kowloon and northern part of Hong Kong Island). Because of Hong Kong's need to rely on its harbor and the associated port facilities for economic development, employment opportunities tend to be located in the urban area surrounding the harbor. Therefore, centralized development has always been very strong. It has been argued by urban planners in Hong Kong that high-density, concentrated development is the most efficient and cost-effective form of utilizing the urban environment (Bristow, 1987). However, considering the fact that most of these areas except the old airport area (which will be vacated for other uses after the airport relocation) have already been overcrowded (Fig. 5), the increased number of residents in these areas is very likely to exceed the residential capacity of these zones and result in poor housing conditions and deteriorating living environment. In fact, the Hong Kong government has recognized the negative effect of overcrowdedness in the urban areas for a long time and vigorously pursued the new town program since the early 1960s with the aim to

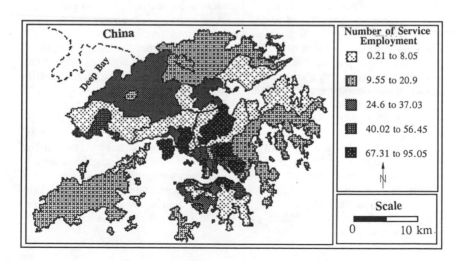

Figure 4. 1986 Distribution of Service Employment (in 1,000)

743

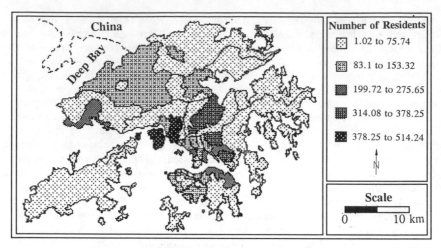

Figure 5. 1986 Distribution of Residents (in 1,000)

alleviate the high population density in these urban areas. However, the modeling results indicate that the relocation of the international airport and the more urban-oriented development strategy will lead to a more centralized development pattern in both population distribution and service employment. Such a development pattern may potentially jeopardize the further development of the new town programs and lead to more overcrowding in the main urban areas. Apparently, the Hong Kong government should have a more cohesive policy to prevent this kind of negative effect caused by the relocation of the Kai Tak International Airport.

CONCLUSIONS

The lack of sophisticated modeling and analytical capabilities has been one of the major weaknesses of applications of GIS in urban planning. This paper demonstrates that urban modeling and GIS can be integrated for urban development simulation, and such an integration is mutually beneficial to both GIS and urban modeling. For urban modeling, GIS can facilitate the derivation of parameters in the urban model and the automatic mapping of modeling results. For GIS, urban modeling extends the GIS analysis from cartographic modeling to mathematical modeling, from simple information management to policy testing and urban development simulation.

However, the model-based GIS approach developed in this paper is not a panacea for all the problems of GIS and urban modeling. Urban modeling is a very complex mathematical process, and currently is integrated with GIS through the "loose coupling" strategy as proposed by Densham (1991). Such an integration usually involves simultaneous use of several different software packages and the implementation of modeling modules using an advanced programming language. This is of course beyond the ability of average GIS users. Future research should focus on the full integration of various urban models with GIS in a more customized, user-friendly manner.

REFERENCES

Batty, M., 1976, *Urban Modeling: Algorithms, Calibrations, and Predictions*. Cambridge, U.K.: Cambridge University Press.

Brail, R.K., 1990, Integrating Urban Information Systems and Spatial Models, *Environment and Planning B*, Vol. 17, pp. 381-94.

Bristow, R., 1987, *Land-Use Planning in Hong Kong: History, Politics and Procedures*. Hong Kong: Oxford University Press.

Clarke, M., 1990, Geographical Information Systems and Model Based Analysis, in Scholten and Stillwell (eds.), *Geographic Information Systems for Urban and Regional Planning*. London: Kluwer Academic, pp. 165-75.

Cripps, E.L., and D.H.S. Foot, 1970, The Urbanization Effects of a Third London Airport, *Environment and Planning*, Vol. 2, pp. 153-92.

Dangermond, J., 1987, The Maturing of GIS and a New Age for Geographic Information Modeling (GIMS), *Proceedings of International Geographic Information Systems (IGIS): A Research Agenda*, Vol. 2, pp. 55-66.

de la Barra, T., 1989, *Integrated Land Use and Transport Modeling: Decision Chains and Hierarchies*. Cambridge, U.K.: Cambridge University Press.

Densham, P.J., 1991, Spatial Decision Systems, in Maguire et al. (eds.), *Geographical Information Systems: Principles and Applications*. London: Longman Press, pp. 403-12.

Echenique, M., 1986, The Practice of Modeling in Developing Countries, in Hutchinson and Batty (eds.), *Advances in Urban System Modelling*. Amsterdam: North-Holland Publishing Company, pp. 275-98.

Foot, D., 1981, *Operational Urban Models*. London: Methuen Press.

Goodchild, M.F., 1991, Spatial Analysis with GIS: Problems and Prospects, *GIS/LIS'91*, Vol. 1, pp. 40-48.

Harris, B., 1989, Beyond Geographic Information Systems: Computers and the Planning Professional, *Journal of American Planning Association*, Vol. 55, pp. 85-90.

Harris, B. and M. Batty, 1992, Locational Models, Geographic Information and Planning Support Systems, *NCGIA Technical Paper 92-1*, National Center for Geographic Information and Analysis, SUNY at Buffalo, Buffalo, New York.

Hong Kong Government, 1989, *Port and Airport Development Strategy (PADS)*, Hong Kong: Hong Kong Government.

Huxold, W.E., 1990, *An Introduction to Urban Geographic Information Systems*. New York: Oxford University Press.

Lo, C.P., 1992, *Hong Kong*. London: Belhaven Press.

Lowry, I., 1964, *A Model of Metropolis*. Santa Monica: Rand Corporation.

Openshaw, S., 1987, Guest Editorial: An Automated Geographical Analysis System, *Environment and Planning A*, Vol. 19, pp. 431-36.

Parrott, R., and F.P. Stutz, 1991, Urban GIS Applications, in Maguire et al. (eds.), *Geographical Information Systems: Principles and Applications*. London: Longman, pp. 247-60.

Piasentin, U., P. Costa, and D.H.S. Foot, 1978, The Venice Problem: An Approach by Urban Modeling, *Regional Studies*, Vol. 12, pp. 579-602.

Putman, S., 1983, *Integrated Urban Models: Policy Analysis of Transportation and Land Use*. London: Pion.

Scholten, H.J. and J.H. Stillwell, 1990, *Geographic Information Systems for Urban and Regional Planning*. London: Kluwer Academic.

Wilson, A.G., 1967, A Statistical Theory of Spatial Distribution Models, *Transportation Research*, Vol. 1, pp. 253-262.

Wilson, A.G., 1974, *Urban and Regional Models in Geography and Planning*. London: John Wiley and Sons.

Worral, L. (ed.), 1990, *Geographic Information Systems: Developments and Applications*. London: Belhaven Press.

A GIS/LIS PUBLIC LAND SURVEY CORNERS CONTROL POINT INVENTORY DATABASE SYSTEM

Larry D. Swenson, Special Projects Coordinator
Minnesota Department of Natural Resources
Bureau of Engineering - Survey Unit
500 Lafayette Road
St. Paul, Mn. 55155-4029
(612)-297-3793

ABSTRACT

Develop a database system to accommodate positions for the most up to date and accurate locations available for all the Public Land Survey System (PLSS) corners in the State of Minnesota. The developed design would incorporate functional features needed by surveyors, cartographers, GIS/LIS developers/users, CAD operators and database managers. The goal being to stimulate wide spread data entry and update of the information in the control point inventory database at the local county level of government were the largest, most up to date set of data presently resides.

INTRODUCTION

The Public Land Survey System (PLSS) is the legal basis for land ownership in Minnesota and most other States within the United States. The PLS corners are an important element in the development of an accurate Land Information System. All users and developers of LIS/GIS will benefit from the use of the most up to date and accurate information.

At least four driving forces came together to get the project started.

First, there has been discussion for decades about the need for an inventory of "known" PLSS monuments. PLSS records can vary from precisely located to no location records available since the orginal survey. To get better understanding of the varying condition of these records, an inventory of the "known" points was needed.

Second, with the expansion of computer mapping via Geographic Information Systems (GIS), an accurate PLSS information set is vital. While one generalized data set is available from U.S. Geological Survey topographic quadrangle maps, better information is available for selected parts of the state. This better information will be expanding over the years and must be brought into a standard format and centralized database.

Third, with the coming technology of Global Positioning Systems (GPS), there is likely to be a greatly expanded influx of geo-referenced surveys both from the private sector and governmental agencies which will expand the PLSS data base with very cost-effective methods.

Fourth, based on the previous three factors, a multi-agency request for funding during the 1992-93 biennium was made to the Legislative Commission on Minnesota Resources (LCMR). While the request was made by the Land Management Information Center (LMIC) of the Department of Administration, the proposal was also supported by the Minnesota Departments of Natural Resources (DNR) and Transportation (MnDOT). The proposal also had the support of the Minnesota Society of Professional Land Surveyors and the Minnesota Association of County Surveyors. LCMR funds were obtained to initiate this effort.

APPROACH

The County Level of government was evaluated as the most logical place to accumulate and perpetuate the most up to date Public Land Survey corner information. Then a centralize database would then be periodically updated from the county level databases.

In order to provide accurate and up to date PLS base layer data for GIS/LIS in Minnesota a database program was developed to provide easy input/output of information plus a set useful tools for the Surveyor to encourage wide spread data entry participation. The database was started with the complete data set of PLS corners as digitized from the USGS Quadrangle Maps. The plan is to upgrade the database at the county level plus any other agencies of major significance in terms of land ownership and/or land surveying accomplishments. A centralized database will be maintained through periodic updates from the counties, thus providing a complete, up to date and accurate common base of PLS information for the developers and users of LIS/GIS. Wide spread participation in the entry of the PLS corner data is the goal of our Statewide Control point Inventory Project.

Process
History of Database

In early 1980's ,Land Management Information Center (LMIC) determines a coordinate value for all PLS section corners in the state by digitizing from the 1:24,000 (85%) and the 1:62,500 (15%) U.S. Geological Survey quad maps. They call the file "GISMO". The obtained values are in Universal Transverse Mercator coordinates in North American Datum of 1927 (NAD 1927). These coordinates were used as the bases for a state wide GIS system.

In 1987 a consultant was hired to fine tune a cost accounting system and develop a computerized file system. This resulted in the entering of some 15,000 drawings and 8,000 projects into the database plus develop a front end for viewing, querying, editing and appending records. This system also includes geoCode information on the various projects.

In 1989, the Minnesota Department of Natural Resources, Bureau of Engineering received a copy of the National Geodetic Survey's (NGS) horizontal control point data file. A database was built to use this data. Unfortunately, not enough information was in the data set to avoid going to Mn/DOT to get hard copies of control point information to use in our surveys.

In 1990, the Bureau of Engineering started to look at ways to keep track of DNR public land survey corner certificates. In doing so the Bureau of Engineering developed the concept for the current Control Point Inventory System. As of Jan. 1, 1992 BOE has 15 three ring binders filled with public land survey government corner certificates. This manual system was getting out of hand with no way to report on information or use the information in our now automated mapping process. We set up a database structure to keep track of the corner certificates tied to our file system database. Many other things were happening at the same time so no implementation of system took place.

In late 1990 the Bureau of Engineering started to look at ways to manage the thousands of survey points and associated data collected each year on hundreds of projects. Although several ideas were considered, no time was available to develop anything of significance.

In February 1991, the Minnesota Department of Natural, Bureau of Engineering received a copy of LMIC's "GISMO" file containing data for all digitized PLS section corners in the State. Many ideas blossomed on ways this data could be used to develop a database structure to accommodate information on corner certificates, PLS section corners, Survey points and land information as reference to the public land survey system. The "GISMO" file was converted to NAD 1983 using NGS's computer program "NADCON" (North American Datum Conversion).

At the same time, the Minnesota DNR Geographic Resources Information Systems Plan prepared by Plangraphics, Inc. was released recommending that the Bureau of Engineering be responsible for the GIS base layers for the Departments GIS.

In March 1991, the ideas, designs and uses of the above systems and data sets were then incorporated into the public land survey corner database concept. Since there were many questions on how to best accomplish our goals, a pilot project was selected. The Bureau of Engineering chose Houston County as the pilot project area to develop and fine tune the concept into a working database.

In May 1991, Land Management Information Center establishes an Advisory Committee for a Legislative Commission on Minnesota Resources (LCMR) funded (from July 91 to June 93) Control Point Inventory Project (PLS corners inventory). Several employees from the Minnesota Department of Natural Resources are members of this committee. The involvement in this committee provided many additional ideas to incorporate into the pilot project.

In October 1991, Lowell Pommerening, a Registered Land Surveyor with the DNR, Bureau of Engineering presented the Houston County pilot project at the Minnesota LIS/GIS Consortium. Members of the Control Point Inventory Committee suggested our pilot project be expanded and used as the pilot phase of the LCMR funded Control Point Inventory System.

In December 1991, Funding was allocated to the DNR, Bureau of Engineering for the purchase of computer equipment, software and personnel time to develop our Houston County pilot project into the Control point inventory system.

Fuvrther enhancements in the system programs, data entry and documentation is prepared in order to pilot the system in five chosen counties.

In June 1992, the control point inventory system software, documentation and training is provided to the five pilot Counties.

September 1992, A scheduled meeting of the Control Point Inventory Advisor Committee and representatives from the five pilot counties will meet to evaluate and discuss funding, enhancements, changes, problems and where do we go from here.

Fall / Winter 1992, prepare a report on the pilot portion of the project and a plan for statewide implementation of Minnesota's Control Point Inventory System.

Overview of The System

The Control Point Inventory is a relational database system. In other words it is composed of a variety of database files which support the main control point system in charge of registering all pertinent point data. This main corner point database offers several applications including: an opening screen for viewing and finding data, a township graphics screen to see where the points are located throughout the township, an edit/update screen for adding or revising data as it becomes available or more accurate, and an output file allowing the user to export the information to CAD, GIS, Database or Ascii files.

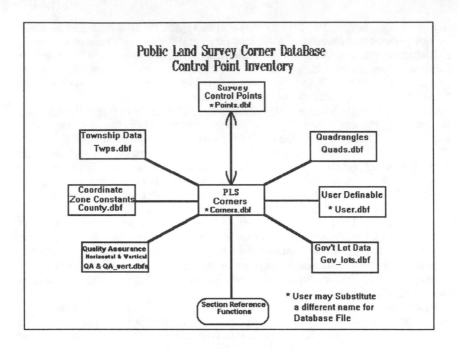

Public Land Survey Corner DataBase
Control Point Inventory

Survey Control Points
★Points.dbf

Township Data
Twps.dbf

Quadrangles
Quads.dbf

Coordinate Zone Constants
County.dbf

PLS Corners
★Corners.dbf

User Definable
* User.dbf

Quality Assurance
Horizontal & Vertical
QA & QA_vert.dbfs

Gov't Lot Data
Gov_lots.dbf

Section Reference Functions

* User may Substitute a different name for Database File

*CORNERS.DBF. . . .Public Land Survey Section Corners Database
This is the main user file which contains information selected from all of the other databases. It produces a conglomeration of information that can be used for general point reference on the screen or paper, as a reference in other database files, to create AutoCAD drawings, and most importantly as a GIS base layer. This file is automatically updated as the other relational files are changed and vise versa (the other database files are altered when CORNERS.DBF information is changed), making the system uniform in information quality/source.

RELATIONAL DATABASES

QUADS.DBF . . . Minnesota's US Geological Survey Quadrangle Maps
This file is used to catalog all USGS quadrangles and references the USGS Quad units to other information sources and quad structures that may be used including DNR Codes, NGS Quad ID, or alternative referencing systems such as that used by LMIC. This file also contains the datum shifts at the four corners of the Quad, to get from NAD 27 to NAD 83 (Same function as the NGS - Nadcon Program).

TWPS.DBF . . . . . Minnesota's Townships
References the county, township and range numbers to its associated township name/s. It also leaves room for comments and notes on individual townships, which may be important for areas of special concern or knowledge.

COUNTY.DBF . . . . Minnesota's Counties
Contains the County Names as well as computational constants necessary to convert coordinate systems (County Coordinates, UTMs, Latitude / Longitude, State Plane

Coordinate Zones) to the 'more accurate'(more detailed/precise scale) County Coordinate system and vise versa. This file is the direct reference for county names.

GOV_LOTS.DBF . . .Government Lots Data

Provides Government Lot data on sections, including the Government Lot number, its acreage, the position within section, and notes of special interest.

QA.DBF . . . . .Coordinates Quality Assurance Codes

Because information is collected from many sources, the database contains a Quality Assurance code to help standardize all point information. This code is given to quantify and relate the 'expected confidence tolerance error level' (accuracy) and to describe the method used in determining the geo-referenced coordinate position of a corner (i.e. GPS, aerial photos, ground survey, digitized). This helps the user to judge the positional accuracy of the individual points shown in the 'CORNERS.DBF' information.

VERT_QA.DBF. . . .Vertical Quality Assurance Codes

Works on the same principle as the QA.DBF, but measures 'expected confidence tolerance levels' in the vertical dimension giving elevation confidence limits.

*POINTS.DBF . . . .Non Public Land Survey Points File

Same structure as CORNERS.DBF with additional information to describe data for non-PLS points. It is a surveyor's "working database" designed to add all non-government corner points from other state/county/city established control monuments which do not necessarily fall at the PLS corners. This will help add to the complexity and precision of the base map as new data becomes available. This can also be use to store Control Points for reference to surveys.

*USER.DBF . . . .User Defined Database

This User defined database may be any type of information that can be linked to the Main database. The user may also define a edit entry screen to view, browse or edit the information in this database.

Corner Reference Functions . . . Public Land Survey Corner Reference

These are programs or functions which references the general topology scheme between sections and corners. This allows the computer to realize that the line between sections (i.e. Sections 6 & 7) is held in common between the section rather than two separate lines.

Database Operational Information and Definitions

Record # - Internal database record numbered chronologically assigned via DataBase software as points are entered.

Active Index - Index for sorting the information/records. This code determines which record the system advances to next using page up and down.

  Choices:*TRSC - Township, Range, Section, Corner *CTRS - County, Township, Range, Section, Corner *QID - NGS Quad Id Number, Number within quad *LL_ID- Every other digit of latitude/longitude *GCDB - Township, Range, Geographic Coordinate Database in a North-South Direction *GCDB2- Township, Range, Geographic Coordinate Database in a East-West Direction *NORTH- Range, X part GCDB, Twp, Y part GCDB *EAST - Twp, Y part GCDB, 55-Range, X part GCDB *PROJECT - Source Project Id Number.

Database In Use - Tells the user the name of the database which is in use.

Key Board Arrow Units - tells the user what units the system will skip when the arrow keys (up, down, right and left) are used.

Record Status - indicates if all corners are displayed (including those replaced) or whether only the updated records are displayed.

Mode of Operation - shows whether the system is in Inquiry or Edit Mode.

Database **Field Information**

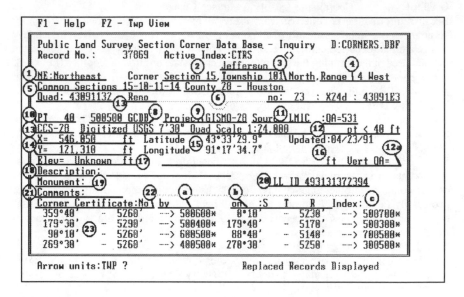

```
 F1 - Help    F2 - Twp View

 Public Land Survey Section Corner Data Base. - Inquiry    D:CORNERS.DBF
 Record No.:    37869    Active Index:CTRS      <>
                    (2)      Jefferson (3)             (4)
(1)NE:Northeast    Corner Section 15,Township 101 North,Range 4 West
(5)Common Sections 15-10-11-14 County 70 - Houston
   Quad: 43091132    Reno         (6)      no:  73  : X74d : 43091E3
              (13)
(10)PT   40 - 500500 GCDB(8)Project:GISMO-20 Sour(11)LMIC  :QA=531
(13)CCS-70  Digitized USGS 7'30" Quad Scale 1:74,000  (12)  pt < 40 ft
(14)X= 546,050    ft Latitude (15)43°33'29.9"    Updated:04/23/91
   Y= 171,310    ft Longitude 91°17'34.7"          (16)        (12a)
   Elev= Unknown  ft(17)                              ft Vert QA=
(18)Description:
   Monument: (19)                        (20)LL ID 493131372394
(21)Comments:         (22)     (a)      (b)
   Corner Certificate:No by         on  :S   T   R   Index: (c)
   359°40'    -  5260'   ---> 500600*   0°10'   -  5230'   ---> 500700*
   179°30' (23) -  5290'   ---> 500400*  179°40'   -  5170'   ---> 500300*
    90°10'    -  5260'   ---> 600500*   88°40'   -  5140'   ---> 700500*
   269°30'    -  5260'   ---> 400500*  270°30'   -  5250'   ---> 300500*

 Arrow units:TWP ?                    Replaced Records Displayed
```

1) <u>Corner : Corner Designation Code</u> - is a two letter code that represents the specific corner point in a section.

Choices:*NE:Northeast *N4:North Quarter *SW:Southeast *E4:East Quarter *NW:Northwest *S4:South Quarter *SW:Southwest *W4:West Quarter *CS:Center Section *CQ:Center Quarter *MC:Meander Corner *WP:Witness Post *MP: Mile Post *AC:Auxiliary Meander

*The full corner designation description follows the colon and is generated by a program. Above this field is a Corner Extension field which displays additional information regarding the corner such as meander corner number or corner of a 40 acre piece. (E4 of NE - East Quarter corner of the Northeast Quarter)

2) <u>Section</u> - Gives section number (1 - 36) in which point falls.

3) <u>Township</u> - Gives the township number in which point falls.

   Choices: Township 26 North to Township 70 North in the 4th Principal
   Meridian & Township 101 North to Township 168 North in the 5th Principal
   Meridian.

4) <u>Range</u> -   Gives the range number in which the point falls.

   Choices: Range 1 West to Range 32 West of the 4th Principal Meridian;
   Range 3 West to Range 51 West of the 5th Principal Meridian & Range 1 East
   to Range 7 East of the 4th Principal Meridian.

5) <u>Common_Sec</u>- References all other section corners held in common with active
   corner. A dash between number indicate that the points are in the same
   township. The plus indicates that the next corner is in a different township.

752

6) County_no - Gives the county number as derived by numbering the counties alphabetically throughout the State. Following the number the county name is displayed. These are also the FIPS codes used in census information.
   Choices: Counties 0 to 87 (0 = statewide)
7) QID - Quad Identification Number - Lists the NGS Quad ID Number followed by the name, DNR ID and USGS ID all from the Quads.dbf linked by the QID.
7A) No - Quad Point Number - All points within each quad are sequentially numbered. Number assignment begins at the southeast corner of the quad and generally proceeds in a northwesterly direction.
8) Geographic Coordinate Database Number - GCDB number is an X-Y indexing system for all points in a township. Each Section line is a even one hundred and each quarter line is an even forty (representing 40 chains). The southwest corner of the township being 100100 and the northest corner of the township being 700700.
9) Project - The project from which the data for a particular point was derived. Project names can be up to eight characters in length and defined by the user. The project 'GISMO-'*county_no* are the project designation for the original digitized points.
   Examples: GISMO-23, CO DATA, DNR, A-92-11
10) PT - Point Number - The point number related to the specific project.
11) Source - Where the information came from or who conducted the project.
    Choices: LMIC, DNR, Mn/DOT, County etc.
12) QA - Quality assurance code - 3 digit quality assurance code relates to the methodology used in deriving the points coordinate position. This relates to the expected statistical tolerance limits for that methodology and/or source of information. The information about this code is displayed on the line below the code and is followed by the expected tolerance.
12A) VERT_QA - One digit quality assurance code relates to the methodology used in deriving the point elevation. This relates to the expected statistical tolerance limits for that methodology and/or source of information.
13) C_version - Coordinate Zone - Coordinate conversion zone code (i.e. CCS-23 which stands for County Coordinate System) The default coordinate system is the NAD 83 datum county coordinates in the county the corner is located. Other Zones UTM-## (Universal Transverse Mercator - Zone 14,15 or 16) SPC-{Z} (State Plane Coordinates - {Z} = N, C or S) , a 27 after zone would use the built in NADCON Conversion to convert to NAD 27 Datum Coordinates and Latitude, Longitude.
14) X= /Y= - X and Y cartesian coordinate values of the point in any of the coordinate systems (County, UTM etc). The default coordinate system is the NAD 83 datum county coordinates from the county in which the corner is located. This default value is what is stored in the database and all other coordinate systems are computed from it.
    Choices: Vary depending on coordinate system and units chosen. Possible Coordinate Systems are: UTM's zones 14, 15, 16 or shifted zones 14S, 15S, 16S; State Plane Coordinates zones North, Central, South; and the default values are the County Coordinate System. Units can be in US Survey feet (ft), meters (m), international feet (if), or Chains (ch).
15) Latitude/Longitude - Latitude and Longitude position.
    Choices: Any value between Approximately 89°15' West to 97°15' West Longitudes and 43°30' North to 49°00' North Latitudes (MN Boundaries).
16) Updated - The date the database record was last updated.

17) <u>Elev - Elevation</u> - Elevation of point if known.
   Choices: Any value between Approx. 602 ft - 2301 ft.
18) <u>Desc - Description</u> - Description of any pertinent landscape features.
   (30 Characters)
19) <u>Monu - Monument</u> - Listing of monument type if in place. (8 characters)
   Choices: *AC - Aluminum Cap *AX - Axle or Shaft *BC - Brass Cap
   *BS - Bridge Spike *CIM - Cast Iron Monument *CB - Concrete Block
   *DM - Dept. Monument *DMS - DNR Survey Marker *FP - Fence Post
   *IP - Iron Pipe *IR - Iron Rod *ND - Nail & Disk *PC - Plastic Cap
   *RRS - Railroad Spike *RH - Redhead Spike *RP - Reference Point
   *SFP - Steel Fence Post *SPK - Spike *ST - Stone *SM - Stone Monument
   *WP - Wood Post.....etc.
20) <u>LL ID</u> - Is the latitude/longitude identification number. The combination of every
   other number of the lat & long, i.e. 47°45'36" West, 94°15'54" North = LL ID
   497441553564. This fields primary function is for indexing the database for quick
   location by Latitude and Longitude.
21) <u>Comments</u> - Any important remarks about site or monuments. (30 characters)
22) <u>Cert - Certification</u> - Tells whether the corner is certified or not. In the database this
   field is a logical field (T or F) on the display screen it will display a Yes or a No.
   Certificate requirements designated by Minnesota Statue 381.12.
   22a) <u>By</u> - Explains by whom the certification was filed.
      Choices: County, DNR, DOT, Private Surveyors RLS#####.
   22b) <u>on</u> - The corner certified as listed on the certificate.
   22b) <u>:S</u> - Section where certified.
   22b) <u>T</u> - Township where certified.
   22b) <u>R</u> - Range where certified.
   22c) <u>Index</u> - The certification index number.
23) Inverse or Note data on the corner. The note data is in a memo field in the database.
It may be edited at anytime by pressing Alt-M which opens a memo window editing
screen. The note data is scrolled through the 4 line display window by pressing the
'M'ore or 'B'ack. The Inverse data is a toggle that displays inversed information to the
two nearest points in all four directions.

Other Features

Other Important features include many Input/Output options that include Ascii data files,
DXF files, and other database files. GeoCoding of points as they are input (X,Y -->
Twp,Rge,Sec,Corner). Automatic fill in of fields you do not know. Coordinate
Conversions to and from UTM's, State Plane, County Coordinate Zones and even Local
Project coordinates to Latitude/Longitude (uses project block conversions) and NAD 27
< == > NAD 83. Future features I hope to include would be Mouse Support, Pull
down Menus, Many more output reports, direct incorporation of data into our ArcInfo
GIS files and Extensive on line Help.

Township Graphic Screen

To allow even easier movement from point to point the township graphics screen may
be accessed by pressing F2. This allows the user to see a visual representation of the
corner position corresponding to the information from the introductory screen. The

display screen is shown below. It allows user to navagate around a township and view the relationship to other points in the township.

Quality Assurance

Quality Assurance Indicator (QA) is a code indicating the accuracy of a corner position. This code is extremely important in order to maintain the integrity of the Data. This code is not absolute in that it represents a methodology of how the coordinate of the point was derived. If the proper procedures were followed for a given code the expected accuracy ranges should be relative to the position. Below is a abridged version of the Quality Assurance Coding.

QA Code	Description	Accuracy Limits
1**	Survey Methods	25 ft
11*	GPS Survey Methods, Static and Kinematic	3 ft
12*	Survey Methods with Computed Error Ellipses (Least Square Adj)	
13*	Standard Survey Methods in NAD 83	3 ft
131	Survey Adjusted in NAD83	2 ft
133	Survey Unadjusted in NAD83	4 ft
15*	Surveys Converted from NAD27	6 ft
151	Survey Converted via NADCON NAD27->NAD83	4 ft
17*	Surveys Converted from Project Datum using GISMO pts	120 ft
171	Survey of >12 pts Converted Project Datum->NAD83	8 ft
173	Survey 6 to 12 pts Converted Project Datum->NAD83	12 ft
175	Survey 2 to 5 pts Converted Project Datum->NAD83	40 ft
3**	Aerial Photography	100 ft

311	Analytical Aerotriangulation	4 ft
33*	Digitized from Rectified Photography	
35*	Digitized from UnRectified Photography	
37*	Ground Position Digitizing using GPS/LORAN-C	300 ft
371	GPS Differential Processing (Code Differential)	30 ft
373	GPS Differential Processing	60 ft
375	GPS(SA Off) - GIS Position	100 ft
377	GPS(SA On) - GIS Position	300 ft
379	LORAN-C with Control Checks	300 ft

| 5** | Digitized from National Map Series | 300 ft |

511	Digitized USGS 3'45" Quad Scale 1:12,000	10 ft
53*	Digitized National Map Series	250 ft
531	Digitized USGS 7'30" Quad Scale 1:24,000	40 ft
533	Digitized USGS 15' Quad Scale 1:62,500	80 ft
535	Digitized USGS 30'X60' Map Scale 1:100,000	120 ft
537	Digitized USGS 30'X60' Map Scale 1:250,000	250 ft
551	Digitized USGS 7'30" Quad Implied Point	300 ft

| 7** | Relation Estimates / Digitized Drawings | 500 ft |

71*	Estimates Based on Survey Points + Measurements	12 ft
73*	Estimates Based on Digitized Points + Measurements	80 ft
75*	Digitized Drawings	500 ft
751	Scale < 1"=600' Reliable Source & Base Points	20 ft
755	Scale < 1"=600' Questionable Map or Base Points	120 ft
759	Cartographic License,Unknown Quality,etc.	500 ft
9**	Unknown	500 ft

SUMMARY

The Control Point Committee's goal for the Public Land Survey Corners Database is simple: to provide the best possible control point information through data sharing and support between county, federal and state offices. The package has been created as a flexible control point referencing system with the ability to store, view, update and export voluminous amounts of accurate point data for use in survey data management, GIS and survey planning on both statewide or localized scale. It offers a unique opportunity for state agencies and county offices to pool information and ease the burden of collecting, updating and imputing control point information for general reference and especially GIS applications. The intent of this system is to provide a user-friendly system which will serve both the GIS and the LIS (Land Information System) user communities.

LESSIONS

Everything takes longer than expected. Everyone has their own goals and work to accomplish thus is can be difficult to incorporate a new system into someone's work place and expect much more than a minimum amount time spent and feedback. A great deal of time can be spent on the special cases that do not fit the general coding schemes as used in a database.

OBJECT-BASED USER INTERFACE STRUCTURES: EXPERIENCES FROM A SOFTWARE PROTOTYPING APPLICATION

Richard Taketa
Department of Geography and Environmental Studies
San Jose State University
San Jose, California 95192

ABSTRACT

Object-based programming techniques provide an effective means to rapidly prototype and develop cartographic applications software. Basic object properties such as inheritance and polymorphism are powerful tools for the software developer. Obtaining the benefits of object programming, however, requires organizing the different interface elements effectively—creating structures that provide flexibility and establishing communication among interface elements to enhance their functionality.

This paper describes the use of object techniques to develop an interactive interface and a user-programming language for a prototype map symbolization application. The interactive interface consisted of a command class for items such as menus, palettes, and dialogs, and a tool class for performing specific application operations. The user-programming language consisted of a language class and a function class. These object classes provided a high degree of flexibility to expand and modify the prototype, while retaining the interface capabilities that are expected in a production application, such as dynamically changing the appearance of interface elements based on changes in the state of the application. Developing the user interface for this prototype application demonstrated the utility of object programming techniques to create highly usable software in an effective manner.

INTRODUCTION

Prototyping an application targeted for a graphical environment invariably involves prototyping the user interface. The relationship between the functionality provided by the application and the means by which a user accesses that functionality is a significant software design consideration. Therefore, the evolution of the application's features and user interface should take place concurrently.

This paper describes the development of user interface software for a prototype application to test ideas for parametric symbols: a form of thematic map symbolization. The prototype demonstrated the feasibility of the parametric symbol concept, and explored ways in which mapping applications using parametric symbols can be organized. In addition, the prototype was designed from the outset to operate in a graphical environment, with user interface concepts considered throughout the prototyping project.

The user interface approach used in the parametric symbol prototyping effort was based on a specific set of goals:

1. *Production oriented.* The user interface should provide the kinds of interaction techniques expected in production software. In addition, concepts should be easily transferred to a commercial application development environment.

2. *Easy to use.* The overall user interface environment should be simple to implement, and should minimize the effort to design and develop the user interface software.

3. *Flexible.* The user interface organization should be easy to modify (e.g. moving menu items within a menu, or from one menu to another).

4. *Extendable.* New application operations and new user interface elements (menus, buttons, dialogs, palettes, etc.) should be easy to add.

Object techniques provided a software development environment that effectively addressed these goals. The prototype's user interface subsystem was organized as a set of object classes, taking advantage of object class hierarchies to create a highly flexible and easy-to-use user interface system. Interactive commands and language functionality were modified and extended easily as the prototyping process proceeded.

BACKGROUND

Much user interface research has focused on the design, creation, and management of the user interface elements. This has included the use of software development tools to assist in the design of menus, icons, and dialogs, some of which involve generation of skeleton code to which application code is added (Schmucker, 1988; Brown and Cunningham, 1989; Olsen, 1992). It has also included the development of software libraries providing user interface tools, and interface approaches supporting the use of such tools. Less effort has been placed on the organization and design of the application software supported by the user interface.

The Macintosh toolbox and Microsoft Windows, two widely-used interface environments, support object-like organizations for user interface elements, such as windows, menus, buttons, and dialogs (Apple Computer, 1985; Microsoft, 1990). The application programming environment is built around a code structure, the event loop, which handles the various interface and system events which occur. The event loop receives an event from the system, then processes each event through the user interface code to execute specific application code. Event processing is based on a hierarchy of information about the events, such as "a mouse-down in the menu bar, selecting the *Reduce* item from the *View* menu." The user interface code processes the kind of event (mouse down), then where on the screen (in the menu bar), then where in the menu bar (View menu), and what item was selected (Reduce). Event processing is then organized as a set of nested case statements (Figure 1).

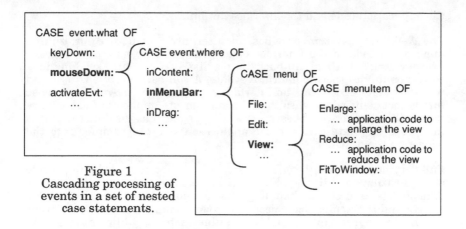

Figure 1
Cascading processing of
events in a set of nested
case statements.

The difficulty with this approach is that the application code structure
tends to mimic the event processing structure (Figure 2). The specific
code to be executed is accessed at the end of the case statement chain.
Any change in the user interface, therefore, necessitates a change in the
case organization. Altering the interface organization can require major
work, with a corresponding potential for coding errors to appear.

User interface management packages, such as Apple's MacApp, provide
mechanisms for handling the interface objects, and the developer has
the choice of associating application code with particular interface
objects, or passing a command number to a procedure incorporating a
case statement. This particular organization provides a way for the
application code to communicate status information back to the interface
objects, and therefore altering the interface objects dynamically, since

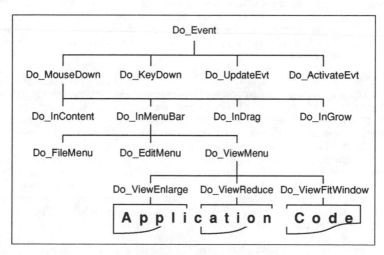

Figure 2
Application code structure can end
up mimicking the event structure.

the application code and the interface are linked.

The X-Window system provides an alternative structure for linking application code to the interface object. Interface objects, or "widgets", are developed with a call-back list: a list of procedure or functions corresponding to specific interface events. A call-back routine is accessed when a widget encounters an interface event, such as moving the mouse into a document window, so that the appropriate application actions are taken. Changes to the appearance of widgets, such as disabling when an inappropriate state exists in the application, are programmed into the widget by the designer.

Fully object-based user interface systems (e.g. Sibert, *et al*, 1986) provide a separation of displayable user interface objects, application data/objects, and objects handling interactions between the displayable objects and the application objects. Such an organization allows different objects to operate in an highly-modular fashion, while maintaining strong linkages to related objects. Altering the different objects becomes easier, increasing flexibility and extendibility, and allowing greater experimentation with the user interface design. These features make an object orientation highly appropriate for prototyping purposes.

THE PARAMETRIC SYMBOL PROTOTYPE

The user interface for the prototype application consisted of an interactive interface subsystem and a parametric language subsystem, both based on object classes. The *interactive interface* subsystem consisted of command and tool objects, while the *parametric language* subsystem consisted of language and function objects. Command objects included standard interface elements, such as menus and menu items, palettes, dialogs, buttons, and scrolling lists. Tool objects were accessed by command objects to operate on the application data. Language objects provided the necessary utilities to process expressions in a parametric description. Function objects were accessed by language objects to perform specific operations in a parametric description. All of these employed object programming features to enhance flexibility and maintain consistency.

Object programming provides a number of features to enhance the software development process (Nierstraz, 1989). Among these, inheritance and polymorphism were particularly relevant to the development of the prototype's user interface subsystem. *Inheritance* allowed common methods to be defined for an object class and its descendant classes. This provided standard interfaces through which the command processor and the language processor could access specific object methods. *Polymorphism* allowed methods in a descendant class to perform unique operations by overriding parent class methods. The combination of these features enabled the user interface software to employ consistent interfaces between subsystems while varying the processing according to the specific object being accessed.

The Interactive Interface Subsystem: Command and Tool Objects

The interactive interface subsystem consisted of two primary object classes: commands and tools. Command objects were standard user interface elements, such as menu items and icon buttons. Tool objects operated on the application's data, performing such tasks as adding and modifying data, and changing the display. Commands and tools operated together to provide interactive control of the application.

Command objects were organized by class, with the highest class, the Command, consisting of data, methods, and, specifically, a command message. The command message was a text string that identified the tool to be invoked when the command object was selected, and optionally provided additional information for the tool. Other command objects, such as menu items, were descendants of the Command class, with unique display and selection methods. All command object classes shared the way in which they accessed tools.

Menus, palettes, and dialogs were collections of command objects. A menu consisted of a list of menu items, a palette consisted of a 2-D array of icon buttons, and a dialog consisted of a variety of buttons (icon buttons, push buttons, check boxes, and radio buttons), label text, editable text, and scrolling lists. These objects provided the means to organize the command objects for presentation to the user (Figure 3).

Command objects were displayed in one of several states. The "enabled" state specified whether or not the user could select that command. If the command object was enabled, then it was shown in its normal display. If the command was disabled, then it was shown in gray and was not available to the user.

A command could also be "selected", indicating that the option or tool represented by the command was in effect at that time. The appearance of selected command objects varied by object class. For example, icon command objects were shown in reverse colors, menu items were check marked, and check boxes were checked.

Changes in a command object's appearance for its enabled and selected

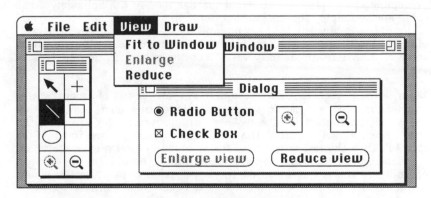

Figure 3
Command objects included menu items, buttons, and text
boxes, and can be grouped in menus, palettes, and dialogs.
A command object's appearance depended on its state:
enabled or disabled, and selected or not selected.

states provided feedback to the user. These states were determined by the tool associated with the command. The command object initiated the status check by sending the tool a SetStatus request. The tool object maintained information about requirements for its operation, and would respond to the SetStatus message by sending SetEnable and SetSelect messages back to the command object. The command object then updated its appearance based on the results of that check.

Command objects were created as part of application initialization. The objects were defined in a resource file, which could be edited and redefined independently of the application code. The resource file included specifications for the type of command object, its initial appearance, and the command message. Appropriate objects were created from this information, and were stored in several lists, such as a menu item list, a command window list, or a dialog command list.

Tool objects performed the work of the application. Selecting a command object invoked the tool associated with that command object. The tool would be found via a lookup procedure using the tool name in the command message. The invocation would either execute the tool immediately, or make the tool the Active Tool, which would select the tool for subsequent events. In either case, tool object execution would change some aspect of the application or its data.

Specific tools were descendants of the Tool class. This general class defined the basic tool methods, but specific tool descendants were created to perform specific actions, such as opening a file or magnifying a view. Instances of these descendant classes were created in the application and accessed as needed. This allowed the command processor to call an abstract tool method as a general interface, as in the EXECUTE method, but the operations differed depending on the specific tool instance being processed (Figure 4).

Tool objects accepted additional information from command objects via the command message, which could further modify the behavior of the tools. For example, the magnification tool was designed to respond differently to key words in the command message such as "by" and "to". "Magnify by 2" caused the tool to alter the current magnification by a factor of 2, whereas "Magnify to .5" set the magnification to 50% of original.

Tool objects set their associated command objects to one of several states (enabled/disabled, selected/not selected), depending on the application context. If the context for the tool was inappropriate (for example, the Line Drawing tool required that a drawing exist before it could be selected), then the tool would send the command object a message to set its enabled/disabled state to "disabled" and to adjust its appearance accordingly. Tools that modified the application data or states would send a message for all command objects to check their statuses.

A tool could also be "selected", and subsequent events would be processed by the tool. For example, the Line Drawing tool could be selected and become the Active Tool. The display for the command object representing the Line Drawing tool would be changed to a "selected" state. A mouse down event would trigger the Line Drawing operation. The mouse

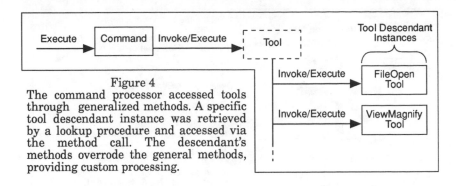

Figure 4

The command processor accessed tools through generalized methods. A specific tool descendant instance was retrieved by a lookup procedure and accessed via the method call. The descendant's methods overrode the general methods, providing custom processing.

position would become the starting point for the line. 'Mouse_stillDown' events would be sent to the tool as long as the mouse button remained down, and the line would be rubber banded, using the starting point as the anchor and the current mouse position as the floating end point. The line would be completed, and made permanent, when the mouse button was released.

Tool objects could be referenced by more than one command object. Different command objects could send different command messages to the tool object, and the tool would vary its execution accordingly. The two view magnification commands "Magnify by 2" and "Magnify to .5" could be sent by different command objects to a single magnification tool. This allowed the magnification operation to be handled by a single tool, but also allowed considerable variation in how the user could specify magnification. More importantly, isolating the operation into a single tool allowed the status of all associated command objects to be set in a consistent manner. For example, when the maximum magnification was reached, all commands that requested an enlargement were disabled (Figure 3).

Tool objects were loaded into a specific application at initialization. A creation function generated the tool instance, initialized its fields, and inserted the instance into the tool list. The tool mechanism allowed tools to be handled in an highly modular manner, so that tools could be combined to form a variety of applications, and new tools could be added easily.

The Parametric Language Subsystem: Language and Function Objects

The second application of object technology in the prototype application's interface involved development of a parametric language to generate parametric symbols. A parametric description was a "program" that created classes of symbols from specific parameters and data. Symbol descriptions were unique to specific symbol types, and were expressed in a programming language designed for this purpose (Figure 5).

The parametric symbol language consisted of two major classes: a language class and a function class. The language object included a mechanism for handling expressions and a function processor. The function class provided the framework for adding specific operations to the language, such as retrieving data from a file. Both the language and

763

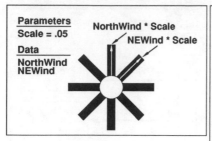

Parameters Scale = .05 **Data** NorthWind NEWind	* Get the parameters and data Scale = GetParameter "Scale" .025 NEWind = GetParameter "Northeast" 0.0 * Calculate intermediate values DeltaNE = NEWind * Scale * Calculate the 45° end points NEAz = Azimuth 45 PNE = Loc 0 0 (Offset NEAz DistP) * Draw the wind rose lines LineWidth 0.05 Line PNE (Loc PNE (Offset NEAz DeltaNE))

Figure 5
Example parametric symbol
description.

function classes were integral to prototyping the parametric symbol capability.

Language Objects. Language objects consisted of a set of methods to handle parametric statements. All statements in a parametric description were converted to token objects (such as variables, constants, operators, and functions) when the description was loaded. Each token maintained information about its token type, as well as the legal operations for that token. The token structure provided for different descendent classes, allowing variation in the language while maintaining consistency. A token manager kept track of the tokens and provided a mechanism for looking up variables, external data, and parameters.

Statement parsing was performed by the language's PARSE method, which was invoked when the symbols required redrawing during a display update. The method was implemented as a recursive descent parser interpreting the lists of tokens that made up a parametric symbol description. It invoked arithmetic operations and invoked the language's function processor to access application-defined functions through calls to their respective EXECUTE methods (Figure 6).

Functions were independent objects linked to the language that created a symbol from a parametric description. For example, a circle might be part of a symbol, and the graphic display of a circle was handled by the CIRCLE function, which interfaced to the graphic system. Functions also provided the interface to the application's parameters and database,

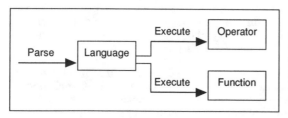

Figure 6
Operators and functions were accessed through their general EXECUTE method, called from the Language object's PARSE method.

as in the GETPARAMETER function.

Functions were descendants of an general Function class, much as tools were descendants of a general Tool class. All functions accessed through a common interface, the EXECUTE method. This allowed new functions to be incorporated into the language easily, since the function processor would access the appropriate function object after having retrieved it at run time. The EXECUTE method was customized to each function, performing the specific tasks for that function. EXECUTE was invoked from the language's PARSE method through the function processor.

Functions were loaded at the initialization of the language much as tools were loaded. An instance of each function was created and inserted into the function table, which could be accessed by the function processor. Functions were highly modular, allowing the parametric language to be extended easily by adding new functions.

DISCUSSION

Organization was important in the design of the prototype's user interface. The user interface subsystem employed narrowly-defined interfaces between object classes by declaring a standard method interface in the general command, tool, and function classes. The method interface was shared among object classes and their descendent classes (inheritance), and descendent methods could override the parent methods to provide unique operations (polymorphism). These two features created a high degree of modularity in the interactive interface and parametric language subsystems.

The modular object structure created an extremely flexible development environment. Command objects could be manipulated easily because they were separated from tool objects. Command objects could be reorganized and changed in form and screen position without requiring changes to the tool code. Tool objects could be developed and loaded in any sequence because command and tool code structures were independent of one another. Finally, tools and functions could easily be used in multiple applications.

The object user interface structure provided a highly extensible environment. Standard interface methods, such as EXECUTE, allowed new commands, tools, and language functions to be added easily. The command and language function processors needed only to look up the correct object instance in a table and call a standard method interface for for the general class. The instance could then override the standard method to perform specific operations.

The object user interface environment was easy to use. The organization and processing of the user interface subsystem were hidden from the application and user interface design process. Little knowledge of the processor's internal operations was needed. Method inheritance eased development even further, since many tool objects could be implemented by simply creating a new EXECUTE method, and inheriting all of other methods. In addition, if a command object or tool object was not interfaced properly, the command object would not be enabled, giving an immediate indication of a problem.

Finally, the interface subsystem was designed to support user interface capabilities expected in a production application. It allowed easy inclusion of new features as the user interface design evolved, it provided dynamic graphical operation and feedback to the user, and it provided direct graphical indications to the developer that command and tool objects had been properly loaded. Yet its modularity also created applications that should be easier to maintain and enhance in the future.

SUMMARY

The parametric symbol prototyping project benefited greatly from the use of object programming techniques. The user interface could have been a cumbersome part of the software to design and develop, given the requirements for a graphical interface. Creating interface components as objects that operate on one another, then creating consistent software interfaces among them, made design and development of the user interface a relatively straightforward task.

REFERENCES

Apple Computer, Inc. (1985). *Inside Macintosh Volume 1*. Reading, MA: Addison-Wesley.

Apple Computer, Inc. (1988). *MacApp® 2.x Manual (interim version)*. Apple Computer, Inc.

Brown, J., and S. Cunningham (1989). *Programming the User Interface*. New York: Wiley.

McMinds, D.L., and B.J. Ellsworth (1990). "Programming with HP OSF/Motif widgets," *Hewlett Packard Journal*, 41 (3), pp. 26-35.

Microsoft Corp. (1990). *Microsoft Windows Guide to Programming*. Redmond, WA: Microsoft Press.

Nierstrasz, O. (1989). "A survey of object-oriented concepts." in Kim, W. and F.H. Lochovsky, eds., *Object Oriented Concepts, Databases, and Applications*. New York: ACM Press, pp. 3-21.

Nye, A., and T. O'Reilly (1990). *X Toolkit Intrinsics Programming Manual*. O'Reilly & Associates, Inc.

Olsen, D.R. (1992). *User Interface Management Systems*. San Mateo, CA: Morgan Kaufman Publishers.

Schmucker, K. (1988). "Using objects to package user interface functionality," *Journal of Object Oriented Programming*, (April/May), pp. 40-45.

Sibert, J., W.D. Hurley, and T.W. Blesser (1986). "An object-oriented user interface system." SIGGRAPH '86, pp. 259-268.

A PERSONAL VISUALIZATION SYSTEM
FOR VISUAL ANALYSIS OF AREA-BASED SPATIAL DATA

Qin Tang
Department of Geography
The Ohio State University
Columbus, OH 43210
(614)847-5592

ABSTRACT

The wide use of digital technology has produced a large amount of spatial data in computer compatible format. Effective methods are needed to explore the data to find interesting patterns or structures that are unknown to us. Current GISs lack the capability of exploring and hypothesis generating. Their strength in spatial data handling can provide a dependable ground for developing new techniques that facilitate the exploration of digital spatial data and provide an effective approach to hypothesis generating. Developing a visualization system that encourages the exploratory analysis of spatial data is one of these techniques. This paper describes the design of a personal visualization system for exploring area-based spatial-temporal data. Conceptual discussion on combining visualization and spatial data analysis is presented. The system features, requirements, components and structure are discussed. The research illustrates how to develop a visualization system through enhancing traditional cartographic methods for visual spatial data analysis.

INTRODUCTION

As the result of modern technology development, a tremendous amount of digital data have been produced. The capability of collecting spatial data is far beyond that of handling and analyzing. To explore the data available, we need to develop effective methods to get insight about the data. Geographic Information Systems(GISs) have proven to be very effective in spatial data handling. Their capabilities of analysis, however, are very limited in the current stage of the development. It is of necessity to explore a variety of ways to enhance the analytical capabilities of GISs. Recent research in many fields demonstrates that visualization can play important roles in understanding huge volumes of data. Applying visualization techniques to spatial data can also enhance the capability of spatial analysis.

Visualization can be considered as both a tool and a process for discovery and understanding. As a tool, it emphasizes technical aspect, that is, how to develop hardware and software to transform numerical data into representative images on the screen. As a process, it focuses on getting insight about the data through computer graphics. Both aspects are important to the significance of the technology. Encouraged by fruitful results in many disciplines, attention on visualization in the spatial domain is dramatically increasing, but most of the efforts are on the conceptual level. Although such efforts are important, it is also necessary to develop experimental systems which demonstrate the capability of visualization to expand research opportunities in spatial realms.

This paper describes the design of a visualization system based on personal computers. The emphasis is on expanding traditional cartographic techniques for the exploratory analysis of spatial-temporal data. Area-based data, e.g., census data, are the empirical data to be explored through the system, though the methodology can be applied to other types of data. This research is sustained by the need to develop a new research approach to spatial data analysis.

A NEW RESEARCH PARADIGM

The Need

The development of technology not only facilitates more efficient scientific discovery, but also has great impact on how research is conducted. The wide use of digital techniques has accumulated enormous amounts of digital spatial data, which represents a significantly different situation in which spatial analysis is performed from that of "quantitative revolution". The traditional data-free, hypothetico-deductive approach to theory development in geography is being challenged. A new perspective of models in geography, focusing on data-driven computer models, is needed (Openshaw, 1989). Exploring spatial data through visualization can play important roles in this new research paradigm.

According to Harvey (1969), there are two routes to scientific explanation. The first route is the inductive route, in which the generalization is derived out of observations. The other one is the deductive route, in which a hypothesis is first derived from prior knowledge or observations and then tested. This route is taken more frequently by geographers because "verification of the hypothesis translates the speculative law (the hypothesis) into an accepted one" (Johnstone, 1987, p 68). Both routes have made significant contributions to scientific progress, but each has its own shortcomings. The inductive method is criticized because "acceptance of the interpretation depends too much on the charisma of the scholar involved and on the unproven representativeness of the case(s) discussed" (Johnstone, 1987, p 66). The shortcomings with the second route, preferred by many researchers because of its "controlled speculation", are the lack of effective ways to generate hypotheses and that it is too theoretical.

The deductive approach also tends to verify what we know rather than to discover what we do not know. When using the deductive method, the procedure takes three steps: generating hypotheses based on prior knowledge, collecting data that are needed to test the hypotheses, and verifying or rejecting the hypotheses utilizing some statistical methods. When facing the huge amount of data, the method becomes less productive because of the lack of knowledge about the data. What is more efficient is to explore the data to find interesting structures that are unknown and then to generate hypotheses based on the insight about the data. The hypothesis is then tested via the deductive route.

Visual Data Analysis

The new paradigm of spatial analysis is philosophically similar to Exploratory Data Analysis, or EDA (Tukey, 1977). EDA is aimed at encouraging hypothesis generation. It detects interesting structures in the data and raises questions about the structures. EDA does not provide solutions to the questions, neither deals with hypothesis testing. As "data detective work," EDA employs a variety of techniques, among which graphic method is the most useful one because graphics are proven to be the best means to reveal the structure in the data.

The development of computer graphics has created a great potential for using graphics in data analysis. Dramatic improvements on hardware and software not only make computer graphics available to many researchers, but also greatly expand the opportunities in data analysis. Compared to manual graphics, computer graphics not only reduce the burden of drawing, but more importantly, they provide new capabilities of exploring data. The most critical one is the interaction between the computer and the analyst. The interaction utilizes the eye-brain system to get deep insight about the data. It is the interaction that allows the analyst explore the data from many different perspectives. When we face uncertainty in the data such a capability is essential to the success of exploration.

The combination of computer graphics and EDA creates a new approach to data analysis - Visual Data Analysis(VDA). VDA is a term that explicitly links visualization and data analysis. Although different views exist on visualization, the most common one is to display information in graphic forms. Such a view separates visualization from analysis. VDA emphasizes that visualization is part of the analysis, rather than an activity coming after analysis (Warner, 1990).

VISUALIZATION AND SPATIAL DATA ANALYSIS

An Analytical View

With the increase of its popularity, visualization has gained a variety of meanings. The basic meaning is to visually portray information. In this sense it has the same meaning as graphics (on paper or computer screen). Such a view on visualization leads our attention to the methods of graphic display rather than to its valuable capability of doing analysis.

In this research, we consider visualization has two essential features. The first one is the use of computer graphics. Although hand-drawn graphics are also very useful, many limitations associated with the medium on which the graphics are drawn make them less effective when huge amounts of data are involved. Computer graphics have the power to process volumes of data. And more importantly, computer graphics provide the capability of interaction between the display and the analyst, which is critical to the discovery and understanding process. The other one is the emphasis on analysis. In this sense the display is the means to reach the end, that is, to get insight about the data from which the display is generated. Such an analytical view on visualization implies that our attention is not on how to generate the display but on how to use the display for understanding our data. In spatial disciplines, we have the tradition of using graphics, especially maps, in research. Maps are not only a means of representing the results of the research, but also a means of conducting the research. Since the computer was introduced to automate the process of map-making, attention has been on the former aspect. The latter remains mostly untouched. It is the focus of this research to develop tools for manipulating the displays, so that the analyst can inspect his data in many ways.

Visualization and GIS

An analytical view on visualization has an impact on spatial data analysis. GISs have become an essential tool in spatial data analysis. Although many views on GIS exist, the definition of four sub-systems (Marble, et al, 1984), that is, acquisition, storage and retrieval, manipulation and analysis, and reporting, reflects the basic structure of a GIS. According to this definition, visualization is part of the reporting sub-system. Its function is to represent the results coming from the manipulation/analysis sub-system. The analytical view on visualization changes the structure. Visualization can find its place both in the manipulation/analysis sub-system and the reporting sub-system. If in the former sub-system it works as an exploratory tool for spatial analysis conducted in the system, while if in the latter sub-system it becomes a tool for presentation of the results.

Features of Visualization for Analysis

With the emphasis on analysis, cartographic display used in visualization has some features that are different from those of traditional cartographic products. The first difference is on the purpose. The purpose of traditional cartographic products is to communicate the information that the map maker has encoded in the map to the user. What is transmitted through the map is the cartographer's point of view on the data mapped. It is known and unchangeable after the map is made. In the case of visualization for data analysis, the purpose is to help the analyst to get insight and better understanding about the data he is investigating. The analyst faces unknown situations. He/she has no clear ideas about what may exist in the data. It is the capability of uncovering unknown structures, relationships, and patterns that makes visualization a powerful research tool.

The second important difference is the dynamic feature of visualization. This feature is supported by the separation of data storage and display. It also comes from computer's capability of instantly changing the image on the screen. With this feature the analyst can do many things that are not possible on an analog map. He/she can interact with the computer to look at the data from different perspectives. He/she can directly access the database to extract accurate information. The user can manipulate one element of the display to cause the change of other elements, so that he/she can investigate the relationship between these elements.

Because of the different purposes, the display used in visual data analysis may be different from a map. On a map, besides the geographic representation itself, there are many other elements, such as the title, the north arrow, and the projection. These elements are usually not necessary for the visual analysis of spatial data. Eliminating these elements can simplify the display and speed up the generation of the display. In addition, thanks to the dynamic capability of the computer, some dynamic representations of spatial phenomena, e.g., animation, are possible to the visualization system, which can enhance our capability of understanding complex phenomena.

Quality vs. Accessibility

The capabilities of a visualization system for analysis depend on many factors. The computer hardware and software have significant impacts on the effectiveness of the visualization system. Powerful capabilities of computing and graphics generating are very desirable. Efficient algorithms for colorful image generation and manipulation can improve system performance to some extent. These factors result in the impression that only a powerful computer environment, for example, a supercomputer and specialized graphic workstations, is suitable for the task of visualization.

It is beyond question that a powerful computer with great computing capability and a high quality graphics workstation can significantly enhance the functionality of visualization for analysis. On the other hand, dependence on it may inhibit many researchers from making use of such a powerful tool in their daily research because many of them do not have access to such an environment. Furthermore, these "requirements" are also raised by emphasizing realistic rendering of the image, which in many cases are not essential to analysis. Therefore, if we can develop a visualization system on personal computers, at the price of reduced display quality that is tolerable for the purpose, visualization will benefit scientific research to a great extent.

Generating informative displays is a fundamental step in visual data analysis. The dynamic features of computer graphics provide a great potential for inventing new representation forms for portraying spatial information. Realizing the potential, however, presents challenges to visualization tool-makers. A more practical approach is to modify existing cartographic methods to fit the needs of visualization, a strategy taken in this research.

THE VISUALIZATION SYSTEM

System Features

The focus of this research is to link visualization and spatial data analysis. The system developed is a tool for visual analysis of spatial-temporal data. Therefore, our attention is on how to use dynamic graphics for detecting interesting patterns and getting insight about the data. Graphics generation becomes an "interior" process which is mostly hidden from the user. It means that most of the tasks of computer map-making are "automatically" done by the system. The user's effort on this aspect is minimized, so that he/she can concentrate on visual analysis of the data.

There are basically two approaches to generating displays in such a system: inventing new representation forms, and modifying existing ones. This research takes the second approach for two reasons. One is that some theoretic and practical guidelines are available for these methods. The other reason is that these conventional methods can be modified and enhanced with the dynamic features of computer graphics to achieve what are not originally possible for these conventional methods.

Guided by this strategy, choropleth mapping is the method that has been modified and implemented in this system. Choropleth mapping is a cartographic method widely used for portraying a statistic surface associated with enumeration units. Using this method requires that the data being explored are based on enumeration units. Therefore, the system is suitable for exploring area-based spatial-temporal data, e.g., census data. Although the display method is limited to choropleth mapping and the data are limited to area-based in this system, the methodology developed in this research is applicable to other display methods and data types. The system can work as an independent system or as a component of an integrated system with a variety of display methods and for many types of spatial data.

System Requirements

Interaction. This is the key factor that determines the success of the system. Interaction provides the powerful combination of the computer's computing and drawing capabilities and the human being's eye-brain system, and is perhaps the most effective way to detect interesting patterns. To get insight about a complex data set the analyst should be able to look into the data from different perspectives. Many displays may be produced before a useful display is generated.

Display Capabilities. Visual data analysis requires more representational powers than traditional cartographic products because of the lack of prior knowledge about the data. Four types of display capabilities are considered to be the most important to the exploration of the area-based data: patterns comparison, relationships, geographic movement, and processes (temporal characteristics). For each of these aspects we use a different display method, which is termed as a visualization module. These modules are critical components of the system.

Good User Interface. A good user interface is critical to insure the effectiveness of the interaction. Besides those common features of a good user interface, that is, user friendly, ease-of-use, and consistent, the interface of a system for visual data analysis must be effective in promoting the involvement of the analyst. The direct manipulation type of graphic user interface is an obvious choice to reach the goal.

Efficient Algorithms. Another important factor that affects the effectiveness of the interaction is the response speed of the system to the analyst input. If it takes too long for the analyst to get the response from the system, he/she may lose his/her concentration on the output, which then makes detecting significant patterns less productive. Furthermore, some of the operations require immediate response to make them work. The response speed is obviously affected to a great extent by the computer hardware, but the methods used to create the display can also have a great impact on the speed. For example, we can reduce the display elements to simplify the graphics. Efficient algorithms for data manipulation, graphic transformation, and interactive control are also of great importance.

System Components and Structure

The personal visualization system consists of the hardware, the software, and the database. Because the personal computer is chosen as the implementation platform in order to provide wide accessibility, it leaves little room to discuss this component of the system. The basic requirements on the hardware include fast computing speed and high resolution graphic devices with as many colors as possible.

The database is what this system explores. A comprehensive database is highly desirable for the exploratory analysis. A complete visualization system should have the capability of manipulating large volume data sets. Due to the experimental nature of the research, however, the system reported here does not have this capability. Because only limited data sets are used to demonstrate the methodology, the database is organized as a simple file system.

The software component, which facilitates the interaction between the analyst and the system, is the most important factor that controls the success of the visualization system. It dictates what we can explore and how we do the work. Although the hardware has great impact on the performance of the system, such as the response speed, it is the software that controls the functionality of the system. Furthermore, the software can also affect the system performance.

The software component consists of a variety of routines, which can be divided into three categories:

The first category of routines performs basic drawing tasks, for instance, drawing polygons, filling polygons, cleaning a window, setting up and changing colors, getting a string from the keyboard, and moving an object. These routines should be as efficient as possible. Some of them can come directly from commercial software for graphics. In the implementation of this system, we utilize a software package called MetaWINDOW, which not only provides basic drawing routines, but also some advanced features for graphics. Experience indicates that such an approach is very cost-effective.

The second category of routines performs the tasks related to data retrieval, manipulation and creation of cartographic displays. These routines are much like those used in computer mapping systems. Because the system only deals with area-based data, only those routines dealing with choropleth mapping are written for the system.

The last category of routines is the key part of the software component. These routines are termed as visualization modules, each of which performs a specific task for the visual analysis of the data. There are five modules available: Interactive Viewing Module, which allows the analyst to extract accurate information from the database and to help him/her interpreting the patterns on the display; Brushing Module, which provides operations for geographic correlation analysis; Alternagraphic Display Module, which gives the analyst more effective ways to compare spatial patterns; Area Masking Module, which provides the analyst a means to examine the pattern of geographic movement associated with individual regions; and Temporal Browsing Module, which allows the analyst to investigate temporal pattern in an animated fashion or in a browsing mode. These modules are described in more details in author's another paper (Tang, 1992). The functions of these modules and the relationships between the modules are illustrated in Fig. 1.

Figure 2 shows the overall system structure. The central part consists of the visualization modules. They control what data are extracted from the database and how to transfer them into the displays.

System Usage

The system runs on a PC 386/486 and requires a VGA monitor. To promote active involvement of the analyst and facilitate effective interaction, the operations are mouse-driven, though input using the keyboard is acceptable.

The interface consists of four windows: map window, message window, main menu window, and pop-up menu/graph window. The display is shown in the map window. The message window gives the analyst instructions and provides some

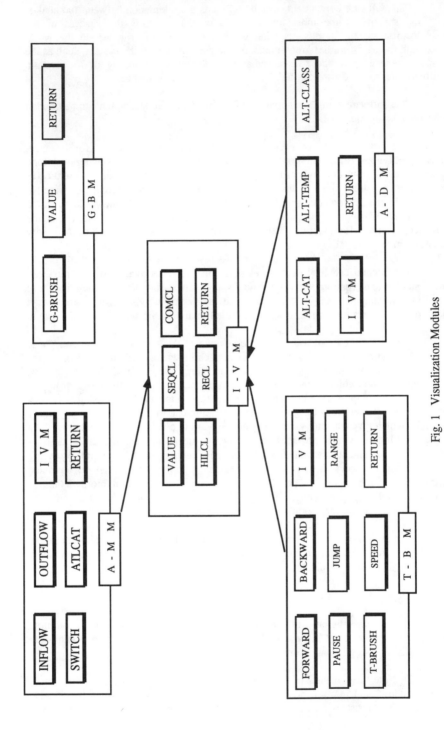

Fig. 1 Visualization Modules

information extracted from the database. The main menu lists the major tasks of the exploration. The pop-up menu/graph window is the place where multiple level submenus and some graphs are displayed.

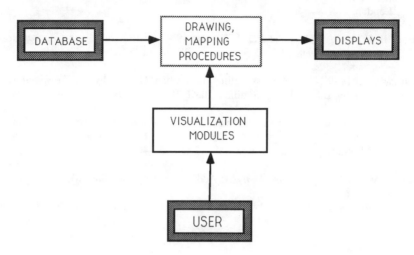

Fig. 2 System Structure

The operation of the system is quite straightforward. Before a main function is selected, a highlight bar indicates which function is going to be chosen. After a main function is picked up, a submenu is popped-up in the pop-up/graph window. Functions in the submenus are identified using "button-style" icons. Some of the sub-functions will generate new displays in the map window. The others will display next level pop-up menu/list again to allow the user to make further selections. Mapping operations are minimized. Simple operation procedures assure that the analyst can concentrate on visual inspection of the displays.

SUMMARY

Effective methods are needed to get better understanding of a huge amount of spatial-temporal data. Visual data analysis is one of the methods that can help the analyst get insight about the data. Visual data analysis combines visualization techniques with exploratory data analysis to encourage detecting unknown structures in the data and generating hypotheses about the findings. Visualization used in this sense puts different requirements on the displays from the maps for traditional cartographic communication.

A visualization system, based on personal computers, is designed to provide the capability of visual analysis of area-based spatial-temporal data. The basic methodology is to enhance choropleth mapping techniques for visual data analysis. Interaction is the most critical feature of the system, which allows the analyst to examine the data from different perspectives. The experience of this research is applicable to other types of spatial-temporal data.

REFERENCES

Harvey, D. 1969, *Explanation in Geography*, Edward Arnold, London.

Johnstone, R. J. 1987, *Geography and Geographers*, 3rd Edition, Edward Arnold, London.

Marble, D. F., H. W. Calkins, and D. J. Peuquet 1984, *Basic Readings in Geographic Information Systems*, SPAD Systems, Williamsvill, NY.

Openshaw, S. 1989, Computer Modelling in Human Geography: in *Remodelling Geography*, Ed. by Bill Maemillan, Basil Blackwell, Oxford, pp. 70-88.

Tang, Q. 1992, From Description to Analysis: An Electronic Atlas for Spatial Data Exploration: <u>Proceedings</u> of ASPRS/ACSM/RT 92 Convention on "Mapping and Monitoring Global Change", Vol. 3, pp 455-463.

Tukey, J. W. 1977, *Exploratory Analysis*, Addison-Wesley, Reading, MA.

Waner, J. 1990, Visual Data Analysis into '90s: <u>Pixel</u>, Vol. 1(1), pp 40-44.

ADAPTIVE RECURSIVE TESSELLATIONS

Paul H.Y. Tsui   and   Allan J. Brimicombe
Centre of Land & Engineering Surveying
Hong Kong Polytechnic
Hung Hom, Kowloon, Hong Kong

ABSTRACT

Quadtree, a recursive tessellation data model can compress the volume of raster data by representing a large area of same characteristic with a larger cell instead of a vast number of small cells. However, the deficiencies of quadtree are the ubiquitous use of square cells and fixed decomposition ratio (1 : 4). These make quadtree a very inflexible data model since the sizes of cells at different levels are fixed once the dimension of the whole area is known. A new Adaptive Recursive Tessellation (ART) data model which allows the use of rectangular cells and variable decomposition ratio is presented. ART offers much greater flexibility to users to cope with the needs of different applications in the aspect of data modelling. A modified two-dimensional run-encoding technique is also implemented on ART to further reduce the storage volume. In order to construct and update an ART, a Tessellation Management System (TMS) has been developed.

INTRODUCTION

In order to speed up the data input process in GIS projects, automated data acquisition techniques, such as remote sensing, scanning and video camera, have been increasingly used in projects. Because most of these techniques provide data in grid format, an efficient data model which can directly store and enable manipulation on raster data becomes crucial. Obviously, the tessellation model is a good candidate but it introduces a very large storage overhead. The emergence of the idea of recursive tessellation provides a solution. The most widely used recursive tessellation model is called *quadtree* (Samet, 1984). It recursively decomposes space into four smaller identical square cells until all cells are homogeneous or the smallest cell level specified by the user (atomic cell) has been reached. However, its ubiquitous use of square and fixed sizes cells at different levels make quadtree very inflexible and may be difficult to fit into a particular application. This paper presents a new *Adaptive Recursive Tessellation* (ART) data model which allows flexible definition of sizes of cells at different levels and which can use rectangular cells. Users can design their own ART models to fit their applications through a *Tessellation Management System* (TMS).

RECURSIVE TESSELLATIONS

Traditionally, there are two regular geometric figures which allow recursive subdivision whilst retaining the same overall shape. They are square and triangle. Squares can be subdivided infinitely with the same shape and orientation. Triangles, however, can be subdivided only with the same shape

777

but different orientation. In fact, rectangles can also be recursively subdivided with the same shape and the same orientation, though omitted in most of the literatures.

Quadtree is a regular recursive tessellation using squares. In the subdivision into four smaller squares, homogeneous cells remain in their largest possible sizes while heterogeneous ones will be further subdivided until an homogeneity criterion is fulfilled. Also, the subdivision is terminated when a cell reaches the smallest defined size (eg. 5m × 5m). Thus quadtree is a variable resolution representation of an image. Another view of quadtree is that in aggregating cells, neighbours with an identical attribute are grouped into larger cells. So the compression efficiency of quadtree greatly depends on the pattern of the spatial phenomena to be represented. If the pattern is scattered or a checkerboard at the smallest cell size, quadtree cannot significantly reduce storage space.

Beside the saving of storage space, quadtree has other advantages. The resolution of the data can be easily lowered by further aggregating cells by adopting an average value for cells now comprising a larger single cell. Thus map generalisation can be achieved. Furthermore, quadtree can be used as a spatial index which can speed up the retrieval of data in large geographic databases. There are nevertheless some unresolved issues. Firstly, the decomposition ratio between each successive level has to be 1 : 4. That is, each cell in the current level must be subdivided into four identical cells in the next lower level. Because of the fixed decomposition ratio and square shape of cell, the size of cell at each level must be generated by geometric progression. Hence, the whole quadtree model is determined by the size of cell of the first level (ie. the area to be represented). The cells generated in this manner may have undesirable sizes, such as 160m or 80m, which cannot meet the needs of users for intuitively useful and more convenient cell sizes, such as 100m or 50m. The shape of cell of quadtree at each level is also square and may not be conveniently coincident with a map series.

DATA STRUCTURES FOR QUADTREE

There have been many data structures developed for quadtree. Topologically, quadtree is usually represented by a tree structure with four branches at each node (Rosenfeld and Samet, 1979). At an early stage, explicit pointers where used to link up cells at different levels in the quadtree. However, pointer-based quadtrees were gradually phased out because of the large storage overhead of pointers. Moreover, in order to access the cells at a lower level, it was necessary to go through the upper levels first. Gargantini (1982) proposed a data structure for quadtree which eliminates the use of pointers. The data structure assigns an unique "locational code" to each maximal homogeneous block based on their positions at each level. Similar data structures are also described in Able (1984) and Able and Smith (1983). The quadtrees encoded in this way are termed *linear quadtrees*. Linear quadtree saves the pointer overhead and provide direct access to each cell through the key.

Lauzon et al. (1985) presented a new data structure for quadtree called *two-dimensional run-encoding* (2DRE). 2DRE run-encodes images in Morton sequence (Morton, 1966). Morton sequencing is a numbering system following the Z-shaped Peano space-filling curve as shown in Figure 1 (Peuquet, 1984). Each atomic cell is assigned a unique Morton key. 2DRE then run-encodes the image following the Morton key order and records the key only whenever there is a change in the attribute of the following cell. In other words, 2DRE reduces storage space by storing consecutive maximal cells of the same value in one record only instead of one record for each maximal cell in linear quadtree.

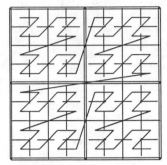

Figure 1   Z-shaped Peano curve for 64 square cells

ADAPTIVE RECURSIVE TESSELLATIONS

In order to solve the problems associated with quadtree, this paper puts forward a new concept of recursive tessellation called *Adaptive Recursive Tessellation* (ART). On the whole, ART follows the divide and conquer principle guiding the decomposition of quadtree. That means heterogeneous areas will be further subdivided until all cells contain one value only while homogeneous area will be represented by maximal cells. Nevertheless, the novel characteristic of the ART model is that it permits the use of rectangular cells and variable decomposition ratios between each successive level. These two points are critical in allowing greater freedom in designing recursive tessellation models. Rectangular cells can increase the ability the data model to have spatial representativeness. For example, rectangular cells can be designed to represent an area covered by map sheets of different scales with this element of spatial organization retained within the data model. Thus, variable decomposition ratios mean that one cell is not necessarily subdivided into four identical cells but into any number of cells provided that they are identical in size, shape and orientation and must fill up the larger cell without gaps or overlaps. In this way, it is also possible to have, at lower levels, intuitively sensible cell size of, for example, 100m, 50m and 10m.

Schematically, ART is composed of levels. A level is a logical layer which comprises one size of cell or *Basic Tessellation Unit* (BTU). BTU is the basic geometric figure used for tessellation at a level. It is a rectangle (or square) as defined by its length and width. The top level of ART model contains the largest BTU and lower the level, the smaller is the BTU. The level comprising the smallest BTU is called atomic cell level and thus the smallest BTU is called the atomic cell. An ART should have at least two levels: the top level and the atomic cell level. Where there is only an atomic cell level in an ART, the ART is a non-recursive tessellation model. It should be noted that BTU in each successive level should be compatible with each other, that is the length and width of the BTU at the current level must comprise of an integer number of those at the next lower level. Finally, the tessellation parameter is defined as the number of cells a BTU is subdivided into in the next lower level and it is also an essential quantity in defining an ART.

From the above specification, quadtree can be defined as an ART with uniform tessellation parameter 4 and the BTUs of lengths being equal to widths at each level. Therefore, quadtree can be said to be a subset of ART.

The data modelling philosophy of ART and quadtree are however very different. In ART, the user first identifies the area of interest and determines what sizes of cells he is going to include in the ART model. Then the set of cells specified by the user is checked for compatibility. If the compatibility criterion cannot be satisfied, addition or elimination of some intermediate levels will need to be considered. On the other hand, in quadtree, once the dimension of the area of interest and the size of atomic cell are known, then sizes of cells and the number of levels will be generated automatically. Obviously, ART takes an user-oriented approach in modelling the data whilst quadtree takes a pure algorithmic approach.

For a clearer explanation of the concept of ART model, consider an example to model the territory of Hong Kong for the urban and regional planning. The cells in the upper three levels can be designed to coincide with the areas covered by the three scales of basic mapping series of Hong Kong (1:20,000, 1:5,000 and 1:1,000) respectively. The cells of the three lowest levels are designed to coincide with the three grid planning units (50m, 10m and 5m) intuitively used by planners. In order to reduce the gap of cell size between 1:1,000 map cell (750m × 600m) and 50m cell, a 150m cell level is introduced in between the two levels. Table 1 summarises the cells used in the ART model and its associated parameters. In this example, the upper part of the ART acts as a map indexing system (eg. for data transfer) whilst the lower part of the model is used for urban and regional planning. This is one of the characteristics of ART model in that a single model can serve two purposes.

Level	Size of BTU	Tessellation Parameters	Remark
7	60,000m × 48,000m		Whole Territory
		16	
6	15,000m × 12,000m		1:20,000 map
		16	
5	3,750m × 3,000m		1:5,000 map
		25	
4	750m × 600m		1:1,000 map
		20	
3	150m × 150m		-
		9	
2	50m × 50m		-
		25	
1	10m × 10m		-
		4	
0	5m × 5m		Atomic cell level

Table 1   Tessellation scheme of the ART model for Hong Kong

DATA STRUCTURE FOR ART

For the ART model to be implemented by computer, it must be encoded in a machine recognizable format (ie. data structure). A data structure which is similar to the two-dimensional run-encoding mentioned above is adopted to encode the ART model. Each atomic cell in ART model is assigned a unique decimal address key according to its position in space. Then the area will be run-encoded two-dimensionally by following the order of the keys. As with 2DRE, a cell's address and its corresponding attribute will be recorded only whenever there is a change in the immediate following cell. The records are then sorted in the ascending order of the keys. In this sense, this data structure is also valid for multiple attribute data.

Unlike the strict definition of the quadtree model, the run-encoding path of ART model varies slightly with different definitions of the ART model as specified by the user. Nevertheless, there is a basic principle guiding the path of run-encoding in ART model. It follows a *row order* path which proceeds in left-right and top-bottom directions at each level and this pattern propagates recursively down the model as shown in Figure 2. Therefore, each atomic cell of the whole area can be run-encoded by repeating the path in each cell at each level. The row order path can facilitate the retrieval of neighbouring cells and has a high repeatability. From Figure 2, it can also be noted that the run-encoding path in each level is slightly different depending on the tessellation parameter specified between two successive levels though it basically follows a repeatable Z-shaped pattern.

Each atomic cell in the ART model has two keys to indicate its position termed the *global* and the *local* address. Global address is a decimal number assigned to each atomic cell following the order of run-encoding. For example, the decimal address of each 5m atomic cell for the whole territory of Hong Kong (60,000m × 48,000m) ranges from 0 to 115,199,999. Figure 3 shows the global addresses of first 100 5m cell of the ART model for Hong Kong. Local address of an atomic cell is a number composed of its positions at all levels. The local position of a cell at a level relative to the immediate upper level is numbered from 0 to n-1 as shown in Figure 2 where n is the tessellation parameter between the two successive levels. If the ART model is specified as a quadtree, the local address must be a base-4 number as there are only four local positions (0, 1, 2 and 3) at each level. Thus the local address for a normal ART model will be a composite-base number. For instance, in Figure 2, the local address of the lower-rightmost cell in the ART model for Hong Kong is $15_{16}15_{16}24_{25}19_{20}8_{9}24_{25}3_{4}$. The global and local addresses of a cell can be converted into each other through the base conversion arithmetic.

Another important issue is the computation of cell address of a point from its coordinates since spatial data is usually expressed in the form of coordinates. The procedure is to calculate the local address of the atomic cell containing the point and the global address is then obtained from the local address. In this computation, the dimension of BTU at each level and tessellation parameters are required. Because this operation will be invoked frequently, the time required for this operation significantly influences the overall efficiency. Thus the factors affecting the time of this operation must be identified. From the point of view of computation, the major factor is the number of levels in an ART model as the greater the number of levels, the larger the number of arithmetic and modular operations that need to be performed. On the other hand, it can be found that the computational efficiency is independent of uniformity of tessellation parameters since irregular tessellation parameters will not increase the number of arithmetic operations performed. For example, in quadtree, even though all tessellation parameters are 4, the computational efficiency will be the same as an ART with irregular tessellation parameters but equal number of levels.

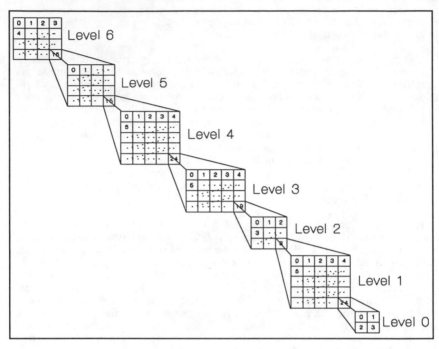

Figure 2   Recursive decomposition of the lower-rightmost cells in the ART
model for the Hong Kong example and the run-encoding pattern for
each level

0	1	4	5	8	9	12	13	16	17	100
2	3	6	7	10	11	14	15	18	19	102
20	21	24	25	28	29	32	33	36	37	120
22	23	26	27	30	31	34	35	38	39	122
40	41	44	45	48	49	52	53	56	57	140
42	43	46	47	50	51	54	55	58	59	142
60	61	64	65	68	69	72	73	76	77	160
62	63	66	67	70	71	74	75	78	79	162
80	81	84	85	88	89	92	93	96	97	180
82	83	86	87	90	91	94	95	98	99	182
300	301	304	305	308	309	312	313	316	317	400

Figure 3   Global addresses of first 100 5m cells in ART model with adjacent
cells

TESSELLATION MANAGEMENT SYSTEM

A *Tessellation Management System* (TMS) is being complied to build and update ART models automatically from both vector and raster data sources. In other words, the TMS works like a Spatial Database Management System (SDBMS). The TMS organizes the spatial database in a systematic hierarchy. At the highest level, the database are divided into different *projects*. Each project forms a sub-directory under which all data and working sub-directories belonging to that project will be created. Under each project, a user can create any number of *blocks*. Each block comprises an ART model which is defined by the size of BTU at each level and the geographical coordinates of the origin of the ART. The TMS will also check for the compatibility and validity of all ART models specified by the user. Within a block, there can be any number of thematic *layers* for different classes of attribute. In other words, all layers in a block are tessellated with the same ART structure and hence the coincidence between different layers can be ensured. Lastly, there is a logical entity termed *frame* which actually is a rectangular window used to index a portion of data of a layer (eg. the area involved in a project). User can create as many frames as they need in a block. Besides, a frame can be applied to different layers within the same block. Frame is quite useful since at most instances, a project will not utilize the whole database and users can quickly reference to the area of interest and hence speed up the retrieval of data.

The TMS can also accept both vector and raster data into its database. Vector data can be the vector exchange files exported from other GISs or CAD systems. Raster data can come from scanner, remote sensing or video camera. No matter the data source, the TMS will transform it into the user-specified ART model. Update function is also crucial in the TMS. This function enables users to input data of higher resolution at different locations into a layer according to the actual needs and the progress of a particular project. Thus, if more detailed data becomes available as a project progresses into another stage, then this new data can be input by the TMS to replace the original less detailed data. Conversely, the TMS is also designed to be able to generalize the data to a lower resolution if necessary by taking the average value of the cells being grouped into a larger cell. Finally, the TMS also provides data storage and retrieval functions for some external modules for spatial analysis. The Tessellation Management System is scheduled for completion by early 1993.

RESULTS OF PRELIMINARY ANALYSIS

In order to study the space efficiency of ART model before the actual implementation of the TMS, a preliminary analysis on the relative space efficiency of the ART and quadtree in the representation of Lamma Island, Hong Kong, was carried out. Both models are used to represent the land area of the Lamma Island. From Figure 4, it can be seen in both models that the larger cells concentrate in the central part of the island while the area near the coastline is mainly represented by small cells. The two images are created by hand digitizing with a vector-based GIS. Then parts of the island represented by both models were manually run-encoded to estimate the effect of 2DRE on the space efficiency of the two models. From Table 2, it can be found that number of cells used by the ART nearly doubles that used by quadtree to represent Lamma Island. In fact, 88% of cells used by the ART are 10m atomic cell. This is because the ART, as designed, can only use 10m and 50m cells to fill up the spaces near the edge of the island. Otherwise, the next

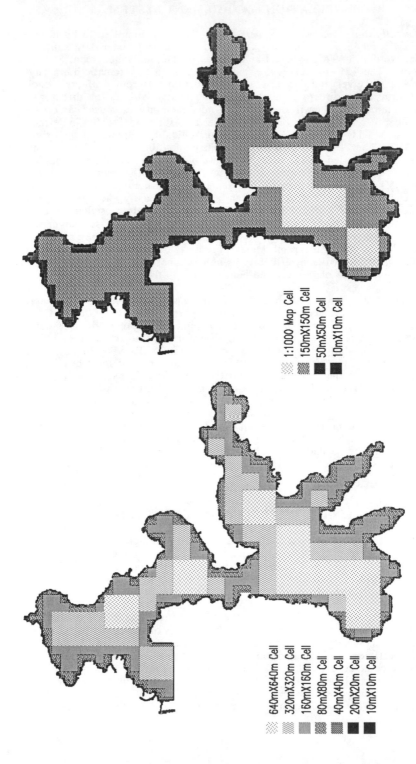

Figure 4 Representation of Lamma Island, Hong Kong, by different sizes of cells in quadtree (left) and an ART model (right)

640mX640m Cell
320mX320m Cell
150mX150m Cell
80mX80m Cell
40mX40m Cell
20mX20m Cell
10mX10m Cell

1:1000 Map Cell
150mX150m Cell
50mX50m Cell
10mX10m Cell

choice in the ART model is 100m cell. On the other hand, quadtree has a greater variety of cells (ie. number of levels) than the ART for Hong Kong. So quadtree can fill up more such spaces with relatively larger cells, such as 10m, 20m and 40m cells, and hence less number of cells are used.

No. of cells used →	Before 2DRE	Estimated % reduction by 2DRE	After 2DRE
ART	10,732	79%	2,254
Quadtree	5,029	55%	2,263

Table 2   Number of cells used to represent Lamma Island in the two models and the estimated % reduction of number of cells by 2DRE on them.

From Table 2, the empirical result shows that the effect of 2DRE on the space efficiency of the two models is quite significant, especially for the ART. After two-dimensionally run-encoding, the number of cells used by quadtree and the ART are nearly equal. This may be due to the common aggregation of large number of continuous atomic cells in the ART model during 2DRE. So 2DRE can be used to compensate the space inefficiency of the ART model in a greater extent.

In this preliminary analysis, a comprehensive comparison on the space efficiencies of both models was made whilst the estimate of computational efficiencies of building up the models was ignored. Nevertheless, based on the arguments stated in the section of "Data Structure for ART" for computation of cell address, we can infer that the computational efficiency of the ART, as designed, is better than that of quadtree since the ART has fewer levels. Obviously, there is a trade-off between space and computational efficiencies in designing an ART model. Therefore, one should get an optimum solution between these two by adjusting the number of levels and using the technique of 2DRE in designing an ART model. The completion of the TMS will allow more rigorous comparison of ART and quadtree and for the investigation of different ART models.

CONCLUSION

ART is proposed as an innovative concept in recursive tessellation. It provides flexibility in the definition of the structure of tessellation in order to adapt the data model to a particular application. It changes the data modelling concept of the traditional recursive tessellation model, quadtree, and allows the participation of users in structuring the data model. In fact, both quadtree and non-recursive tessellations with grids and rectangles are a special case of ART. In designing an ART which can both fit into a particular application and be space efficient, it is necessary to consider the overall structure of ART beside the mere inclusion of those cells desirable for the application. In particular, the variety of sizes of cells of an ART model is a critical factor affecting the space efficiency. In case there exists a trade-off between the space efficiency and the model suitability for an application, 2DRE is an efficient and practical technique in reducing the storage requirement of ART model. Although the conceptual modelling advantages of ART have been stressed, initial results show potential for computational efficiency.

ACKNOWLEDGEMENTS

The authors gratefully acknowledge the support of the Hong Kong Universities and Polytechnics Grants Committee through research grant 340/935 and the Hong Kong Polytechnic Research Degrees Committee.

REFERENCES

Able, D.J., 1984, "A B$^+$-Tree structure for large quadtrees". *Computer Vision, Graphics and Image Processing 27*, pp. 19-31.

Able, D.J. and Smith, J.L., 1983, "A data structure and algorithm based on a linear key for a rectangle retrieval problem". *Computer Vision, Graphics and Image Processing 24*, 1, pp. 1-13.

Gargantini, I., 1982, "An effective way to represent quadtrees". *Communications of the ACM 25*, 12, pp. 905-910.

Lauzon, J.P.; Mark, D.M.; Kikuchi, L. and Guevara, J.A., 1985, "Two-dimensional run-encoding for quadtree representation". *Computer Vision, Graphics and Image Processing 30*, pp. 56-69.

Morton, G., 1966, "A computer oriented geodetic data base and a new technique in file sequencing". IBM Canada Ltd., unpublished report.

Peuquet, D., 1984, "A conceptual framework and comparison of spatial data models". *Cartographica 21*, 4, pp. 66-113.

Rosenfeld, A. and Samet, H., 1979, "Tree structure for region representation". *Auto-carto 4*, Vol. 1, pp. 108-118.

Samet, H., 1984, "The quadtree and related hierarchical data structures". *Computing Surveys 16*, 2, pp. 187-260.

INFORMATION ENGINEERING AND GIS APPLICATIONS

Nancy von Meyer
Vice President
Fairview Industries
8616 Fairway Place
Middleton, Wisconsin 53562

ABSTRACT

Traditional geographic information system (GIS) development relies on user interviews, data flow diagrams, and functional requirement documents to guide programming and physical table design. While these approaches lead to relatively rapid application development, they do not promote complete documentation and system and data integration.

Information engineering is a set of procedures for developing integrated systems. It uses "data-driven' principles and results in complete and standardized documentation. There are many different types of information engineering, each type is called a methodology. All information engineering methodologies share some basic components.

INTRODUCTION

GIS is growing up. A technology that once existed in isolated areas of corporations and scattered public agencies is penetrating to board rooms and public policy offices. Many have speculated on the reasons for this growth (Dangermond, Parker, Harlow, and others). This paper focuses on one of the technical implications of the growth: GIS as an integration technology. GIS is more than another way to manage spatial data. If properly designed, it can provide the basis for data and organization integration.

Historically, GIS design and development has been driven by single applications on stand alone platforms. As examples, a planning administrator may develop a system for growth management, a soil scientist may develop a system for erosion control, and land records staff may used GIS to organize and modernize deeds and plats. Often GIS design has started after the software and hardware was purchased or as an add-on to an existing platform. Another indication of GIS development independence is that the developments were fairly advanced before linkages to external databases and related software were necessary.

Williamson and Hunter (1989, p.14) have observed that a major limitation of existing development practices is that they rarely address the incorporation of large-scale institutional changes into the process. Because GIS incorporates institutional, technology, and data related activities, it presents significant difficulties during implementation on a corporate-wide scale. They

787

suggest that descriptions of the conceptual framework of the GIS should be built before implementation.

One of the newest tools in information system development is information engineering. The goal of information engineering is to capture the data, technology, and institutional components of an organization in an enterprise model. It strives to provide an integrated structure for all systems that is in harmony with the goals and missions of the organization. This new discipline has emerged almost in parallel with the relational database management technology. James Martin and Clive Finkelstein have been credited with developing the first information engineering principles at IBM in the 1970's. (Martin, 1989; Finkelstein, 1989). Most of this early development was taking place in non-spatial computer systems for banking, hospitals, and schools (Bedard, 1989, page 44).

Information engineering can be used in two modes for GIS development, system design and organizational structuring. In the system design mode it provides for active end-user participation in system design. Rather than having programmers and database developers build systems in isolation and then deliver the final product, it is a means for users to guide and contribute to system construction. But the real strength of information engineering is in how it can help re-shape the organization by re-engineering work and optimizing the information resource. In organizations that use spatial data, GIS is a full partner in the re-shaping equation.

This paper presents an overview of what information engineering is and a few examples of how it has been used with GIS in organizations for both design and re-shaping the organization.

WHAT IS INFORMATION ENGINEERING?

Information engineering has been defined by Martin (1989) as a structured methodology for defining, developing, and implementing information systems which are derived directly from the enterprise's objectives. This means that the system development does not occur for single applications, but instead is tied to the organization's mission. The relationship of systems to the mission occurs as follows:

Mission Every corporation or organization has a mission or overall purpose. For example a utility's mission might be to provide electric, gas, and water service to current and future customers in a timely, efficient, and environmentally safe manner. A public agency's mission might be to manage resources to the maximum benefit of the consuming public within the financial constraints of good government.

Organizations Organizations are assembled to meet missions. They are the human, technological, and physical resources working for a common purpose, i.e. the mission. The structure of organizations are expressed in organizational charts.

Information Information is required by organizations to meet their assigned mission or purpose. Information can generally be classified in three levels: planning, analysis, and implementation. The priority of an organization's purpose influences the priority of all information needs.

Functions   Functions apply information to activities. Typically, functions are where an organization does its work. The levels of functions generally parallel the levels of information.

Data   Data are the smallest pieces of information used in an organization. Typically, data are not thought of as information until they are applied to a function to solve a problem or meet a need.

Many authors describe information engineering as "data-driven" because it stabilizes and integrates the data required for functions.

As with any structured method, information engineering occurs in phases. Every flavor of information engineering provides slightly different names for the levels and sometimes varying numbers of phases, but in general there are three levels: planning, analysis, and implementation. Implementation may also be called physical design and construction. These levels are often represented as a pyramid, with the planning or strategic level at the top and the implementation at the bottom.

At the planning level, the corporate mission is described in terms of high level information models. The strategic activities and information requirements are prioritized from the mission. In some cases this high level modeling can provide insight to executives and leaders on how to structure the organization. Some methodologies call this level of modeling information strategic planning or enterprise modeling. It is used to define where the analysis should begin.

At the analysis level, a portion of the enterprise model is examined in more detail. Business experts meet to describe the function they do and the information they need to meet the corporate goals. These are captured in process and data models. Each analysis project either builds on or combines with the results of the preceding efforts.

The design and construction or implementation level is where the models are turned into physical systems. This where the prototypes and pilot projects occur. The system is actually built in this level.

If information engineering is being used in the design support mode, its goals can be achieved with the modeling efforts at the various levels. To achieve organizational structuring support further analysis of the modeling results are required. Four matrices used in this analysis are:

Function - Data Matrix

The function - data matrix relates functions to their required data elements. This matrix is sometimes called the Create, Read, Update, Delete

(CRUD) matrix. Data elements are defined in terms of whether a function creates, reads, updates, or deletes a data element. A data element or data entity can not be read, updated, or deleted until it is created. It is preferable to avoid multiple functions creating the same entity.

Function-Information Matrix

In this matrix functions are mapped against their information needs. This helps determine if multiple functions have similar information needs. Colloquially, this is the "while you are out there" analysis. This matrix also helps verify that every function has all the information it needs.

Data -Information

The data to information relationship verifies that all the pieces are present for the whole. That is, are all data elements present for every information need and is every data element defined completely in terms of the information needs it serves. For example, if an information need for parcel location contains township, range, and section and tax billing has another information need for township and range, and section are both information needs accessing the same, complete data for township, range and section.

Information - Organization Matrix

This matrix helps analyze the corporate organization. It is where the re-engineering or analyzing work flows can be done. By matching organizations to information it is possible to optimize both information and organizations within the context of the corporate mission.

GIS AND INFORMATION ENGINEERING

More and more agencies and corporations are beginning to use information engineering principles in GIS development. Two major areas that have seen GIS and information engineering are utilities and in government landownership applications.

Utility developments in GIS have typically started from AM/FM or SCADA (supervisory control and data acquisition) systems. Benjamin (1992) and Bachmann and Kindrachuk (1992) have documented the early investment in technology and the move to broader based system development in utilities.

> Those computer systems (AM/FM and SCADA) had been developed one by one for separate relatively narrow purposes: energy management, energy accounting, distribution monitoring and control, load management, circuit mapping, filed work management or anything else a computer could speed up or organize. But across the board computerization requires integration, and by the late 1980's, utilities were recognizing how powerful these systems could be if they could work together across a company-wide network. All too often they couldn't. (Benjamin)

Utilities were part of the first wave of major corporations to implement large computer systems. They regularly incorporated technological innovation into their business practices, for example, to improve customer services. However, the full benefits of these systems, through increased productivity, better decision support, or improved competitive positioning, are yet to be realized. (Bachmann and Kindrachuk)

In some cases utilities have used information engineering methods to achieve integration benefits. These have ranged from high level organization models to specific application data models. Duke Power Company has used information engineering methods to define the corporate business functions and information needs. Puget Sound Power and Light has used information engineering methods to develop a facilities management system, which unifies the corporate mapping functions. Wisconsin Power and Light is using information engineering to support a corporate-wide information integration system. "We want to integrate our corporate data so we can reduce the amount of redundant data we are collecting and maintaining. We want to collect and maintain it in one spot, but make it available to all people who need it. We also want to re-engineering the work order process." (Arc/News, 1992).

Because so much of the data that utilities use is tied to geography and because they typically have such large volumes of data, it is not surprising that so many are moving towards an information engineering - GIS mix. The increasingly competitive environment in energy sales and management is also contributing to the move to more carefully designed systems.

Public sector based landownership systems are also driven by the large and complex data sets. However, for the landownership applications, the extremely long data life, sometimes exceeding 200 years, and the legal complexities surrounding land title, has led to the need for information engineering principles. "The object of policy, legislation, and administration is managing use of land and, therefore rights to land. ... A system cannot be simple graphic; it must integrate the physical record with activities that change it and the history of these activities." (Parr, 1992)

The Bureau of Land Management is using information engineering concepts to build an enterprise model of the land holdings aspect of the agency. The ALMRS (Automated Land Records and Mineral System) and GCDB (Geographic Coordinate Data Base) projects have used data and process modeling to specify system requirements and to guide data conversion. (von Meyer and Scruggs, 1991).

Parr (1992) has used information engineering principles to describe a common information architecture for local government land use planning. A review of the data demonstrated the need to examine the geographic data together with the non-graphic data to get a full picture of the county's needs. "Capturing physical changes in boundaries is no more important than capturing the reasons for those changes and concomitant changes in government responsibilities. Even in the subdivision process that changes basic parcel records, graphic representation is a fraction of the information that is collected." (Parr, 1992, page 6). Information engineering provided the tools to combine the record and geographic component of land use management in one system.

The US Army Corps of Engineers is another agency that has embraced the information engineering concepts and applied them to real estate related problems. A large encyclopedia is being constructed to documented existing real estate systems in the Corps. Using an automated information engineering encyclopedia some of the Corps' goals are to: support an integrated repository of corporate meta-data, facilitate development, maintenance, and technical support of applications and information systems, and to provide a smooth transition from present systems to an automated distributed data architecture.

CONCLUSIONS

Information engineering is a powerful tool that can help GIS installations reap the benefits of geographic data integration. To date, very few organizations have used GIS to its full integration potential. This may be due to the absence of rigorous development methods. But certainly the time has come to look at information systems and GIS as partners in building corporate information infrastructures.

The partnership is not without problems. Bedard (1989) has identified some generic issues that have not been fully addressed. How and when to make the inventory and analysis of geographic data, including lineage? How are the special issues of geographic data like generalization, spatial meta-data, and topologic relationships introduced in business models? How should we model spatial processes? How and when should base reference systems and accuracy requirements be included in information engineering? How should cartographic data dictionaries be constructed?

Some solutions to these issues are evolving as more and more organizations take on information engineering principles with GIS projects. However, more complete answers will require the GIS community to begin to think more globally in terms of the enterprise role of GIS and the traditional information systems community to being to recognize the revolutionary change presented by GIS technology.

REFERENCES

Arc/News, (1992), "Wisconsin Power and Light Selects ARC/Info Software for Corporatewide Information Integration System," Spring issue.

Bachmann, C. and Kindrachuk, E.P., (1992), "Utility Enterprise Information Generates Productivity Gains," GIS World, April, pages 94-96.

Bedard, Y., (1989), "Information Engineering for the Development of Spatial Information Systems: A Research Agenda," Urban and Regional Information Systems Annual Conference Proceedings, Boston, Massachusetts, Volume 4, pages 43-53.

Benjamin, S., (1992), "Riding the Fourth Wave," Electric Perspectives, July/August, pages 22-30.

Finkelstein, C., (1989), <u>An Introduction to Information Engineering</u>, Addison Wesley, New York, New York, 393 pages.

Martin, J. (1989), <u>Information Engineering - Books 1, 2, and 3</u>, Prentice Hall, Englewood Cliffs, New Jersey.

Parr, D., (1992), "Land Data in Local-Government Automation," <u>Surveying and Land Information Systems,</u> Volume 52, Number 1, March, pages 5-12.

von Meyer, N., and Scruggs, R., (1991), "Cadastral Survey Data in a Land Information System," <u>Surveying and Land Information Systems</u>, March, pages 49-52.

Williamson, I.P., and Hunter, G.J., "The Importance of Conceptual Modelling in the Design of land nad Geographic Information Systems," <u>Urban and Regional Information Systems Proceedings of Annual Meeting</u>, Boston, Massachusetts, Volume 2, pages 7-15

MODELING POTENTIAL NATURAL VEGETATION
FROM A TOPOGRAPHIC GRADIENT
IN THE SOUTHERN SIERRA NEVADA, CALIFORNIA

Richard E. Walker, David M. Stoms and Frank W. Davis
Department of Geography, University of California
Santa Barbara, California 93106-4060
Internet: walker@castor.geog.ucsb.edu
(805) 893-7044

Jan van Wagtendonk, Yosemite Research Center
Yosemite National Park
El Portal, California 95318
(209) 372-0465

ABSTRACT

In this paper, we describe the application of a vegetation model to predicting cover types over a 610,000 hectare study area of the southern Sierra Nevada. Topographic relief in the area spans over 4000 meters, from grasslands of the Central Valley to alpine communities on the high peaks. Elevational and topographic moisture gradients, were derived from the Fresno East 1:250,000 scale DEM. Vankat's (1982) ecological model was used as a classifier to assign the 100m DEM pixels to the most probable cover type. Generally, the model provides a good first approximation of vegetation cover over this large, heterogeneous region, both in thematic accuracy and in the realistic appearance of vegetation polygon shape.

INTRODUCTION

Characterizing and mapping natural vegetation have long been the subject of study. Küchler and Zonneveld (1988) provide a cartographically-oriented synthesis of techniques and guidelines which have evolved over the years. More recently, with advances in computers and their data handling and storage capabilities, new tools are being developed which can facilitate depicting the distribution of vegetation across a landscape (e.g. Kessell 1990, Twery et al. 1991). Mapping natural vegetation can now be done with the assistance of computer-stored, spatially co-registered environmental information used as inputs to models of vegetation. Maps created by this process can be as taxonomically and structurally precise as the input data can support, and can be tailored to a specific region.

This paper summarizes an approach we used to model the native vegetation for a mountainous region in California. Taking a previously published vegetation gradient diagram (Vankat 1982), we produced a synthetic image to simulate "environmental space" in the form of a two dimensional look-up table. We then used the table to encode a digital elevation model (DEM) of the region according to its most probable potential vegetation. The results of this preliminary work provide insights into the level of accuracy attainable using such simple data and techniques, and suggest other environmental parameters which would help refine and improve the natural vegetation output model.

BACKGROUND

R. H. Whittaker, an ecologist specializing in plant-environment relationships, began a practice of depicting the continuum of vegetational changes and environmental gradients using diagrams with temperature (or elevation) on the vertical axis and moisture availability (topo-moisture) on the horizontal axis (Whittaker 1960). This technique has since been used to portray vegetation gradients in a number of other studies (e.g. Peet 1978, Rundel *et al.* 1978, Parker 1989). Vankat (1982) applied this method to model the relationships between elevation and moisture and the dominant vegetation types for Sequoia National Park, California. While omitting some important environmental influences on plant distributions, the gradient diagrams are useful generalizations in that they limit the possibilities of the species that can potentially inhabit a given site.

Digital elevation grids (DEMs) can be used to derive many different microclimatic and hydrologic variables, including temperature regime, slope, aspect, and slope position. DEMs have been used to calculate site energy regime, runoff potential, siting of facilities and other related applications (e.g. Stefanovic and Wiersema 1985, Dubayah et al. 1990, Brown 1991).

Numerous studies have employed DEMs for regional vegetation mapping and modeling. Kessell (1990) used an approach similar to ours to obtain vegetation maps for large natural areas in Australia. Strahler *et al.* (1981) and Franklin *et al.* (1986) used mainly elevation information to predict forest tree species occurrences in a timberland mapping effort. Davis and Goetz (1990), in their study of the distribution of Coast Live Oak (*Quercus agrifolia*) in an area of southern California, used DEM data to characterize several energy-related environmental variables. Agee and Kertis (1986) employed the DEMs as a primary data source in their project mapping cover types of the North Cascades in Washington state. Cibula and Nyquist (1987) used similar methods to map Olympic National Park.

APPLICATION OF DEM TO GRADIENT MODELING
OF POTENTIAL VEGETATION

Study Site and Data

The slope west of the Sierra Nevada crest in the Mineral King one degree latitude by one degree longitude cell, or latilong, was selected as the site for this analysis (figure 1). Topographic relief spans over 4,000 meters, from grasslands in the Central Valley to alpine communities on higher peaks such as Mt. Whitney. Besides the fragmented terrain of the Sierra Nevada range, this region is dissected by two extremely deep drainages of the Kern and Kings Rivers. Thus, there are rapid changes in vegetation from canyon bottom to crest within quite short geographic distances. The

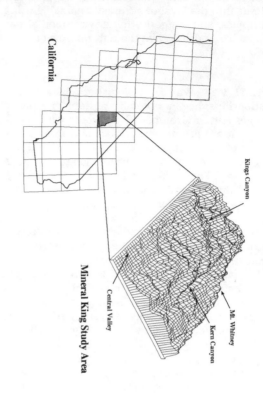

Figure 1: The Study Area

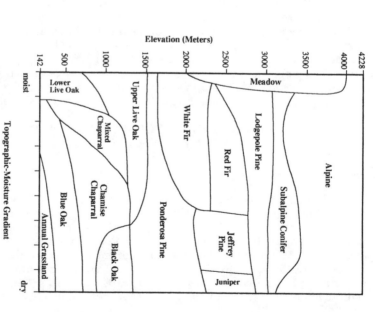

Figure 2: The vegetation of Sequoia National Park, California, modified from Vankat (1982).

study site contains most of Sequoia and Kings Canyon National Parks, and portions of the Sequoia and Sierra National Forests. Because of the steep environmental gradients (arid to moist; hot to cold; frost-free to 3 month growing season) and the resulting variety of vegetation types, the Mineral King latilong should be especially sensitive to misclassification of vegetation polygons.

Vankat (1982) divided the natural vegetation of Sequoia National Park into 15 units. We extended the elevation at the lower extreme to include areas outside the Park and dominated by annual grasses (figure 2). We then modeled potential vegetation based on its association with regional topographic moisture and elevational gradients using the Fresno East 1:250,000 scale DEM (U. S.Geological Survey 1987).

Topo-moisture, as used by Vankat (1982), was based on slope aspect scaled to the eight cardinal directions plus flat surfaces, ranked by relative solar exposure (e.g., southwest aspect was the driest). We did not include ridges and riparian areas of Vankat's diagram as these were not readily derived from the DEM data. Each 10 degree increase in slope angle, in accordance with Vankat's original model, equated to a shift of one unit toward the xeric end of the topo-moisture axis. Because of noise in the aspect and slope class maps, the topo-moisture map was smoothed with a 5 x 5 pixel moving filter before classification. The two gradient maps were then transformed into a UTM projection with 100 meter square grid cells (i.e., 1 ha) and masked by the study area boundary (figures 3 and 4).

Encoding the Topo-moisture Elevation Model

The next step was to produce the look-up table with which to map the relationship between environmental site characteristics and the vegetation types. First, the diagram was digitized into a synthetic image, and each vegetation type was encoded with a unique DN value. This image was then geometrically scaled to match in the number of rows the dynamic range of the elevation, and in the number of columns the range of values from wet to dry of the moisture index image. After reflecting this synthetic image about its horizontal axis, we then had a raster to serve as a look-up table. The final step in the production of the potential vegetation image was simply to read the topo-moisture and elevation values at each cell and use them to index the location of the vegetation type in the image/look-up table.

RESULTS

The resultant potential vegetation image, with a minimum mapping unit (MMU) of 1 ha, had a complex pattern, with 4,850 patches, averaging 125 ha and ranging from 1 ha to one polygon 117,000 ha in size. Elevation is clearly the dominant influence, and in a few areas a simple contour-encoding of the classified terrain was evident. However, in most areas the topo-moisture regime as we modeled it also strongly influenced the type prediction. These sites typically had a complex mosaic of several types, obviously determined mostly by slope-aspect. The ecological model did not include water or bare rock habitats as these are not necessarily related

797

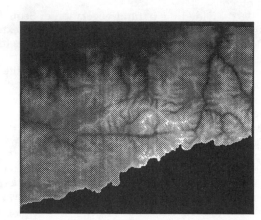

Figure 3: The digital elevation model (DEM) of the study region. Elevation ranged from 100m to over 4000m in the region.

Figure 4: The topo-moisture gradient image as derived from the DEM. The gradient was coded from 1 to 10, where lower values represented more xeric conditions.

798

to topographic gradients, nor did it include urban and agricultural development or early successional stage vegetation.

We compared our terrain-based potential vegetation coverage with the most floristically and spatially detailed vegetation maps of the region, produced during the 1930's by the VTM survey (Wieslander 1935), and based upon enlarged 1:125,000 scale USGS topographic maps. This map was believed to have the most accurate floristic information available for the region, and served as a surrogate for ground truth. Only the areas within Sequoia National Park and a few other lower elevation areas to the west were mapped in the VTM survey, however.

We selected 100 randomly placed points which fell into the areas mapped in the VTM survey to be used as a test of the agreement between our predicted vegetation and the species depicted in the VTM map of the points. As the vegetation classifications differ between the maps, cross-walks were developed to facilitate the inter-map comparisons. The agreement between our potential vegetation coverage and the VTM map of actual vegetation is summarized in table 1. Our potential vegetation map was correct in 44 out of the 100 points. In addition, 10 more points were within a single pixel from the correct type. With a positional error buffer of 0 to 3 pixels (0m to 300m), the percent accuracy rises to 63.

The table of predicted (our coverage) versus actual (VTM maps) vegetation (table 1) reveals some biases which appeared in our model. For example, in four of the locations where our model predicted Lodgepole Pine Forest to occur there was in actuality Jeffrey Pine Forest. In general, it appears that the model has somewhat of a bias of predicting vegetation types in areas lower or drier than where they are present.

DISCUSSION

Models relating vegetation to environmental variables can serve a number of purposes. In addition to their use in producing maps of current vegetation, the models can be manipulated according to projections of future climates, with the objective of predicting the probable effects of the changes on the vegetation. Another advantage of such models is the ease with which they can be refined and updated as new and more accurate information becomes available. For example, we could try moving the boundaries of the Vankat model in the lookup table and observe whether our results improve. Manual techniques are much more labor-intensive, and map updates are infrequent due to the high cost involved.

In conjunction with remotely sensed data, such as NDVI measurements, vegetation-environment models can be used to establish prior probabilities of "what should occur" on a given segment of the landscape (e.g. Strahler et al. 1981, Franklin et al. 1986). Templates of this nature complement digital satellite data well, as taxonomic details are largely not extractable from the remote sensing data alone. Conversely, where the models tend to predict only the vegetation potential of a site, remotely sensed data assist to identify what is actually present at that location.

Remotely sensed measurements might also indicate more definitively the moisture regime at each cell, through indirect means such as an index

Table 1. Predicted vegetation type versus VTM mapped vegetation type. Rows give predicted vegetation from our model; VTM is given in the columns. The model was exactly correct in 44% of cases, and generally correct (within 3 pixels error buffer) in 63% of the cases. The types codes in the table are given below.

	1	2	3	4	5	6	7	8	9	10	11	12	13	14	15	16
1	2	5														
2	1	6			1											
3	1															
4					1		1									
5		2	1													
6			1		1											
7							2									
8							3	1		2						
9							2									
10							1	1		2	1			1		
11								1			3					
12							1	1		3	1					
13						1				4	2		1	2		1
14											1	1	2	2		2
15																
16												1		5	1	25

1 - Annual Grassland
2 - Blue Oak Woodland
3 - Lowland Live Oak Woodland
4 - Mixed Chaparral
5 - Chamise Chaparral
6 - Black Oak Woodland
7 - Upland Live Oak Woodland
8 - Ponderosa Pine Forest
9 - White Fir Forest
10 - Jeffrey Pine Forest
11 - Red Fir Forest
12 - Juniper Woodland
13 - Lodgepole Pine Forest
14 - Subalpine Forest
15 - Subalpine Meadow
16 - Alpine Fell-fields

of leaf area. Dense broadleaf vegetation can be detected and often indicates greater site evapotranspiration and thus more mesic conditions. By means such as spectral mixture analysis, soil and leaf reflectance may be separated, and cells of high soil or rock exposure can be inferred as more xeric in their general condition. At the least, rock exposure occurs where there is no canopy or soil to obscure it, and represents lack of soil and no water holding capacity.

Vankat's diagram was not originally presented as a definitive deterministic model of the vegetation along the Sequoia National Park environmental gradient. In fact, the author noted that rather than being separated by distinct boundaries, most types intergrade along continua to a greater or lesser degree. In its original version, the distances separating the dashes used as borders were used to indicate the distinctiveness or ambiguity of the vegetation transitions. Thus, to improve our model, type overlap could be incorporated and a probabilistic rather than deterministic output would be desirable.

Stephenson (1988) showed the importance of soil depth upon species occurrences in the Sequoia Park region, and offered a different set of diagrams of potential vegetation for the higher elevations. With soils information for the region, the landscape could be segmented and the Stephenson diagrams employed to provide a potentially more accurate vegetation model.

CONCLUSIONS

The work presented in this paper only begins to address the complex task of constructing accurate vegetation models from gradient-type information. Though not able to account for such factors as hillslope position and soil moisture availability on a site basis, a very simple model in the form of a two dimensional look-up table was correct about half the time in a region of the southern Sierra Nevada. Models such as this type are of strong interest, since their simplicity and ease with which they can be altered to conform to new information (or projected climatic changes) make them powerful tools for the ecologist or land managers.

In the near future we will pursue this line of research in Yosemite National Park, using an approach which integrates potential vegetation models with digital remotely sensed data as well as other environmental variables which have important bearing on plant distributions. We will also look into refining such a model for creating potential vegetation maps for other segments of the Sierra Nevada ecoregion, to support work being done related to "gap analysis" -- locating important and unprotected sites of biodiversity in California.

ACKNOWLEDGMENTS

Ken McGwire contributed valuable input related to this work. This research was sponsored in part by NASA grant NAGW-1743 awarded to Jack Estes at the University of California, Santa Barbara; and by a grant from the National Fish and Wildlife Foundation.

REFERENCES

Agee, J. K., and J. Kertis, 1986. "Vegetation Cover Types of the North Cascades", Report CPSU/UW 86-2, National Park Service Cooperative Park Studies Unit, College of Forest Resources, University of Washington, Seattle, Washington.

Barbour, M. G., and J. Major, eds., 1977. Terrestrial Vegetation of California. Wiley-Interscience, N. Y.

Brown, D. G., 1991. "Topoclimatic models of an alpine environment using digital elevation models within a GIS", in *Proceedings of GIS/LIS'91*, Atlanta, Georgia, October 28-November 1, 1991, pp. 835-844.

Cibula, W. G. and M. O. Nyquist, 1987. "Use of topographic and climatological models in a geographical data base to improve Landsat MSS classification for Olympic National Park", *Photogrammetric Engineering and Remote Sensing*, **53**: 67-75.

Davis, F. W. and S. Goetz, 1990. "Modeling vegetation pattern using digital terrain data",*Landscape Ecology*, **4**: 69-80.

Dubayah, R., J. Dozier and F. W. Davis, 1990. Topographic distribution of clear sky radiation over the Konza Prairie, Kansas", *Water Resources Research*, **26**:629-690.

Franklin, J., T. L. Logan, C. E. Woodcock, and A. H. Strahler, 1986. "Coniferous forest classification and inventory using Landsat and digital terrain data", *IEEE Transactions on Geoscience and Remote Sensing*, **GE-24**: 139-149.

Kessell, S. R., 1990. "An Australian geographical information and modelling system for natural area management", *International Journal of Geographical Information Systems*, **4**: 333-362.

Küchler, A. W., and I. S. Zonneveld, eds., 1988. *Vegetation Mapping*. Kluwer Academic Publishers, Dordrecht, The Netherlands.

Parker, A. J., 1989. "Forest/environment relationships in Yosemite National Park, California, USA", *Vegetatio*, **82**: 41-54.

Peet, R. K., 1978. Forest vegetation of the Colorado Front Range: patterns of species diversity. *Vegetatio*, **37**:65-78.

Rundel, P. W., D. J. Parsons, and D. T. Gordon, 1977. "Montane and subalpine vegetation of the Sierra Nevada and Cascade Ranges", In: M. G. Barbour and J. Major, eds., Terrestrial Vegetation of California, p. 559-599, Wiley-Interscience, N.Y.

Stefanovic, P., and G. Wiersema, 1985. "Insolation from digital elevation models for mountain habitat evaluation", *ITC Journal*, pp. 177-186.

Stephenson, N. L., 1988. *Climatic control of vegetation distribution: The role of the water balance with examples from North America and Sequoia National Park, California.* Ph.D. Dissertation, Cornell University.

Strahler, A. H., J. Franklin, C. E. Woodcock and T. L. Logan, 1981. *FOCIS: A Forest Classification and Inventory System Using Landsat and Digital Terrain Data*, Report NFAP-255, USDA Forest Service, Houston, TX.

Twery, M. J., G. A. Elmes, and C. B. Yuill, 1991. "Scientific exploration with an intelligent GIS: predicting species composition from topography". *AI Applications*, 5: 45-53.

U. S. Geological Survey, 1987. *Digital Elevation Models, Data Users Guide Number 5*, Reston, Virginia.

Vankat, J. L., 1982. "A gradient perspective on the vegetation of Sequoia National Park, California", *Madroño*, 29: 200-214.

Whittaker, R. H., 1960. "Vegetation of the Siskiyou Mountains, Oregon and California", *Ecological Monographs*, 30:279-338.

Wieslander, A. E., 1935. "First steps of the forest survey in California", *Journal of Forestry*, 33:877-884.

Incorporating a Neural Network into GIS for Agricultural Land Suitability Analysis

Fangju Wang
School of Engineering and
Department of Computing and Information Science
University of Guelph
Guelph, Ontario, Canada N1G 2W1

ABSTRACT

Agricultural land suitability assessment involves analyzing physiographic characteristics to classify tracts of land in terms of their suitability. The currently used methods, including the method developed by the United Nations Food and Agriculture Organization (FAO) and the statistical pattern recognition method, have limitations which may lead to inaccurate results.

This paper presents a new approach in which the techniques of neural network and geographical information system (GIS) are integrated for land suitability analysis. Neural network is an effective tool for pattern analysis. By transforming the land suitability assessment problem into a pattern analysis problem, neural network can be used to achieve more accurate results. Incorporating a neural network into a GIS may strengthen the GIS's ability for data analysis.

INTRODUCTION

Information about agricultural land suitability is very useful for rural and regional planners and farm managers. In areas with dense population and limited arable land, such information is especially important for utilizing the scarce land resources in the most appropriate manner.

Agricultural land suitability assessment involves analyzing measurable quantitative physiographic characteristics to classify tracts of land in terms of their suitability. The suitability is assessed by the presence, absence or level of specified climatic requirements (temperature and rainfall), internal soil requirements (soil temperature, soil moisture, soil aeration, natural soil fertility, soil depth, soil texture, salinity), and external soil requirements (slope, flooding, and accessibility). For an area, we usually need to assess its "absolute" suitability and "relative" suitability. The former refers to its suitability for individual crops. The latter refers to its suitability with respect to a group of crops: which is the most suitable crop, which is the second most suitable crop, and so on.

Accurate suitability assessment requires collecting and processing a large variety of land characteristic data. Geographical information systems (GISs) may largely facilitate the assessment. In a GIS, land characteristic data can be stored in different data layers. By overlaying the layers, land characteristic data can be collected for the regions to be assessed. A GIS may thus provide an efficient means for data collection. Furthermore, to obtain accurate land suitability information from the data collected, a GIS must be equipped with an appropriate assessment

804

method. However, most of the commonly used methods have limitations which may lead to inaccurate results.

In this paper, a new assessment method is described, and some experimental results are presented and analyzed. This method is based on the techniques of artificial neural network. It may overcome some of the limitations of the currently used methods and provide more accurate analysis. Before introducing the new method, two of the currently used methods – the method developed by the United Nations Food and Agriculture Organization (FAO) (1976) and the statistical pattern recognition method – are discussed.

THE MOST LIMITING CHARACTERISTIC METHOD

The method developed by FAO typifies the most commonly used assessment methods. The main constituents of the FAO method are a collection of assessment criteria and a rating rule. For a given crop, the criteria classify the values of each land characteristic into different suitability classes, and the rating rule rates an area according to the area's most limiting land characteristic. Such methods are referred to as *most-limiting-characteristic* methods in this paper.

Let us take a variation on the FAO method as an example, which was developed by the Indonesian Center for Soil Research in conjunction with the FAO (CSR/FAO 1983). In this method, 15 land characteristics are taken into consideration and 4 suitability classes are defined: $S1$ (highly suitable), $S2$ (moderately suitable), $S3$ (marginally suitable), and N (not suitable). In the appendix, the assessment criteria for wetland rice are listed. When assessing suitability for an area, each of its characteristic data is rated according to the criteria, then the class of the most limiting characteristic is assigned as the area's suitability class.

A major limitation of the CSR/FAO rating rule is that a single characteristic may have too large an influence on the rated suitability and thus the *most-limiting-characteristic* rating rule may quite often generate inaccurate suitability information. For example, we have three areas: Area 1, Area 2 and Area 3. The rooting conditions of the three areas are all rated as $S3$, but most of the remaining characteristics of Area 1 are rated as $S1$, most of the characteristics of Area 2 are rated as $S2$ and most of the characteristics of Area 3 are rated as $S3$. According to the CSR/FAO rating rule, all the three areas are rated as $S3$. However, the three areas actually have quite different suitability. In practical situations, the most limiting land characteristic is unlikely to influence suitability in such a decisive way, unless the characteristic is extreme, such as very high (or very low) temperature.

An additional limitation is that the CSR/FAO rating rule can hardly provide accurate information about relative suitability, especially when an area's absolute suitability classes for different crops are the same. In many planning tasks, the information about relative suitability is very important to decide the most appropriate land use.

A STATISTICAL PATTERN RECOGNITION METHOD

The core components of the statistical pattern recognition-based assessment method include vector representation of land characteristic data and feature space partitioning.

In pattern recognition, a pattern (or object) can be considered as a vector $\mathbf{x}$:

$$\mathbf{x} = (x_1, x_2, ..., x_n)^T$$

805

where n is the number of features (or attributes) which are useful for pattern recognition, x_i $(1 \leq i \leq n)$ is the value of the ith feature, and T denotes vector transpose. When a geographical area is represented as a vector, the elements are values of the area's land physiographic characteristics. All the vectors form a n-dimensional feature space. Each feature defines a dimension. An effective approach for pattern recognition is *feature space partitioning*. In this method, the feature space is partitioned into regions each of which is associated with a pattern class. A pattern belongs to a class if and only if it resides within the corresponding region.

Feature space can be partitioned by using representative vectors to preresent classes and a distance metric to measure similarity. It is assumed that we have c predefined pattern classes. μ_i, the representative vector of class i $(1 \leq i \leq c)$ is of the form

$$\mu_i = (\mu_{i1}, \mu_{i2}, \ldots, \mu_{in})^T. \tag{1}$$

Euclidean distance is among the most commonly used distance metrics for measuring similarity between vectors. The Euclidean distance between vector $\mathbf{x}$ and μ_i is

$$d_E(\mathbf{x}, \mu_i) = (\mathbf{x} - \mu_i)^T (\mathbf{x} - \mu_i). \tag{2}$$

The smaller the distance, the more similar $\mathbf{x}$ is to μ_i in terms of the features. In statistical pattern recognition, the more similar a pattern is to μ_i, the more likely it should be classified into class i.

A decision rule can be formulated as

$$\mathbf{x} \in i \quad \text{iff} \quad d_E(\mathbf{x}, \mu_i) < d_E(\mathbf{x}, \mu_j) \quad \forall j \neq i. \tag{3}$$

The vectors (points) equidistant to class i and class j, i.e., the $\mathbf{x}$'s satisfying $d_E(\mathbf{x}, \mu_i) = d_E(\mathbf{x}, \mu_j)$, form the decision surface which separates the two classes. All the decision surfaces partition the feature space.

It can be observed from Equation (2) and (3) that all the features make equal contributions in classifying a pattern. A statistical pattern recognition method enables us to assess an area's suitability based on all of its land characteristics instead of the most limiting one. By means of a similarity measurement, it can provide more accurate information about relative suitability (Wang et al, 1990).

However, the statistical pattern recognition method has the following shortcomings: (1) It can hardly handle the situations where a single characteristic does have a decisive influence on an area's suitability. (2) It can hardly be applied to features of unordered values, for example, soil texture, because it is meaningless to measure a distance between two soil types.

ARTIFICIAL NEURAL NETWORK

The techniques of neural network can be used to overcome the limitations of the two methods, and produce more desirable results.

Network Topology and Processing Functions

A neural network interconnects processing units in imitation of the neurons in the human brain. A neural network can be used as a pattern recognition tool. It can be taught to classify patterns in a desired way (Schalkoff, 1992).

The elementary processing units in a neural network are called *nodes*. The nodes are organized in *layers*. The neural networks developed in this research are *feedforward, back propagation neural networks.* In such a neural network, there is one *input* layer, one *output* layer, and zero or more *hidden* layers between. Usually only the nodes in neighbouring layers are connected. Data are input at the input layer, move *forward* in one direction through successive layers, and are output at the output layer. When a neural network is used for pattern recognition, each output node corresponds to a predefined class. For a pattern presented to the network, each output node produces a value. The pattern is classified into the class which corresponds to the node whose value is in a prespecified range. The input pattern is represented as a vector. An input node takes an element of the vector. Figure 1 illustrates a three-layer neural network.

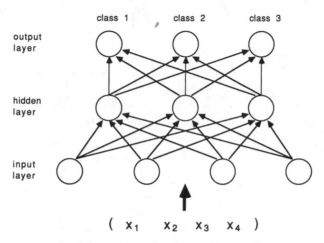

Figure 1: A three-layer neural network.

The connections between nodes serve as data paths. Each connection is associated with a *connection strength*. All the nodes, except the input nodes, performs the same two functions: collecting the activation of the nodes in the previous layer and generating an activation (see Figure 2). An input node takes input data and pass them as its own activation to the nodes in the successive layer. The following is the collection function and activation function used in this research.

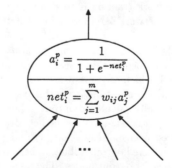

$$a_i^p = \frac{1}{1 + e^{-net_i^p}}$$

$$net_i^p = \sum_{j=1}^{m} w_{ij} a_j^p$$

Figure 2: The processing functions of a node.

The collection function of node i for pattern p is

$$net_i^p = \sum_{j=1}^{m} w_{ij} a_j^p \qquad (4)$$

where m is the number of nodes in the previous layer, w_{ij} is the strength of the connection from node j in the previous layer to node i (see Figure 3), a_j^p is the activation of node j for pattern p. (Superscript p denotes "for pattern p".)

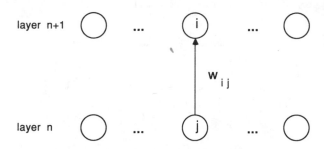

Figure 3: The connection between two nodes.

The activation of node i for pattern p is

$$a_i^p = \frac{1}{1 + e^{-net_i^p}}. \qquad (5)$$

The activation of a hidden node is passed to nodes in the successive layer, and the activation of an output node is an output value.

Before using a neural network for pattern recognition, we have to teach it how to classify patterns. This process is called *training*. A trained network *recalls* what it learned in the training process to classify unknown patterns presented to it.

Network Training and Recall

The task of neural network training is to determine appropriate connection strengths in order that the network can classify input pattern vectors into desired classes.

There are two methods for network training: *supervised* training and *unsupervised* training. The former is used in this research. For supervised training, a set of *training data* are needed, which are "input – desired output" pairs.

In a training process, the connection strengths are adjusted to minimize the differences between the actually calculated output values and the desired output values. When a pattern vector is presented to the network, the actually calculated output values are produced by using Equation (4) and (5) at each nodes. The differences between the calculated output values and the desired output values are used to calculate *node errors* for output nodes. The node errors are used to adjust connection strengths between the last hidden layer and the output layer. Then, the output node errors are used to calculate node errors of the last hidden layer. The

808

node errors of this layer are used to adjust connection strengths between the second last hidden layer and the last hidden layer. The same procedure is applied until the connection strengths between the input and the first hidden layer are adjusted. Then another training pair is presented to the network. After all the training pairs have been presented, total output error or maximum output error is examined. If the total (or maximum) output error is less than a threshold, the training is completed and the network is ready for pattern classification. Otherwise, the network is trained again using the same training pairs.

The node error of output node i for pattern p is

$$\gamma_i^p = a_i^p(1 - a_i^p)(t_i^p - a_i^p) \tag{6}$$

where t_i^p is the desired output value of node i for pattern p and a_i^p is activation, that is, actually calculated output value of node i for pattern p.

The node error of hidden node j for pattern p is

$$\gamma_j^p = a_j^p(1 - a_j^p)\sum_{i=1}^{m}\gamma_i^p w_{ij} \tag{7}$$

where m is the number of nodes in the successive layer, γ_i^p is the node error of node i in the successive layer for pattern p and w_{ij} is connection strength between node j and node i.

The update rule for w_{ij} is

$$\Delta^p w_{ij} = \epsilon(\gamma_i^p a_j^p) \tag{8}$$

where ϵ is the *learning rate* which can be used to control the speed of training.

The training algorithm is based on *gradient descent*. In training a network, the objective is to minimize error

$$E^p = \sum_{i=1}^{c}(t_i^p - a_i^p)^2. \tag{9}$$

To adjust connection strength w_{ij}, gradient $\partial E/\partial w_{ij}$ is calculated. The strength adjustment Δw_{ij} is set proportional to $-\partial E/\partial w_{ij}$. For details of the gradient descent method see reference [3].

Since the information about errors propagates backward in adjusting the connection strengths, such a network is called a *back propagation neural network*.

The process for a trained neural network to classify unknown patterns is referred to as *recall*. When an unknown pattern is presented to the network, output values are produced by applying Equation (4) and (5) at each nodes. The pattern is classified into a class if the corresponding output node produces a value which is within a prespecified range.

NEURAL NETWORKS FOR LAND SUITABILITY ASSESSMENT

In this research, a group of neural networks were developed for absolute suitability assessment and a network was developed for relative suitability assessment.

A Neural Network for Absolute Suitability Assessment

To assess absolute suitability, a neural network is created for each crop. The neural networks for absolute suitability *evaluate* land characteristics according to the CSR/FAO criteria. They were trained to *rate* suitability according to the following rule: If all the characteristics of an area are rated as suitable (i.e., as $S1$, $S2$ or $S3$), rate the area based on its overall land quality; If some characteristics are rated as unsuitable, rate the area as unsuitable.

The networks for different crops are almost the same in network topology and training process. The difference is in training data selection. Not loss of generality, only the neural network for wetland rice is described.

Network topology

In creating a neural network, the first step is to determine network topology, that is, the number of layers and the number of nodes in each layer. The neural network for wetland rice has three layers: an input layer, an output layer, and a hidden layer.

The number of input nodes are determined by the length of input vectors. The length depends on the data representation scheme used for coding the input. A commonly used coding approach is to code the input by a binary coding scheme.

Three types of land characteristic values are involved in land suitability assessment: numerical values, ordered textual symbols, and unordered textual symbols. To make the data suitable for neural network classification, textual symbols in the vector elements were converted into numerical values. To ensure that all the characteristics may make equal contributions, the values were normalized into the same range.

In this research, the ordered textual symbols were converted into integer numbers in the range between 0 and 15, and then the numbers were converted into binary numbers. For example, "very low", "low", "medium", "high" and "very high" were converted into 2, 5, 8, 11, and 14, and then converted into binary numbers 0010, 0101, 1000, 1011, and 1110. The above numbers were selected because any two of them had at least two binary digits which are different. The unordered soil types were converted in a similar way: Each soil type was represented by an integer number and then the number was converted into a binary number. The numerical characteristic values were first normalized into the range from 0 to 15, converted into integers, and then converted into binary numbers. After the conversion, an area is represented as a vector with each element being a binary digit. The number of input nodes is selected equal to the number of binary digits. The following illustrates the conversion process:

$$
\begin{bmatrix} \dots \\ medium \\ \dots \end{bmatrix} \rightarrow \begin{bmatrix} \dots \\ 8 \\ \dots \end{bmatrix} \rightarrow \begin{bmatrix} \dots \\ 1000 \\ \dots \end{bmatrix} \rightarrow \begin{bmatrix} \dots \\ 1 \\ 0 \\ 0 \\ 0 \\ \dots \end{bmatrix}
$$

As mentioned before, the number of output nodes is equal to the number of predefined classes. However, in developing a network for wetland rice, the number of output nodes cannot be simply chosen as 4. One reason is that annual average temperature and surface soil pH both have two value ranges classified into each class, except for $S1$. For example, annual average temperature ranges $30°C - 32°C$

and $24°C - 22°C$ are both classified into $S2$, and surface soil pH ranges $7.1 - 8.0$ and $5.4 - 4.5$ are both classified into $S2$. Therefore, we need four subclasses for $S2$. Each subclass is for a combination of the two characteristics. Similarly, we need four subclasses for $S3$ and four subclasses for N. Thus the minimum number of output nodes is 13 $(= 1 + 4 \times 3)$. since some special cases must be taken into consideration, more subclasses are needed for N. For example, if an area has soil texture of gravels or sands, it is unsuitable for wetland rice, even the remaining characteristics are very good. In this case, the classification should be based on the soil texture only, instead of the average land quality. To classify such areas into class N, a subclass of N can be defined. Constructed in this way, a neural network can be easier to train and classification results are more accurate.

Training data preparation

The training data for the wetland rice neural network were obtained from the table of assessment criteria. The data were prepared in the following steps: (1) substituting each value range [A, B] in the table by its central value $(A+B)/2$, and substituting each inequality, for example, "$< C$" or "$\leq C$", by C and one or more possible extreme values which are less than C, (2) for each class or subclass, using the values in the table to form all the possible land characteristic vectors, (3) converting the vector element values into binary, and (4) attaching a class name to each of the vectors to form "land characteristics – desired class" pairs.

Recall

When an unknown vector is presented to a trained network, the network classifies it into the most likely class according to its overall land quality or a single land characteristic.

It can be observed from Equation (5) that the range for output values is $[0,1]$. A threshold θ $(0 \leq \theta \leq 1)$ is needed. An output value within $[\theta, 1]$ indicates that the input pattern should be classified into the corresponding class. If $1 - \theta$ is too small, some areas may be left unclassified.

A Neural Network for Relative Suitability assessment

The neural network for relative suitability assessment is created in a similar way. It has four output nodes which corresponds to the four crops involved: wetland rice, soybean, sugar cane and pasture. The training data for a crop is obtained from the $S1$ column of the crop's assessment criterion table. (The data from other columns, for example $S2$, can also be used. But the $S1$ column data generate the most reasonable results.)

When a land characteristic vector is presented to the network, each output node produces a value. The relative suitability is ranked according to the output values. The crop corresponding to the node which produces the highest value is the most suitable one, the crop corresponds to the node which produces the second highest value is the second most suitable one, and so on. This is based on the fact that the value of an output node is proportional to the similarity between the input vector and the $S1$ characteristics of the corresponding crop.

Implementation

The neural networks were coded in the C programming language. Each network is associated with three C programs. They are used for training, recall, and interpreting classification results, respectively. The connection strengths determined by the training program are stored in a data file. The recall program reads in the strengths before it performs suitability assessment.

The central data structures of a neural network are three one-dimensional arrays and two two-dimensional arrays of floating numbers. Each of the one-dimensional arrays is for a network layer. The ith element of such an array is used to store activation values of node i in the corresponding layer. Each of the two-dimensional arrays is for connections between two layers. The element with subscript $[i][j]$ is used to store connection strength w_{ij}.

The training program applies the algorithm described above to determine connection strengths. The recall program uses Equations (4) and (5) to produce output values. The interpreting program translates the output values from the recall program into absolute or relative suitability classes.

The GIS was created by using ARC/INFO[1] software tools. The geographical data were stored in coverages. ("Coverage" is an ARC/INFO terminology for data layer.) In an assessment task, the data required for suitability assessment were collected by overlaying the coverages, and then transferred to the neural networks.

The neural networks were incorporated by using a macro (ARC/INFO program unit). The macro includes commands for calling the C programs, and also command sequences for coverage overlay, data transfer, and result display. The macro was designed to work in an interactive way. It allows the user to select different neural networks to work with. Data communication between the GIS and neural networks is achieved by using two data files. One file is used to transfer the land characteristic data to the neural networks and the other to transfer the assessment results back to the GIS for display and analysis.

RESULTS AND ANALYSIS

Experimental results have been obtained by applying the neural networks to practical data.

Study Area

Data used in this paper were collected from the lower part of the Cimanuk watershed area on the north coast of west Java, Indonesia.

Data on physiographic characteristics central to land suitability classification in this area were digitized as five separate coverages: temperature, rainfall, soil characteristics, soil salinity and slope.

For assessing land suitability, the five coverages were overlaid. The composited coverage consists of 642 areas. Each area is associated with 13 land characteristic values. In the following, the results of assessing absolute suitability for wetland rice and relative suitability for the four crops are presented and analyzed.

Absolute Suitability

The neural network method may assess absolute suitability based on overall land quality if all the characteristics are considered suitable. It may also assess the suitability based on a single characteristic if the characteristic has a decisive influence. In the following discussion, four areas are taken as examples to illustrate this point. Their characteristics and the characteristics' suitability classes are

---

[1] ARC/INFO is a product of the Environmental Systems Research Institute, Inc.

listed in Table 1. The areas' suitability classes rated by using different methods are listed in Table 2.

area#	AAT	AAR	NDM	SRC	ST	RD	CEC	pH	TN	P_2O_5	K_2O	SAL	SLP
219	27.0	1500	5.51	W	MACL	125	VH	4.25	VL	M	VL	5.55	6
	S1	S1	S1	S3	S3	S1	S1	S3	S3	S3	S3	S3	S3
309	27.0	1500	5.51	P	STCL	125	VH	4.75	L	L	L	4.46	0
	S1	S1	S1	S2	S2	S1	S1	S2	S2	S3	S2	S2	S1
326	27.0	1750	5.51	MW	SICL	125	VH	6.75	M	L	H	0.0	0
	S1	S1	S1	S1	S2	S1	S1	S1	S1	S3	S1	S1	S1
8	27.0	1750	5.51	VP	STCL	125	VH	7.75	M	M	VH	15.0	0
	S1	S1	S1	S1	S2	S1	S1	S2	S1	S3	S1	N	S1

Symbols are explained in alphabetic order:

AAR – Average annual rainfall	AAT – Annual average temp.	CEC – Cation exchange capacity
H – High	L – Low	LO – Loam
M – Medium	MACL – Massive clay	NDM – Number of dry months
P – Poorly drained	RD – Rooting depth	SAL – Salinity
SICL – Silty clay	SLP – Slope	SP – Somewhat poorly drained
SRC – Soil drainage class	ST – Soil texture	STCL – Structured clay
TN – Total nitrogen	VH – Very high	VL – Very Low
W – Well drained		

Table 1: Characteristics of the selected areas.

area#	class by neural network	class by the CSR/FAO method	class by the statistical PR method
219	S3	S3	S3
309	S2	S3	S2
326	S1	S3	S1
8	N	N	S2

Table 2: Suitability classes assessed by different methods.

Area 219, 309, and 326 all have medium or low availability of surface P_2O_5 which are rated as $S3$. The remaining characteristics of the three areas are different. The majority of the characteristics of Area 219 are rated as S3, those of Area 309 are rated as S2, and those of Area 326 are rated as S1. Thus the three areas have different actual suitability for wetland rice in terms of their overall land quality: Area 326 is the best and Area 309 is better than Area 219. However, the CSR/FAO rating rule indiscriminately rates all the three as S3 simply because their most limiting characteristics (P_2O_5 availability) are rated as S3. Obviously, this rating is inaccurate. The neural network rates them as S3, S2 and S1 respectively, which are more consistent with their overall quality.

Most land characteristics of Area 8 are rated as $S1$ or $S2$. When the area was assessed in terms of its average quality (by the statistical pattern recognition method) it was rated as $S2$. However, this area has a very high subsoil salinity value (15 mmhos/cm) which makes the area unsuitable for wetland rice. This area thus should be rated as N even if the remaining characteristics are very good. The neural network did classify it into the right class.

Relative Suitability

The neural network gives each area a rank of wetland rice, soybean, sugarcane, and pasture to indicate the relative suitability. Reasonable results have been obtained. For example, most of the areas with very high soil salinity values have the highest output values in class "sugar cane". Sugar cane has the strongest resistance to soil salinity among the four crops. If an area has a very high salinity value but its remaining characteristic values are suitable for several crops (including sugarcane), the most suitable crop for this area should be sugarcane. This is commensurate with the results from the neural network analysis. For example, the neural network correctly ranked sugar cane as the most suitable crop for Area 8 which has very high salinity. Most of the the areas with very low K_2O have the highest output values in class "soybean". Soybean has lower requirement on K_2O than other crops. When the remaining characteristics are suitable for several crops, it is optimal to use an area with a very low K_2O to grow soybean. This is also consistent with the experimental results. For example, soybean was ranked as the most suitable crop for Area 219 which has very low K_2O availability.

In addition to identifying the most suitable crop, the neural network method ranks the crops in terms of their suitability for an area. This enables us to make optimum multiple purpose use of the valuable land resource.

CONCLUDING REMARKS

The experimental results obtained in this research have shown that the techniques of neural network may provide effective tools for agricultural land suitability assessment in a GIS context. A neural network can be constructed and trained to assess the suitability in a way desired by the analyst. For example, the neural networks developed in this research for absolute suitability may combine the desired features of both the CSR/FAO method and the statistical pattern recognition method and apply the rules in right situations. In this sense, the neural network techniques are more effective and flexible for land suitability assessment.

ACKNOWLEDGEMENTS

This work was supported by the Natural Sciences and Engineering Research Council (NSERC) of Canada research grants. The author would like thank Dr. G. B. Hall and Mr. Subaryono for the testing data.

REFERENCES

[1] CSR/FAO, 1983, *Reconnaissance and land resource surveys 1: 250,000 Scale atlas format procedures*, Bogor, Indonesia, Centre for Soil Research, Ministry of Agriculture, Government of Indonesia (United Nations Development Program and Food and Agriculture Organization).

[2] FAO, 1976, *A framework for land evaluation*, FAO Soils Bulletin No. 32, FAO Rome and ILRI, Wangeningen, Publication No. 22.

[3] Schalkoff, R, 1992, *Pattern Recognition: Statistical, Structural and Neural Approaches*, New York, NY: John Wiley & Sons, Inc..

[4] Wang, F., G.B. Hall, and Subaryono, 1990, 'Fuzzy information representation and processing in conventional GIS software: database design and application', *International Journal of Geographical Information Systems*, Vol. 4, No. 3, 261-283.

APPENDIX

Assessment criteria for Wetland rice

Land Characteristics grouped by Land Qualities	Land Suitability Ratings			
	S1	S2	S3	N
t - Temperature Regime				
1. Annual Average Temp.	25-29	30-32	33-35	> 35
(°C		24-22	21-18	< 18
w - Water Availability				
1. Dry months (<75mm)	0-3	3.1-9	9.1-9.5	> 9.5
2. Average Annual rainfall	> 1500	1200-1500	800-1200	< 800
(mm)				
r - Rooting Conditions				
1. Soil drainage class	somewhat poor, moderately well	very poor, poor	well	somewhat excessive, excessive
2. Soil texture (surface)	sandy clay loam, silt loam, silt, clay loam	sandy loam, loam, silty clay loam, silty clay, structured clay	loamy sand, massive clay	gravels, sands
3. Rooting depth (cm)	> 50	41-50	20-40	< 20
f - Nutrient Retention				
1. CEC me / 100g soil	≥ medium	low	very low	
2. pH (surface soil)	5.5-7.0	7.1-8.0	8.1-8.5	> 8.5
		5.4-4.5	4.6-4.0	< 4.0
n - Nutrient Availability				
1. Total N	≥ medium	low	very low	
2. Available P_2O_5	very high	high	medium-low	very low
3. Available K_2O	≥ medium	low	very low	
x - Toxicity				
1. Salinity mmhos/cm	< 3	3.1-5	5.1-8	> 8
s - Terrain				
1. Slope %	0-3	3-5	5-8	> 8
2. Surface stoniness	0			≥ 1
3. Rock outcrops	0		1	≥ 2

Source: CSR/FAO (1983)

CONVERSION AND INTEGRATION – CHALLENGES FOR GIS IN ELECTRIC UTILITIES

Caroline Williams, Senior Software Engineer
Computervision GIS,
Hohlstrasse 534,
8048 Zurich, Switzerland.
41 1 432 21 41

ABSTRACT

This paper will focus on the issues encountered in the design and implementation of data conversion and subsequent data collection processes as part of a large electric utility application. The project was recently completed as a turn–key system using the application development facility of an existing GIS.

The issues of data conversion included the challenge of converting a large volume of existing electrical distribution data of a utility that serves millions of customers. The difficulties of integrating the schemas of the source and target models became evident. Error and data loss were guarded against with effort; it was a task that would be reduced by an industry standard. In this respect, the role of standards in the exchange of information is pertinent, and will be discussed with attention to the specific difficulties encountered.

The integrity of the data after conversion and during capture adheres to a data model designed for electrical networks. The data model includes high, medium and low tension networks. Special concerns for electrical utilities, such as connectivity and voltage level, are further quality controlled by the application processes. Emphasis will be centered on the special requirements of maintaining electrical networks.

INTRODUCTION

The topic of this paper is the design and development of part of a system recently developed for Union Fenosa, one of Spain's largest electricity distribution companies. The project was based on SYSTEM 9 using it's application development facilities. Fenosa is one of Spain's largest utilities serving over 2 million customers and one of the world's most progressive. They are progressive because they're part of a recent trend of utilities to re–engineer their processes using GIS. Many utilities understand that their existing systems have a strong demand for spatially related data and are looking for integrated and long term solutions for their work flow rather than to simply automate and speed up manual processes (Weber,

1992). The first step of the project was to establish functional specifications for the system and define the data model. The data model is of underlying importance to the system's processes and will be discussed first. The next two sections will describe the task of transferring data from one system to SYSTEM 9 and the processes of data update, respectively.

THE DATA MODEL

Views

A view restricts a data model to the information that is relevant to the viewer (Data, 1986) In the case of Union Fenosa there are four views. The positional view shows the location of cables and stations. The detailed view shows the location of cables in conduits and the layout of the conduits. In this view, cables are seen differently. These two views are different since the data has different relevance to the viewer. In the detailed view the position of a cable is only important with respect to the conduit. In the positional view the locational of the cable is important with respect to other criteria, such as location of substations.

Two views which will not be described here are the orthogonal and urban views. The orthogonal view models the schematics of an electrical network. The urban view contains buildings and roads. The urban view is used as background data only.

Modelling the Views

The conceptual model of SYSTEM9 is feature oriented and relational. Relations exist between features and their primitives. It is also possible to model logical relations using composite feature classes. As an example, the simple feature classes house, shed and laneway can be logically grouped together as a composite class, allotment. The four views of the electrical system were modelled using composite classes with relations existing between the views. In this way the different views can be combined in one model.

Positional View

The data model designed for this project includes the generation, transmission, distribution, and utilization levels of operation of an electrical system. Feature classes were created for two classes of substations in the transmission system and two classes of lines, high voltage and subtransmission. In the distribution part of the system there were, again, modelled two classes of lines (or feeder) primary and secondary and two classes of centers. Further entities were modelled in classes such as switches (divided by voltage level), poles, and connections. The overlap of the operational divisions were modelled using composite classes, where, for example, the transmission substation as a composite class can logically reference a high voltage line coming into the substation and a subtransmission line coming out. This part of the data model, concerning the lines and stations, is called the positional view (see Figure 1).

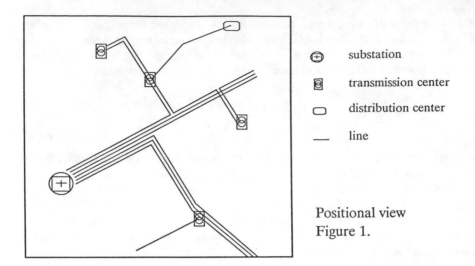

substation

transmission center

distribution center

line

Positional view
Figure 1.

Detailed View

The detailed view models the positions of cables in conduits with cables and the cross–section of the conduit as node features. The conduit itself is a linear feature. The perimeter of substations or centers are modelled as surface features (see Figure 2). In the positional view, substations and centers are node features. The views are integrated using a composite feature class that references both, in the diagram it is Cable_Conduit (see Figure 3).

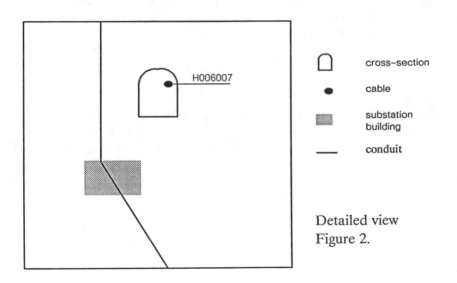

H006007

cross–section

cable

substation
building

conduit

Detailed view
Figure 2.

Integrated Data Model
Combining these four views of an electrical system in one data model using composite features minimises data duplication. The composite features logically reference different views of the same entity and keep the common attributes. For the sake of simplicity, the data model is generalized here to include only one class of substation, one class of line in the positional view and the corresponding classes in the detailed view (see Figure 3).

The Data Model

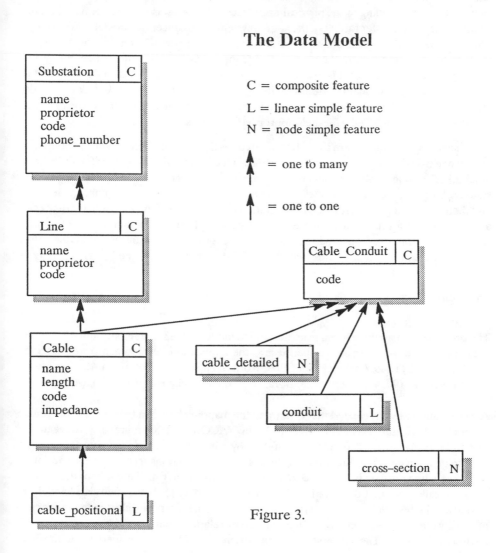

Figure 3.

819

Schema Mapping

After the definition of the data model the next major objective of this project was to take graphical and attribute data from an existing system and transfer the data to SYSTEM9. Looking at the three levels for data transfer suggested by Marini, physical, structural and semantic, it was discovered that the most difficulty would be at the semantic level(Marini, 1989). The understanding of the physical and structural levels was satisfactory based on the documented data transfer format of the source system . The conceptual models of the source and target systems differed. Both were feature oriented, however, the source system encoded features in layers. As well, in the source model, there was a combination of geometry and topology based features. That is, features could be points or nodes. In the target model the feature would be a node feature. Additionally, in the source system, text and labels could exist independently. In the target model, text is integrated as attributes of features. The text and labels should to be merged with their appropriate features.

In the source system there were a number of relations that were modelled using attribute values. In one case, it is a reference attribute whose value is the value of another feature's code attribute. The code attribute should be maintained in the target data model for use by other systems, however, the reference attribute can be eliminated by the use of composite features and their implicit logical relations. These relations are stored and maintained internally and automatically in a relational database. To further complicate the situation these relations were stored in an external database and were not part of the data transfer files. Other cases were also based on attribute value and in SYSTEM9 could be replaced by composite relations.

These differences in the models had to be overcome in the schema mapping process.

The Loading Process

The processes to map the schema from the source model to the target model and transfer the data consisted of programs developed using the application development tools of SYSTEM9, the SYSTEM9 Analytical Tool Box, Korn Shell scripts, LEX and YACC programs, and the SYSTEM9 Data Interchange Librararies, the Interchange Input (ii) (see Figure 4).

A specification file was created that contained the mapping between the layers in the source system and the feature classes in SYSTEM9. The YACC and LEX programs were created to recognize special tokens defined to identify the layer information and the feature classes and what to do upon encountering these tokens. As well, the programs recognized the special tokens that define the data in the source data file. Upon encountering these tokens the ii_ functions are called to input the data into SYSTEM9 according to the defined data model. Included also in the specification file is the feature class of the composite class of the case one to one simple to composite relations. There are other relations that have to be loaded that will be described further. The data is stored in an intermediate file and then loaded into the database. At this point, all geometry has been loaded in the database in the appropriate feature classes. Composite relations have been formed if they were specified in the specification file. Some attributes have also been loaded.

The attribute relations in the source data model can be replaced by SYSTEM9 composite features. This is done by a series of scripts, programs using ATB functions, and ATB tools atb_pf_composite_upload and atb_pf_upload that create SYSTEM9 composite features and transfer the attribute information. This data is loaded from the external database.

After these processes are carried out all the possible data that can be loaded has been loaded. What remains is data in the external database that is not intended to be loaded into the SYSTEM9 target database. This is data that has no graphical representation and will remain in an external database. This data is maintained by the turn-key system and is beyond the scope of this paper. Other data that is not loaded has errors and has been rejected. The types of errors will be further described in the next sections.

Data Loss
Data that was not loaded is data that failed to meet the topological requirements for the SYSTEM9 data model. For example, lines with less than two unique points are rejected. Polygons with less than 4 points are rejected. Composite features were formed using attributes from the source data to make logical relations in the target data. In some cases, data was lost due to missing attribute data in the source. In this way bad data is filtered out and not integrated into the target database.

Data Integrity
After the data conversion process, data that was not topologically connected became connected due to snapping procedures in the loading. Lines were connected by a common node. This is important for connectivity in the electrical distribution network.

Data Transfer Standard
In this case, a data transfer standard would eliminate the difficulties of mapping the schemas. The choice of the YACC and LEX tools to parse specification files for the data transfer simplified this schema mapping process.

The choice of transferring data was largely preferrable to capturing data as was done by San Diego Gas & Electric in California whose re-engineering was started as early as 1977. In their case, data conversion required 100 digitizers and 30 support staff for four years (GIS WORLD, April 1992). In these times, data is transferred more often than collected, naturally, when the choice exists. The evolution of GIS systems allows for this and a data transfer standard would simplify the process. The evolution of transfer standards, requires, as evolution does, time. In the meantime, data transfer is possible using existing tools and data formats.

821

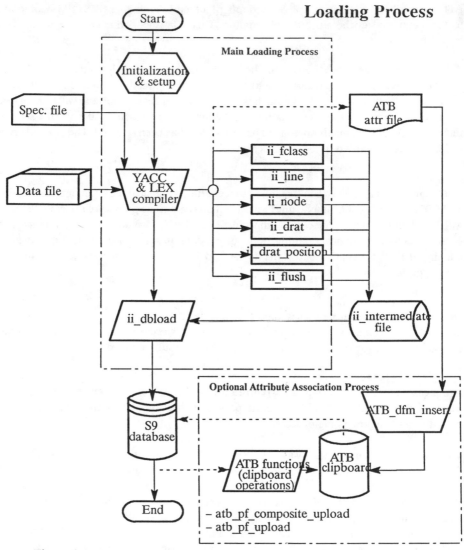

Loading Process

Figure 4.

Once data is loaded into the target system it's manipulation does not stop. Data can be queried, changed, added to or deleted. This task is ongoing and is carried out using a data capture and edit application that maintains data based on the data model.

DATA CAPTURE AND EDIT

Quality Control

For the purposes of other applications and systems sharing the data of the electrical network it is necessary for the data capture and edit procedure to maintain certain important qualities of the data. One quality is the connectivity and another is the correct voltage level. These things can be controlled by the application and the data model and will be discussed here. The importance of this quality control is realized when the data is used by the connectivity processes. This data is found in the positional view.

Data Collection

To control the connectivity of the electrical network in this electrical distribution application it was necessary that the user never interrupt a network without continuing connectivity. A center to be added should be connected to an existing line. This continues to make possible the flow of electricity since a topological connection is made. Making a topological connection to a line splits the line into two and makes a logical relation with a common node. With no application control a user could add a center unconnected.

Application control was possible using a macro language called Application Control Language (ACL). Normally, a user of the capture and edit application would have a choice of creating new points or connecting. This is communicated by which button of the input mouse is used for input. A GREEN button implies a new coordinate is being digitized. A BLUE button is a request for a connection to an existing object.

As part of the application development environment ACL scripts control the user's actions. In the case of this turn–key project, the ACL script controlled the button code by necessity. A BLUE button code was always enforced when the operation required a connect.

To further maintain the electrical distribution network it is necessary to maintain the correct voltage levels and the correct relations of the composite features according to the data model. To do this the ACL scripts controlled the current and valid feature classes and their entry. First, this was done by letting the user select the current voltage level. In this data model there are four. They are the four levels of operation, generation, transmission, distribution and utilization. Based on this level of operation, only certain feature classes are valid for entry. The ACL script controls this by only allowing these feture classes to be captured.

Data Editing

By the same token editing operations are controlled by the ACL scripts. It is not possible to connect incompatible feature classes if they don't "overlap" in the data model. The meaning of "overlap" in this context is that of the operation levels of electrical systems. That is, the generation level overlaps with the transmission level at the substation. It is not possible to remove objects when it causes incompatibilities in the relations.

CONCLUSIONS

This project was developed with the intent of providing an integrated environment for the electrical distribution utility. The data was loaded from data files and transferred to match the target data model. A data transfer standard would simplify this to some extent, however, some additional data transfer processes would still be necessary. The application processes were designed to control the data according to the data model and it's relations. The data will be further used for connectivity applications. The implementation of this project is part of a trend towards integrated systems and shared data.

REFERENCES

Date, C.J, 1986, An Introduction to Database Systems, Addison–Wesley Publishing Company, Inc., Reading, Massachusetts.

Marini, Mauro., 1989, "Multilevel Approach to Geographical Data Transfer", GIS/LIS '89 Proceedings , Orlando, Florida, Vol. 2, pp. 478–484.

Pansini, Anthony J., 1992, Electrical Distribution Engineering, The Fairmont Press, Inc., Lilburn, Georgia.

Vaughn, Fred D., 1992, Utility Cuts Costs, Speeds Service with AM/FM: GIS WORLD, April 1992, Vol. 5, No. 3, pp. 62–66.

Weber, Stan P. , 1992., GIS Strategies for Utilities; A GIS Must Be Designed to Integrate with Existing Systems: GIS EUROPE, May 1992, Vol. 1, No. 4, pp. 42–43.

OBJECT-ORIENTED MODELS OF SPATIOTEMPORAL INFORMATION

MICHAEL F. WORBOYS
Department of Computer Science
Keele University
Keele, Staffs ST5 5BG UK
email: michael@cs.keele.ac.uk

ABSTRACT

Geographic information has three components: spatial, attribute and temporal. An essential requirement of future geographic information systems (GIS) will be the ability to handle the temporal dimension. This requirement will involve the modelling and management of spatiotemporally referenced information. Object-oriented database technology is set to play an important role in the management of GIS since relational technology has limitations with regard to geographi information, due to the complex structure of spatiotemporal data types.

This paper describes the author's research into an object-oriented model for spatiotemporally referenced information. The approach extends earlier work on a generic model for spatially referenced objects. The object-oriented approach to data modelling is described, followed by a discussion of an object model for spatiotemporal information.

GENERIC MODELS OF GEOGRAPHIC DATA

For any type of information system there are essentially two distinct kinds of data modelling activity: the first is specific and relates to the customization of a particular system; the second is generic and involves the formation of models which can be used for many systems. The formation of specific models and customization of specific systems are the tasks of those engaged in the design and implementation of systems for specific applications. Such modelling will be more efficient if it can draw on generic models. The construction of such generic models, using the object-oriented approach, is the subject of this paper.

High-level conceptual models of spatial information can be conveniently divided into layer-based and object-based models. A *layer-based model* treats spatial information as being the distribution of a set of attributes over a region. A *spatial framework* is a covering of the given region with a finite tessellation. The tessellation used in a spatial framework is often regular (but this is not a requirement), and will often be a grid of squares (Ordnance Survey Grid or latitude/longitude) or triangles. Measurements of, say, topographical altitude, with respect to this framework constitute a single layer of our model. Another layer may consist of the variation of temperatures over the same region. In general, there will be many layers in our database, each with respect to the same underlying spatial framework. The layer-based approach can be extended to spatiotemporal geographic information

825

by extending the layer concept to layers based upon an underlying spatiotemporal framework. However, this is not a direction which is followed in this paper.

An *object-based* model structures geographic information as collections of entities, which inhabit a spatial (or, more fully, a spatiotemporal) framework. In this case, the reference is from the entity set to the spatial framework. This contrasts with layer-based modelling, where the reference is reversed. The spatial framework may be provided by the well-known arc-node structure in the case of *vector-based* implementations, such as ARC/INFO. This paper discusses a more general object-oriented approach to such models.

OBJECT MODELLING FOR GIS

Object-oriented modelling concepts

The concepts and terminology of the object-oriented approach are beginning to stabilize. (For a recent attempt at term definition in this area, see X3/SPARC OODBTG 1991). The main ideas are listed below. This paper is only concerned with the object-oriented approach as it applies to the conceptual modelling of systems. Thus are excluded object-oriented database management systems (OODBMS) and object-oriented programming languages (OOPLAS).

An *object* is a self-contained, uniquely identifiable conceptual entity. Objects have behaviour which may impinge upon other objects. An example of such an object is the county of Staffordshire in the UK. An object will support a set of *operations*, the action of which may depend upon the current state of the object and which may have an effect on its future behaviour. An operation on the object Staffordshire might be to update its county boundary. This operation may depend upon the current state of the areal reference of the object and will certainly change its future behaviour (for example, its spatial relationship with other counties).

Many objects may support similar collections of operations (*protocols*). Such objects may be grouped into *object classes*. Thus 'Staffordshire', 'Lancashire' and 'Cheshire' are all names of members of the object class UK-county. Classes fall into a natural partial order based upon their level of generality. Thus, class UK-county is a special case of class UK-administrative-region. We say that UK-county is a *subclass* of UK-administrative-region, or that UK-administrative region is a *superclass* of UK-county. The set of classes under this ordering forms an *inheritance hierarchy*. Classes which are subclasses inherit all the behaviour and state from their superclasses and they add their own behaviour and state.

Another important structuring principle for object modelling is *aggregation*. An example should illustrate this idea. The class coordinate-pair is the aggregate of classes x-coordinate and y-coordinate. The final object modelling construct which we introduce here is *grouping*. An example of a grouped object class is UK-county, where each member of the class may be considered as a collection of objects of type UK-ward.

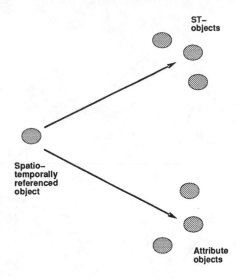

Figure 1: A spatiotemporally referenced object

Object modelling of geographic information

There is a growing body of work (for example Egenhofer and Frank 1989, Worboys *et al* 1990) on the application of object-oriented ideas to the modelling of GIS. Much of this work assumes that geographic objects have a purely spatial reference. However, the work described in this paper is based upon the view that to model geographic space, one must take account of the temporal dimension. Objects in geographic space have references to spatiotemporal (ST-) and non-spatiotemporal objects. (See figure 1). Thus, for example, a land parcel references an area existing in an interval of time as well as its reference number, classification, etc. A town may be modelled as a composite object which is the aggregate of units, each of which references an ST-object defining its extent in space-time and also other objects defining its areal magnitude, population, etc.

We are making the assumption here that a geographic object may be modelled (approximated) as an aggregate of units, each of which references an ST-object which is its extent in space-time, and also references other non-spatiotemporal attributes. The units are assumed to be homogeneous in their non-spatiotemporal attributes. That is, these attributes are assumed to be constant throughout the spatiotemporal extents.

In fact, a geographic object may be the aggregate of units each of which may have several spatiotemporal references. For example, a road unit may reference an area through time and also a line through time (being its centre-line), as well as name, classification, etc.

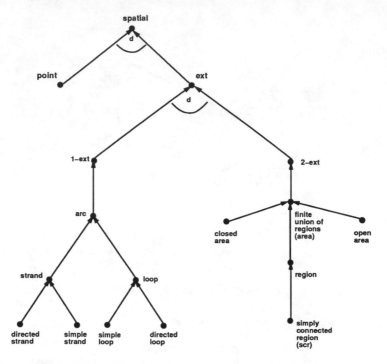

Figure 2: The inheritance hierarchy of spatial object classes

GENERIC OBJECT MODEL FOR PLANAR DATA

Spatial object classes

A discussion of the object model for geographic objects developed in this work rests on an understanding of the model of purely spatial objects. We assume, for simplicity, that all the spatial objects to be considered are embedded in the coordinatized euclidean plane, $\Re^2$, with the usual topology. Figure 2 shows the inheritance hierarchy of spatial object classes given by the author in (Worboys 1992). The class spatial has as disjoint subclasses point and ext. Class point has instance variables, x-coordinate and y-coordinate. Class ext is intended to model any extended objects embedded in two dimensions, and consists of sets of points. Class ext has as disjoint subclasses 1-ext and 2-ext of one and two-dimensional extents, respectively.

The most general class of one-dimensional objects considered here is arc. The intuitive notion of an arc may be grasped by imagining a finite piece of string thrown onto a table. It may cross over itself a finite number of times and may or may not have its ends touching. The class strand is a subclass of arc where each member does not have joined ends. The class loop is a subclass of arc where each member has its ends touching, and therefore is closed up. Simple arcs are those which do not cross themselves. An arc may have a direction or orientation.

The most general two-dimensional object class considered here is the class,

828

Table 1: Classification of spatial relationships

Spatial operations				
Group	Operator	Operand	Operand	Resultant
Set-oriented	equals	spatial	spatial	Boolean
	member	point	ext	Boolean
	subset	ext	ext	Boolean
	intersection	ext	ext	ext
	union	ext	ext	ext
	difference	ext	ext	ext
	cardinality	set(spatial)		cardinal
Topological	interior	area		open area
	closure	area		closed area
	boundary	area		1-ext
	components	area		set(region)
	extremes	strand		set(point)
	begin	directed strand		point
	end	directed strand		point
	inside	point	simple loop	Boolean
	clockwise	oriented loop		Boolean
Metric	distance	point	point	$\Re$
	length	arc		$\Re$
	perimeter	region		$\Re$
Euclidean	bearing	point	point	$[0, 2\pi)$
	area	area		$\Re$

area, members of which are disjoint finite unions of regions. A *region* is a homeo-morphic image of a disc embedded in the plane with a finite number of holes, cuts (arcs subtracted from the interior) and punctures (points subtracted from the interior), and which may include all, some or none of its boundary. There are the specialized classes, closed area and open area, whose members contain all of or none of their boundaries, respectively. A subclass of region is simply connected region, whose members are regions with no holes. A full discussion, including formal definitions, of this material is given in (Worboys 1992).

Spatial operations

Table 1 shows operations which are defined upon the spatial object classes which may be taken as a base set. The operators are divided into four groups: set-oriented, topological, metric and euclidean, each one dependent upon a richer sub-structure of the cartesian plane than those which precede it.

Spatial objects may be treated as pure sets (collections of objects). The set-oriented operations which result are equals, member, subset, intersection, union, difference and cardinality. Topological properties of the embedding of the spatial objects in the plane are needed for the second group of operators. Operators interior, closure and boundary of an area are defined in the usual topological manner. The operator extremes returns the pair of end points for each strand. Operators begin and end are similar operators, but apply only to directed strands. The Boolean operator inside tests whether a point is inside a loop, and clockwise

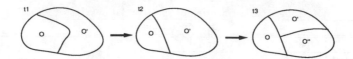

Figure 3: Change of regional structure through time

determines the orientation of a directed loop. Metric properties of the space allow the definition of operators in the third subclass. The final subclass of operators derives from strictly euclidean properties of the plane, in particular, from the measurement of angle.

GENERIC SPATIOTEMPORAL OBJECT MODEL

Spatiotemporal Objects

Several authors (for example, Langran 1989 and 1992, Langran and Chrisman 1988, Vrana 1989, Worboys 1990) have recognized the importance of the temporal dimension in the modelling of geographic information. Our model extends the model of the last section and is based upon some further simplifying assumptions. Firstly, the underlying spatial framework is assumed to be embedded in a Euclidean plane. Secondly, time is represented as one-dimensional, linearly ordered (we do not allow branching time for the purposes of this work) and uni-directional. The time dimension is taken to be orthogonal to the two spatial dimensions. Finally, changes to the spatial configuration of objects only take place at finitely many time points.

The extension to ST–objects is accomplished in this special case by representing temporal extension as a length in a direction orthogonal to the spatial plane. Since changes to the spatial extent of objects occur only at finite discrete intervals, we may view the underlying ST–objects as finite collections of disjoint right prisms (ST–atoms) whose bases are the spatial extents and whose heights represent their temporal extents. Thus, an ST–object is a finite collection of ST–atoms, where each ST–atom is a right prism. Formally, an ST–atom is an ordered pair $< S, T >$, where S is a spatial object and T is an extended temporal interval.

Figure 3 shows an example of the application of these concepts to the modelling of changing administrative boundaries. At time t_1, there are two areal objects, O and O'. At time t_2, the spatial extent of O is reduced and that of O' is correspondingly enlarged. At time t_3, the spatial extent of O' is reduced to accommodate a new areal object O''. The ST–objects O, O' and O'' may be decomposed into ST–atoms as follows (see Figure 4):

> O is represented by the collection of two ST–atoms $< S_1, [t_1, t_2] >$
> and $< S_2, [t_2, t_\infty[>$.
> O' is represented by the collection of three ST–atoms $< T_1, [t_1, t_2] >$,
> $< T_2, [t_2, t_3] >$ and $< T_3, [t_3, t_\infty[>$.
> O'' is represented by the single ST–atom $< U, [t_3, t_\infty[>$.

S_1 and S_2 represent the initial and modified spatial extents of object O. T_1, T_2 and T_3 represent the three stages of spatial extension that object O' undergoes. U

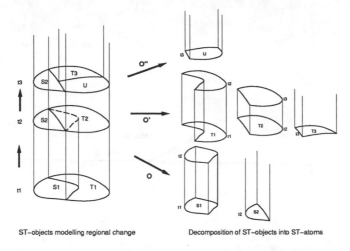

Figure 4: Modelling variation in regional boundaries

represents the spatial extent of O''. A point of time indefinitely far into the future is represented by t_∞.

Spatiotemporal Operations

Having described the fundamental spatiotemporal object as composed of a finite collection of ST-atoms, the next requirement is to identify and model the ways in which such objects can interrelate. We propose two projection operators, π_S and π_T, which project the set of ST–atoms onto the sets of spatial objects and time intervals, respectively. Thus, a purely spatial relationship between ST–atoms can be decomposed into an S–projection and some known spatial operators, while a purely temporal relationships can be decomposed into a T–projection and known temporal operators. With respect to mixed spatiotemporal relationships, in our model, where the temporal dimension is orthogonal to the spatial dimensions, any spatiotemporal relationship between ST–atoms can always be decomposed into the cartesian product of separate spatial and temporal relationships (figure 5). Purely spatial relations between objects have been discussed earlier. Relationships between temporal intervals have been fully analyzed by Allen (1983, 1986).

Example 1: Suppose we have the two ST–objects, O and O' with the following associated sequence of events:

Time	Event
t_1	O created
t_2	O' created spatially disjoint from O.
t_3	O' expanded to share a common boundary with O
t_4	O' dropped
t_5	O dropped

If we assume that the spatial extent of O is unchanged throughout its existence, then O may be represented by the single ST–atom $A = < S, [t_1, t_5] >$. Let O' be

831

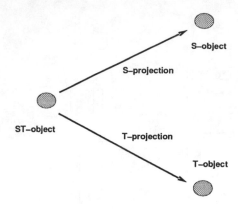

S-object

S-projection

ST-object

T-projection

T-object

Figure 5: Projection mappings

represented by the pair of ST–atoms, $A_1' =< S_1', [t_2, t_4] >$ and $A_2' =< S_2', [t_3, t_4] >$. Then, the total spatial extent of O' during the interval $[t_3, t_4]$ is given by $S_1' \cup S_2'$. This is shown in figure 6.

That O and O' have at some time shared a common boundary can be seen by observing that:

$\pi_S(A)$ and $\pi_S(A_2')$ share a common boundary (are adjacent);
$\pi_T(A)$ and $\pi_T(A_2')$ are not disjoint as sets.

Thus, the spatiotemporal relationship between O and O' is expressible in terms of spatial and temporal relationships between the projections of their constituent ST–atoms.

Example 2: Imagine a road and its relationship to a county. At time t_1 the road is created, but does not run into the county. At time t_2 the road is extended into the county. At time t_3 the county area is extended to include a further portion of the road. Suppose further that at the required level of generalization, the road is represented by set of atoms, each a vertical sheet and the county by a set of prism atoms as above. A representation of is given in figure 7.

The following questions are posed.

1. Has the road ever passed through the county?

2. Does this road pass through land which has ever belonged to this county? (S-project and take S-union of county and road atoms. Is the intersection non-disjoint?).

3. Does this road pass through land which has always belonged to this county? (Intersect, T-project, does interval equal whole time?)

In order to answer question 1, it is necessary to take the intersection of the union of atoms representing the road with the union of atoms representing the county. If this intersection is empty, then the answer is 'no'. For question 2, firstly take the spatial

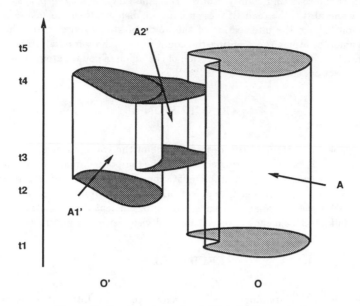

Figure 6: Spatiotemporal relations in Example 1

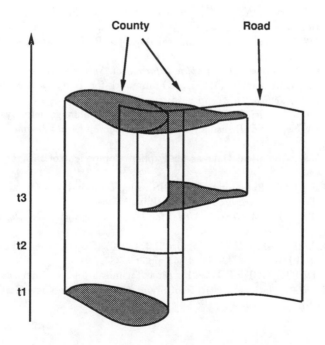

Figure 7: Spatiotemporal relations in Example 2

projection of the road and county atoms. Take the union of spatial projections of the county and also of the road. If these unions are disjoint, then the answer is 'no'. For question 3, take the intersection of the union of atoms representing the road with the union of atoms representing the county. Take the temporal projection of this intersection. If the temporal projection of the union of county atoms is a subset of this intersection, then the answer is 'yes'.

CONCLUSION

This paper has attempted to describe some of the concepts involved in modelling geographic information taking account of the temporal dimension. The author is currently constructing a pilot implementation of this model using the proprietary system Smallworld GIS and its object-oriented programming language Magik. Major issues for the future are the construction of efficient access methods, query languages and effective means of presenting and visualizing the information.

REFERENCES

Allen J.F. (1983). Maintaining Knowledge about Temporal Intervals, *Communications of ACM*, **26**(11), pp.832–843.

Allen J.F. (1986). Towards a General Theory of Action and Time, *Artificial Intelligence*, **23**, pp.123–154.

Egenhofer, M.J., Frank, A. (1989). Object-oriented modeling in GIS: Inheritance and propagation, *Proceedings Auto-Carto 9*, Baltimore, Maryland, US, pp.588–598.

Langran G. (1989). A Review of Temporal Database Research and its use in GIS, *IJGIS*, **3**(3), pp.215–232.

Langran G. (1992). *Time in Geographic Information Systems*, Taylor and Francis.

Langran G. and Chrisman N. (1988). A Framework for Temporal Geographic Information, *Cartographica*, **25**(3), pp.1–14.

Maiocchi R. and Pernici B. (1991). Temporal Data Management Systems: A Comparative View, *IEEE Transactions on Knowledge and Data Engineering*, **3**(4), pp.504–524.

Vrana R. (1989). Historical Data as an Explicit Component of LIS, *IJGIS*, **3**(1), pp.33–49.

Worboys M.F.(1990). Reasoning about GIS using Temporal and Dynamic Logics, *Proceedings Temporal Workshop*, NCGIA, Univ. Maine.

Worboys, M.F. (1992). *A Generic Model for Planar Geographic Objects, IJGIS*, to appear.

Worboys M.F., Hearnshaw H.M., Maguire D.J. (1990). Object-Oriented Data Modelling for Spatial Databases, *IJGIS*, **4** (4), pp.369–383.

X3/SPARC/DBSSG/OODBTG (1991). Final Technical Report, *American National Standards Institute*, National Institute of Standards and Technology, Gaithersburg, MD 20899, USA.

INTEGRATING SOIL LOSS STANDARDS AND GEOGRAPHIC INFORMATION SYSTEMS IN THE MANAGEMENT OF OFF-HIGHWAY VEHICLE PARKS

Richard D. Wright, David McKinsey, and Barbara Bell
Department of Geography
San Diego State University
San Diego, CA 92182

ABSTRACT

The California Department of Parks and Recreation is responsible for managing seven off-highway recreational vehicle parks. Federal and State laws require that the parks' natural resources be protected for the benefit of present and future users. To assist in meeting this mandate the Department of Geography at San Diego State University and the Off-Highway Vehicle (OHV) Division of the California Department of Parks and Recreation have cooperatively developed an environmental geographic information system. The primary purpose of the GIS is to facilitate the monitoring of vehicular impacts on the parks' wildlife and soils. To deal with the management of the soils, State Parks, through consultation with various Federal agencies has adopted a generic soil loss model and standards that allow areas to be set aside for rehabilitation if they fall below soil loss levels. In this study we show how the soil loss model can be linked with a GIS. The linked system is employed to red flag existing problem areas and to predict potential soil hazards. An evaluation of the integrated system is provided.

INTRODUCTION

Soil erosion models are employed by many organizations with responsibilities for managing land and water resources. A prime example is the erosion hazard rating (EHR) system adopted by the California Department of Parks and Recreation for managing soils in its off-highway recreational vehicle parks. Although EHR was not developed as a computer-based system, it has characteristics which allow it to be linked with a GIS. In this paper we describe the State of California Off-Highway Vehicle (OHV) Division GIS and the development, components, and uses of the EHR system. We show how EHR is being integrated with a GIS and conclude with an evaluation of the linkage between EHR and GIS.

DEVELOPMENT OF THE GIS

San Diego State University's Department of Geography and the California Department of Parks and Recreation's Off-Highway Vehicle Division are jointly preparing a GIS for managing the State's seven OHV parks. The parks comprise a total area of more than 65,000 acres and range in size from 220 acres to more than 35,000 acres. ARC/INFO software, loaded on Sun and VAX platforms, is employed for preparing and using the database. Other ARC/INFO related modules such as TIN, NETWORK, GRID and ARCVIEW are also used as the need dictates. The database consists largely of environmental and trails layers since the GIS is primarily intended to be a tool for monitoring vehicular impacts on the parks' wildlife and soils. Primary sources for the database include OHV Division master planning maps ranging in scale from 1:24,000 to 1:6,000 and color photographs at the scale of 1:6,000. Photography is flown every two years to provide reasonably current data about the parks' conditions.

DEVELOPMENT OF THE EROSION HAZARD RATING SYSTEM

As required by state legislation, a Soil Loss Standards Committee (SLSC) of soils experts from state and federal agencies was formed in 1988 to develop generic soil loss standards and soil conservation guidelines for managing OHV parks (California Department of Parks and Recreation 1991). Existing soil erosion models such as the universal soil loss equation, which provide a quantitative measure of soil loss per unit area, were deemed inadequate because they are not easily translated into soil loss standards applicable to the rehabilitation of OHV areas and trails. Instead, the committee adopted the erosion hazard rating (EHR) system which was being evaluated concurrently by the California Soil Survey Committee (CSSC) as a possible standard for National Cooperative Soil Survey participating agencies in California. EHR was subsequently approved by CSSC in 1989 and accepted by the Off-Highway Motor Vehicle Recreation Commission in 1991. It is a fairly sophisticated rating system based on ordinal scaling of nominal and ratio data, not a highly precise mathematical model.

APPLICATION OF THE EHR SYSTEM

The EHR system has a wide variety of potential applications. For example, EHR can be used:
(1) to measure the impacts of OHV activities on soils and surface configuration. Areas rated as highly eroded can be closed temporarily for rehabilitation, or in extreme cases, closed permanently.
(2) to monitor the effects of a revegetation effort in a disturbed area with the objective being to ascertain if an acceptable EHR level has been reached. This could result in an area being reopened for recreational activities.
(3) to model the impact or potential impact of vegetation loss resulting from a fire. This could be useful in making certain that the potential erosion that could result from a prescribed burn program falls below some predetermined threshold value. The amount and location of acreage to be burned within a given watershed over a given time period could be a function of the EHR rating.
(4) to site a recreational trail in a location that will not produce an unacceptable amount of erosion. One could also determine the erosion rating for a single proposed trail, for several proposed alternatives, or for one or more existing trails.
(5) to measure the erosion rating component of an environmental sensitivity model. The model could be used to identify environmentally constrained areas in which human intrusion should be eliminated or at least kept to a minimum.

(6) to provide an overall assessment of soil hazards in a park unit to identify specific areas that need the attention of the managers in order to rectify serious soil erosion situations.

(7) to create "what if" scenarios such as: Determining the impact of changes in the factors which comprise the soil hazard rating system. For example, what would be the effects on the soil erosion rating of an area given a decrease in vegetation cover, an increase in the intensity of precipitation events, or an alteration in slope length. Modeling these types of possible changes allows park managers to consider a wide range of possible alternatives to solving erosion problems in their parks.

COMPONENTS OF THE EHR SYSTEM

The EHR consists of four major components: soil erodibility, runoff production, runoff energy, and soil cover. With the exception of runoff energy, all components consist of two or more factors.

Component 1: Soil Erodibility
The soil erodibility factor provides a rank ordering of soils according to the transportability of detached soil particles. Two variables--soil texture and slope steepness--are combined on a 4-point ordinal scale (See Table 1). An adjustment of +1 is made for soils having a sodium content that decreases aggregate stability and -1 for special situations, e.g. high iron content, that result in increased aggregate stability.

Table 1. Soil Texture

Texture Class	Slope steepness			
	0-15%	16-30%	31-45%	46 and above
sand	1	1	2	2
loamy sand	1	2	3	3
sandy loam	2	2	3	3
sandy clay loam	2	2	3	3
sandy clay	1	1	1	1
clay	1	1	1	1
clay loam	2	2	2	2
loam	3	3	3	3
silty clay	2	2	2	2
silty clay loam	3	3	3	3
silt loam	4	4	4	4
silt	4	4	4	4

Component 2: Runoff Production
Runoff production is a function of precipitation amount, water movement in the soil, runoff from adjacent and intermingled areas, and uniform slope length. These four factors are summed to form a measure of the runoff production component.

Precipitation. The precipitation factor is represented by the amount of precipitation for a 6-hour event over a 2-year period. Precipitation amounts are classed and ranked on a 5-point ordinal scale (See Table 2).

Water movement in the Soil. The movement of water in the soil is a composite factor that takes into account the interrelated subfactors of surface soil infiltration, permeability of the subsoil, and the presence (and depth) or absence of a layer that restricts the vertical movement of water. An 8-point rating scale is applied (See Table 2).

Table 2. Runoff Production

Precipitation

inches	< 1.0	1.0-1.7	1.8-2.2	2.3-2.7	> 2.7
rating	1	2	3	4	5

Water Movement in the Soil

Infiltration	Permeability	RL depth (in)	Rating
rapid	any	> 40	1
rapid	mod	20-40	2
rapid	mod	< 20	3
mod	any	> 40	4
rapid of mod	slow	< 20	6
slow	any	any	8

Runoff From Adjacent and Intermingled Areas

Amount	low	moderate	high
Rating	0	2	5

Uniform Slope Length

Length	< 25	25-50	> 50
Rating	1	4	6

Runoff From Adjacent and Intermingled Areas. Impervious or nearly impervious surfaces adjacent to or within an area have an impact on the amount of surface runoff. Areas are rated as low, moderate, and high in terms of the percentage of surface porosity of adjacent and imbedded areas. The adjectival descriptions low, moderate, and high are converted to the integers 0, 2, and 5 respectively.

Uniform Slope Length. The length of uniform slopes between breaks in the terrain influences the runoff production. Length of uniform slope in feet is classed as less that 25, 25 to 50 and greater than 50 and rated as 1, 3, and 6, respectively.

Component 3: Runoff Energy

Runoff energy is indicated by slope steepness. The rating for this component is determined by computing slope as a percentage and then rounding it off to the nearest hundredth, keeping two decimal points.

Component 4: Soil Cover

The soil cover component considers the quality, quantity, and distribution of vegetation and its impact on erosion. This component consists of two factors--quality and quantity of vegetation and distribution of vegetation. The two factors are added to provide a soil cover rating.

838

<u>Quantity and Quality of Cover</u>. This factor is measured by computing the percentages of ground cover and shrub and/or tree canopy. The two are combined in a rating that ranges from 0 to 5 (See Table 3).

<u>Cover Distribution</u>. This factor is included to take into account significant variations in the continuity of the cover. Areas having a uniform cover are assigned the value 0 and those having a patchy distribution receive the value 1.

Table 3. Cover Quality

Percent Shrub and or Tree Canopy	Percent Ground Cover					
	0-10	11-30	31-50	51-70	71-90	> 90
0-10	5	4	3	2	1	0
11-30	4	4	3	2	1	0
31-50	4	3	3	2	1	0
51-70	3	3	3	2	1	0
71-90	3	3	2	2	1	0
> 90	3	2	2	1	0	0

THE EROSION HAZARD RATING

The erosion hazard rating is obtained by taking the product of the ratings for the four components. The numeric ratings are classed into the adjectival categories low, moderate, high and very high.

<u>Low EHR - less than 4.</u> Accelerated erosion is not likely to be a significant problem in areas having a low EHR.

<u>Moderate EHR - 4 to 12.</u> Accelerated erosion is likely to be a significant problem during periods of intense precipitation activity.

<u>High EHR - 13 to 29.</u> Accelerated erosion will occur in most years. Erosion control measures should be undertaken.

<u>Very High EHR - More Than 29.</u> Accelerated erosion is a significant problem even during periods of below average precipitation intensity. Erosion control measures are essential for areas having a very high EHR.

INTEGRATION OF EHR AND GIS

<u>Hungry Valley SVRA</u>.
<u>Hungry Valley SVRA</u>, a 19,000 acre park located north of Los Angeles near Tejon Pass in the Transverse Ranges, is the test area for this paper (Fredrich et al 1991). This SVRA has been an official OHV park since 1978, but has been used for OHV activities since the early 1970's. Prior to 1970, portions of it were grazed extensively by livestock. Hungry Valley SVRA is

a topographically and biologically diverse landscape. Elevations range from over 5,300 feet in the northwest to less than 3,000 feet in the southeast. Annual precipitation ranges from about 10 to 15 inches, with at least 70 percent of the total occurring during the period November to April. The diverse vegetation includes pinyon pine, oak, cottonwood, red willow, California juniper, Great Basin sagebrush, rabbitbrush, and several native grass species.

Data Employed

To adapt EHR to a GIS required many data layers, most of which were available in the existing ARC/INFO database. Additional data to augment the soils layer were obtained from the Soil Conservation Service Survey. The following data layers have been extracted from the database:

Soils (texture, permeability, infiltration, depth to a restrictive layer)
Vegetation (percent cover, type, distribution).
Slope, Roads and Trails, Barren Areas and Rock Outcrops

The above vector data have been converted to raster form and processed using ESRI's GRID module.

EHR - GIS Integration

The steps followed in processing EHR using the GRID module are represented in Figure 1.

Soil Erodibility. The first step is to create a map of soil erodibility. Using definitions from the Soil Conservation Service descriptions, soils are classed into soil texture groups. The slope data are classed into equal interval categories, each with a range of 15 percent. These two data sets are combined and ratings are assigned in accordance with Table 1. Since none of the soil types in the area has significant amounts of iron or sodium the aggregate stability adjustments are not used.

Runoff Production. To calculate runoff production the climate must be considered. Precipitation values are extracted from the NOAA six-hour precipitation map of California. The value for the Hungry Valley area is about two inches. From Table 2 a weight of 2 is assigned for the precipitation variable. The soil descriptions from SCS are used to create a lookup table which replicates the water movement weights of Table 2. The weights for water movement in the soil are assigned from this lookup table. To determine runoff from adjacent areas, data for roads, trails, barren areas, rock outcrops, and areas with a weight of 6 or higher for water movement in the soil are unioned. The data is then binary coded as impervious and porous. A 3X3 filter adding the nine cells and assigning the sum to the central cell is passed over the map. The sum for each is divided by nine to obtain a percentage. The percentages are compared to the runoff portion of Table 2 to obtain a rating for this variable. For the uniform slope factor the maximum weight is assigned. This is because the cell size of 25 meters is larger than the 50-foot slope value in the uniform slope section of Table 2. These four factors are summed and divided by three, resulting in a weight for this step.

Runoff Energy. For this component, a slope map consisting of categories, each with a range of one percent is created. These percentages are rounded off to two places to produce weights.

Figure 1. Erosion Hazard Rating - GIS Integration

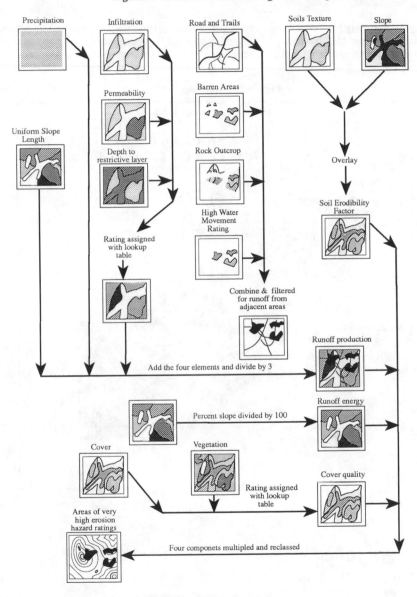

Precipitation

Infiltration

Road and Trails

Soils Texture

Slope

Permeability

Barren Areas

Uniform Slope
Length

Depth to
restrictive layer

Rock Outcrop

Overlay

Soil Erodibility
Factor

High Water
Movement
Rating

Rating assigned
with lookup
table

Combine & filtered
for runoff from
adjacent areas

Add the four elements and divide by 3

Runoff production

Percent slope divided by 100

Runoff energy

Cover

Vegetation

Rating assigned
with lookup
table

Cover quality

Areas of very
high erosion
hazard ratings

Four componets multipled and reclassed

Soil Cover. The percentages of canopy and ground cover are estimated for each vegetation polygon. The roads and trails are added to this layer as barren areas. The weights are assigned from Table 3. The distribution of cover is determined from the scattered and very scattered vegetation categories. These distinctions are made through field surveys and from aerial photographs. This factor is added to above weights to produce a total for soil cover.

Final Product. As mentioned previously, the final soil erosion hazard rating is created by multiplying the weights of the four components and then classing them into four categories (see figure 2).

Figure 2. Final Map

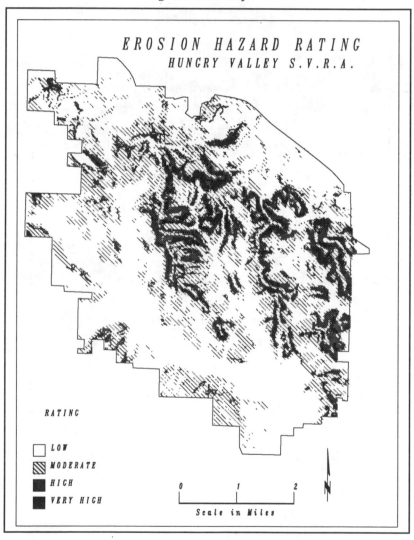

SUMMARY AND CONCLUSIONS

In summary, the EHR system can be integrated with a GIS to facilitate the process of monitoring erosion resulting from motorized vehicle activity in areas devoted to off-highway vehicles. The GIS modeling done in this research has limitations, most of which relate to:

1) data which are too coarse in relation to the data requirements of some of the small area rehabilitation and conservation tasks, and

2) the limitations of maps and aerial photographs for providing the type of data needed to quantify certain factors in the EHR model.

In view of these limitations, a two phased approach for using EHR with a GIS is recommended. In the first phase the combined EHR/GIS is employed to generate an overall picture of erosion hazards in a given park and to identify specific areas that require closer examination. In the second phase, fieldwork involving GPS and other techniques can be applied on a site specific basis to obtain more refined data appropriate to larger scale needs. Research work on the EHR model is continuing to further refine its integration with GIS technology.

REFERENCES

California Department of Parks and Recreation, Off-Highway Motor Vehicle Recreation Division. 1991, Soil Conservation Guidelines/Standards for Off-Highway Vehicle Recreation Management. Sacramento, California. 78pp.

Fredrich, B., Salazar, D., Wright R. andWoodward,R., 1991. "Using a Geographic Information System," Fremontia, vol. 19:2, pp. 10-14

United States Department of Agriculture. Soil Conservation Service. 1969, Soil Survey. Antelope Valley Area California.

TRAFFIC FLOW MODELING WITH VISUALIZATION TOOLS

Demin Xiong
Department of Geography, The Ohio State University
103 Bricker Hall, 190 North Oval Mall
Columbus, OH 43210

ABSTRACT

Recently, transportation studies in the area of Intelligent Vehicle/Highway Systems (IVHS) have been directed mainly toward the development of new types of traffic flow models and the technology necessary to communicate with and guide drivers. The present research, however, is conducted from a GIS perspective that places emphasis upon the development of methods to integrate modeling and visualization techniques in order to analyze traffic flows in more effective ways. This paper first introduces a conceptual framework which can be used to integrate various procedures for effective and efficient traffic flow modeling and visualization. Then techniques to carry on model calculation with user-equilibrium modeling as a specific example will be presented. Finally, visualization tools that can be used to analyze and display various aspects of the modeled traffic flow data such as shortest paths, travel time, flow compositions on a network in user-equilibrium will be described. Ultimately, this research is expected to be beneficial to IVHS development and certainly to road network analysis in general.

INTRODUCTION AND BACKGROUND

Intelligent Vehicle/Highway Systems (IVHS) are currently an important focus in transportation research. The key feature related to the operations of IVHS is that transportation operations and services are managed primarily based on the provision of accurate, real-time traffic flow data, which are to be collected and transmitted in real time (Euler 1990). Using this real-time data, traffic flow models can be used to predict future traffic flow patterns, to anticipate areas of possible traffic congestion and to test alternative strategies for traffic control or route guidance. Obviously, traffic flow models can play a key role in IVHS, and provision of sufficient capacity to perform real-time model calculation and the capability to carry on real-time data examination and interpretation will be critical. The potential of using visualization tools in the examination and interpretation of volumes of data generated from complex modeling and simulation has been demonstrated in various disciplines, but has not been fully explored for data examination and interpretation in traffic flow modeling. Even though the present research is not intended to provide a systematic treatment at all the problems involved in real-time traffic flow modeling, it emphasizes the introduction of visualization techniques to form a modeling framework that is suitable for efficient model calculation and data analysis in traffic flow modeling. The development of detailed procedures to undertake model calculation, and data analysis with visualization tools are also important aspects of this research.

The value of visualization techniques for complex data analysis and interpretation has been demonstrated in various disciplines. Grotjahn and Chervin (1984) experimented with animated graphics to present meteorological data and found that these graphics provide a very efficient way for the analysis and understanding of their data. Farrell (1987) discovered that when digital simulation and computation become more complex, large volumes of data, often several million bytes of multidimensional data, for each study can be generated, thus simple graphic methods do not have sufficient capacity to convey necessary information of the data, so that the use of color, three-dimensional, dynamic or animated displays is required. In transportation, efforts have also been made to develop graphical tools to display traffic related data. The earliest computer generated graphics used in transportation were in the Chicago Area Transportation Study (Tobler 1987). In recent years, the application of GIS technology has begun to provide a systematical treatment of transportation data management, analysis and display. Many GIS packages such as TransCAD and ARC/INFO feature a combination of mature capabilities of data analysis and data display. Other interesting research which is related to the graphical representations of network data can also be found in the efforts to develop computer programs for mapping migration data (Tobler 1987), displaying location-allocation solutions (Allard and Hodgson 1987), and visualizing computer networks (Becker et al 1990).

Nevertheless, traffic flow modeling and visualization have generally been treated as quite separate processes in previous research. In order to visualize the output from traffic flow models, data generally need to be transferred forward and backward between graphical software and the modeling program. In some cases, even though data transformation is not necessary, modeling process still can not be visualized. The separation between model calculation and data visualization will not allow model calculation and data visualization to be executed simultaneously and makes the full investigation of the modeled data impossible in many cases. The lack of visualization tools suitable for exploring modeled traffic flow data represents another problem that has not been addressed adequately in previous research. Available visual tools for flow data presentations are limited in general to displaying existing data with a clearly defined spatial coordinates and topological relationships. For traffic flow modeling, data related to the different aspects of traffic flows may not be readily available, and it may not be possible to make it readily available because of the limitations on storage space in a computer. To address these problems, a conceptual framework which is intended to couple modeling and visualization procedures into an integral system to achieve computational efficiency is proposed, and then detailed procedures to perform model calculation and data visualization based on a user-equilibrium (UE) problem are presented.

FRAMEWORK FOR MODELING AND VISUALIZATION

Traditional traffic flow modeling in many instances takes a batch-processing mode in which modeling procedures with all the parameters and input data are determined before being submitted for calculation. Upon completion of the calculation, output files containing the results of the calculation will be created, then investigation of the output data follows. Such a practice, on the one hand, can seriously delay the delivery of the modeling result. On the other hand, searching for a better result over a set of alternatives or experiment with various procedures and different input data or different parameters for comparative explanations is generally discouraged because of the large work load involved in analysis of the output data. The main idea of the proposed framework for traffic flow modeling shown in Figure 1 is twofold. First, data involved in a modeling procedure should be graphically visualized so that information related to the data can be conveyed to modelers effectively and efficiently. Second, modelers should have the means to interact with the modeling process so that

845

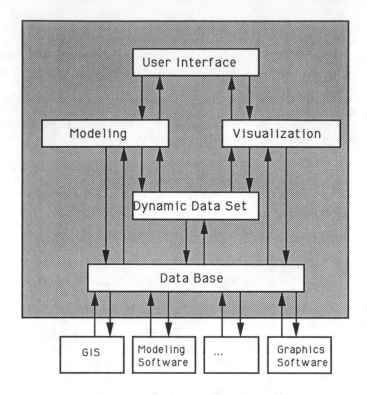

Figure 1   Framework for Traffic Flow Modeling and Visualization

modeling procedures, input data and formulation parameters can be changed or modified instantaneously as modelers observe the proceeding modeling process with visualization tools. In addiction to those two aspects, it is realized that data, as the input to feed the models, data, as processed in modeling process, and data, as the output of the models should be organized and managed in a manner that will be suitable for efficient modeling and visualization purposes.   In the following, each component in the proposed framework is described briefly.

User Interface

The user interface not only provides the basic means for interacting with the modeling and visualization process, but also assists in establishing an user-friendly environment. With an appropriate user interface, modelers will have the flexibility to change or modify some or all of their input data, to select different parameters and to choose alternative modeling procedures. For investigation of output data, interactive graphics will be especially handy. The investigator can control and respond to graphical display so that data are displayed in a way suitable to convey the various complex aspects of the situation. When appropriate menus are provided, analysts can point to an object or select appropriate menu items by means of input devices such as a mouse or a key on the keyboard in their manipulation of the modeling and visualization process.

Modeling and Analysis

Because of the complexity of human travel behavior, analytical models must be used extensively in traffic flow modeling. The most important function of the modeling and analysis module proposed here is to provide procedures to carry on the necessary calculations in the traffic modeling process. In the preliminary stages of this research, basic procedures such as calculation of shortest paths, all-or-nothing assignments, etc, have been developed, and a program to calculate user equilibrium using Convex Combination Method has been constructed. As long as traffic flow modeling has to largely depend on analytical tools, the modeling and analysis module will remain to be a key component in the proposed framework. The design of the user interface, visualization tools and data management functions to a great extent should reflect the design of the modeling and analysis module.

Visualization

A modeling process may not be completed until its result is carefully investigated and interpreted. Graphical display in general can provide a concise tool to convey information for the investigation and interpretation of the output data. Further, it is noted that output data in traffic flow modeling often involves working with complicated network topology in addition to the network geometry. Therefore, as indicated earlier, tools developed for general visualization purposes may not be useful for the purpose of modeled traffic flow data visualization. For example, in order to visualize the distribution of traffic flows over a network generated by a O-D pair, a visual tool has to have the ability to identify all the possible paths and the flow volumes assigned to these paths connecting this O-D pair, then it must render an appropriate graphical display to represent each possible path including the direction and magnitude of flows. In addition to the requirement that visualization tools should be precise enough to satisfy a special need, visualization tools should be systematically designed so that they can be used to explore various aspects of the data for different purposes.

Dynamic Data Set

Many calculations in traffic flow modeling demand intensive computational effort, and involve large volumes of input and output, thus appropriate strategies for data management are needed to ensure that the potential of the computer can be fully exploited. The idea of building a dynamic data set is to keep a set of data in the fast memory of a computer because data in the fast memory can be retrieved much faster than from a disk or from other types of mass storage devices. In structuring the dynamic data set, it is important to have an estimate of the sizes of different data classes that will be involved in the modeling and visualization process, as well as an estimate of the frequency with which each kind of data will need to be retrieved. Then a priority order may be established to determine which kind of data may be held in the fast memory, and what kind of data may be temporarily loaded into the fast memory. In general, network attributes and topology are necessarily included in the dynamic data set because they will be used in modeling and visualization operations with a very high frequency.

<u>Database</u>

The dynamic data set provides a solution for fast retrieval and storage of the data. However, a database is still necessary in order to integrate and to manage various data to feed the modeling and visualization procedures, and a database can be a means for necessary manipulation and storage of the output data that are generated from model calculations or data analysis. Also with a database, traffic flow modeling and visualization can take advantage of the existing GIS, transportation software, and visualization systems by sharing data with this existing software. For example, the traffic flow modeling and visualization software could share a database with a GIS or utilize a database from a GIS directly. To do so, GIS functions can be used to digitize networks, to edit attribute data and to do necessary analysis.

The framework presented above, which contrasts with the traditional modeling practice by incorporating integrated capabilities of analytical methods, visualization tools, interactive techniques and data management functions, will allow traffic flow modeling to be carried out more effectively and efficiently. The provision of the visualization tools will increase our ability to comprehend the complex modeled traffic flow models, and by using the interactive techniques, a modeler can control or modify modeling and visualization process so that errors or biases can be reduced, alternatives can be compared and new ideas can be generated. Through effective data organization and management, overall efficiency for data storage and computational speed can be achieved.

USER EQUILIBRIUM MODELING

Traffic flow modeling involves a series of procedures including trip generation, trip distribution, modal split and trip assignment. This research will focus on the static user-equilibrium (UE) problem. The existing UE formulations and solution strategies that are utilized may be put into the framework shown in Figure 1 in order to form an effective modeling scheme. Even though the static UE problem is a relatively simple modeling problem, compared to dynamic and integrated modeling problems, it can serve as an example to demonstrate the concept proposed in this paper.

In transportation research, the user-equilibrium problem has been studied extensively, and there exist well established procedures and algorithms to carry on the derivation of user-equilibrium solutions. The Convex Combinations Method (CCM) that is used in the present study is one of the most widely used methods, which not only has an efficient solution procedure, but also can create a converged UE flow pattern. The detailed mathematical formulation upon which the CCM rests can be found in Sheffi (1985). Figure 2 (a) (also see Sheffi, 1985) shows the procedure used to calculate the UE flow pattern. The first step, as shown in Figure 2 (a), is actually to find a feasible solution which will minimize the total travel cost on an empty network, that is done with an all-or-nothing assignment with free-flow travel cost on each link. From step 2, an iterative scheme is established. At step 2, the travel cost of each link over the network will be updated with respect to the flow assigned to that link in previous iteration, then an all-or-nothing assignment is carried out to generate an auxiliary flow pattern which conceptually will minimize an objective function set in the convex combination program. Step 3 finds an optimal move size by deriving an solution to another objective function. Then at step 4, a new flow pattern is calculated by applying the move size to both the previous flow pattern and the auxiliary flow

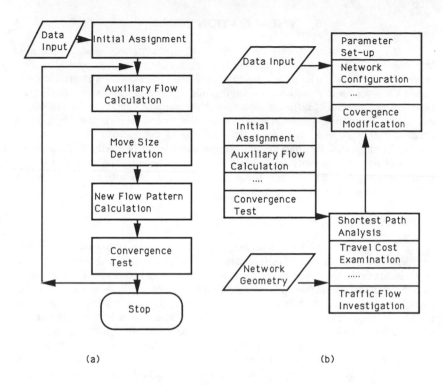

(a)   (b)

Figure 2   UE Modeling Procedures

pattern. In last step, convergence with respect to the UE flow pattern are tested. If the program has met the specified convergence criteria, flows assigned in the last iteration are taken as an approximation of the UE flows; otherwise, the program returns to step 2 and starts another iteration.

To incorporate a user interface and graphical tools, however, modifications of the above modeling procedure are required. First, an interactive mechanism has to be built into the modeling process. It is realized that if an interactive mechanism is built to determine whether the convergence criteria should be changed or whether the model calculation should be stopped for each iteration, the modeler's work load will be increased substantially and computation time to complete the whole calculation will be increased. A practical choice then is to place an interactive procedure at the beginning of the model calculation so that when a model calculation is started, it will not be constantly interrupted although modelers still have the means to control the modeling process. A second modification is required for incorporating graphic tools. With the same concern as in adding the user interface to the modeling process, the utilization of visualization tools should not interrupt model calculations. Therefore a method similar to post-processing is proposed here, that is, graphical display is rendered only after a cycle of the UE calculation is completed. With these two modifications, the modeling procedure will be structured as seen in Figure 2 (b).

VISUALIZATION TOOLS

To portray data with appropriate graphical representations is one of the most important goals for data visualization, but not the only one. As in traffic flow modeling, data from model calculations in many cases have complex interaction with network topology and multi-attribute aggregation. To provide adequate visualization tools for investigation of different aspects of these complex data, both the procedures that render appropriate graphical displays and the procedures that can "dig" out or synthesize necessary data to support the rendering of the displays are required and need to be integrated. Keeping this idea in our mind in the following discussion, we will focus our attention specifically on the design of visualization tools that can be used to investigate and to interpret the modeled UE flow data with respect to shortest path, travel time and traffic flow assignment.

Visualization of the Shortest Paths

Shortest path analysis has many implications for the examination and interpretation of the modeled UE flow data. For instance, unusual shortest paths may be an indicator for an inappropriate selection of modeling procedures or of improper choices of model parameters. With shortest paths in user-equilibrium, road users can find alternative paths which can minimize their travel times. A procedure that can be used to calculate and to display shortest paths in UE is shown in Figure 3. Notice that flows assigned for each iteration are involved in the calculation of shortest paths and all the resultant shortest paths are required to be merged into a single layer to eliminate possible overlays of shortest paths calculated from different iterations. The validity of this method can be verified by examining UE modeling procedures, which however is beyond the scope of this short paper. However, two points do need to be clarified. First, under extreme conditions, a shortest path calculated from the assigned

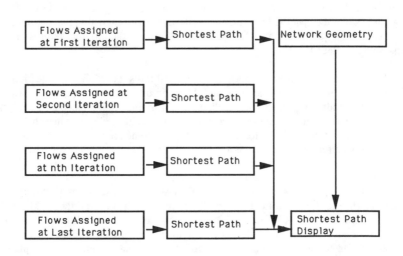

Figure 3   Procedure for Shortest Path Analyisis in UE

850

flows in an iteration may not turn out to be a shortest path. In this case, the flow assigned to that path needs to be checked. If the actual flow assigned to this path turns out to be zero, this path should not be considered as a shortest path. Second, a shortest path calculated with the flow assignment of an iteration may not be unique among the shortest paths calculated with the flow assignments of other iterations so that shortest paths calculated with different flow assignments of different iterations need to be merged into a single layer so as to eliminate those shortest paths which are not unique. To render a graphical display of the shortest paths, each link on the network is assigned an ID number so that the geometric data of a shortest path can be retrieved whenever a shortest path is calculated. Actually, in many cases we may need to highlight the shortest paths while all other links are also displayed. To do so, the location of a shortest path relative to other links can be easily identified. The direction of a shortest path can be indicated by using particular labels at origin and destination or by displaying the paths with the order from origin to destination. With an appropriate design for the graphical interface, and the above shortest path calculation and display procedures, an user can instantly select an O-D pair to find out the shortest paths between them. By observing various displays, the user may obtain some useful evidence above the validity of the model or have a good understanding of how people will select their routes to travel on the network.

Visualization of Travel Costs

Travel cost analysis constitutes another important aspect in the examination and interpretation of the modeled UE flow data. Whether the essence of the actual conditions with respect to the travel cost on a network is captured by a selected model or not can be intuitively verified by rendering various graphical displays based upon travel cost data. The analysis of travel cost based on a modeled flow pattern can provide various estimations of travel cost on the network. Procedures shown in Figure 4 can be used to display travel cost data on the network. From Figure 4, it is known

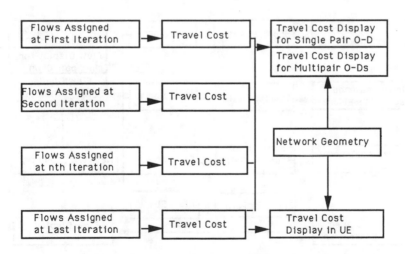

Figure 4   Procedure for Travel Cost Analysis in UE

851

that link travel cost can be calculated directly from the UE traffic flow pattern of the network, which is actually the assigned flows at the last iteration. To display travel cost with respect to its flow direction on a link, a band with its width proportional to its travel cost is placed in the side corresponding to the side that is ruled by traffic regulation (left side for the United States). The analysis and display of travel cost for single O-D pair and multiple O-D pairs all require shortest path calculations between defined origin(s) and destination(s), then the cost along the calculated path(s) is computed with the UE flows, e.g. the assigned flows at the last iteration.

Visualization of Traffic Flows

To some extent, the essential part of data examination and interpretation in traffic flow modeling is the analysis of the assigned traffic flows. On the one hand, traffic flow is an important element that not only can be used to make a good sense of whether the modeled traffic flow pattern is close to the actual one, but also needs to be referenced in project design, traffic control and route guidance, etc. On the other hand, traffic flow is the basis upon which the analysis of other aspects of the network can be carried out. For example, both the shortest path and link travel cost have to be calculated from traffic flow data. Figure 5 outlines various procedures needed to carry on the assigned traffic flow analysis which include procedures for the analysis of overall flow assignment, flow assignment from one origin to one destination, flow assignment from one origin to multiple destinations, flow assignment from multiple origins to one destination and also the composition of a flow on a path with respect to where the flow comes from and where the flow terminate. A key technique in these procedures, however, is the method to calculate the share of the UE flow pattern from the assigned flows at each iteration, which requires that the actual assignment taking place in model calculation can be restored. Because the assigned flows for each iteration are already recorded, the restoration of the flow assignment can be

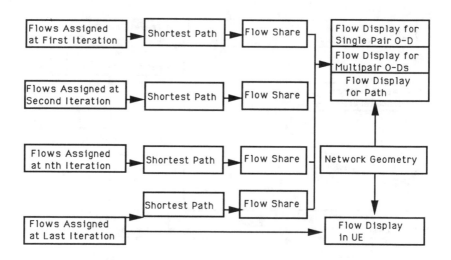

Figure 5   Procedure for Traffic Flow Analysis in UE

852

accomplished with much less calculation effort than that the actual assignment in model calculation needs. For graphical representation of traffic flows, the bandwidth method can also be used to portray their volumes, directions, etc.

CONCLUSION

In this study, a prototype program with the applications of the concepts and techniques presented above has been constructed. Based on a preliminary experiment with this program, it is concluded that these concepts and techniques can provide not only a general framework and detailed procedures to carry on model calculation and data visualizations in traffic flow modeling, but also provide the ability to achieve high computational efficiency in model calculation and data visualization, which will be the key to the applications in IVHS or in other related real-time transportation operations. Even though this study still has a limited scope in exploring the potential of the technology provided by various disciplines including computer graphics, traffic flow modeling and GIS for traffic flow modeling and visualization, it has demonstrated that the introduction of visualization tools can offer promise for complex data investigation in real-time or near real-time so as to make real-time traffic flow modeling a useful tool in IVHS.

ACKNOWLEDGMENTS

This research is conducted under the guidance of Professor Duane Marble. Professor Duane Marble and Professor Morton O'Kelly offered many invaluable comments and suggestions for the paper.

REFERENCES

Allard, L. and M.J. Hodgson, 1987, Interactive Graphics for Mapping Location-Allocation Solutions: The American Cartographer, Vol. 14, No. 1, pp. 49-60.

Becker, R.A. et al. 1990, Dynamic Graphics for Network Visualization: Proceedings of the First IEEE Conference on Visualization '90, San Francisco, California.

Euler, G.W. 1990, Intelligent Vehicle/Highway Systems: Definitions and Applications: ITE Journal, Vol. 60, No. 11, pp. 17-22.

Farrell, E.J. 1987, Visual Interpretation of Complex Data: IBM Systems Journal, Vol. 26, No. 2, pp. 174-199.

Grotjahn, R. and R.M. Chervin, 1984, Animated Graphics in Meteorological Research and Presentations: Bulletin of American Meteorological Society, Vol. 65, No. 11, pp. 1201-1220.

Tobler, W.R. 1987, Experiments in Migration Mapping by Computer: The American Cartographer, Vol. 14, No. 2, pp. 155-163.

ANIMATION AND GEOGRAPHIC INFORMATION SYSTEM FOR EVALUATION OF PROJECT PERFORMANCE

Lee-Chang Yang, Deputy Director
Performance Evaluation Department
Council for Economic Planning and Development
Executive Yuan
87 Nanking East Road, Section 2
Taipei, Taiwan 10408, R. O. C.
Tel: 886-2-5624490   Fax: 886-2-5413600

ABSTRACT

Council for Economic Planning and Development is in charge of economic planning and development of the whole nation. Major projects are subjected to strict evaluation to ensure that they can achieve their objectives. This paper will discuss how visualization of project performance can be performed with GIS and animation. Thus, project performances can be reviewed both in spatial and time domains. Traditional approaches were implemented to evaluate project performance providing tabular information and statistical charts. A self-developed GIS was used to depict the environment in which all projects occurred and to overlay related information for spatial analysis. Animation is performed to indicate relative project performances   and performance of a single project changing in a period of time. Terrain animation is implemented for viewing results of spatial analysis and performance changes. All results can be obtained by mouse clicks in a windowed environment.

INTRODUCTION

Council for Economic Planning and Development is in charge of economic planning and development of the whole Republic of China. There are 47 important projects with a budget exceeding 120 billions U.S. dollars under strict monitoring now. Project performance is evaluated on its progress report and expenditure relative to its approved budget. The most important issue is to bring every single project following its schedule. For one single project which is behind its schedule what management prescription should be taken is also enforced.

All the 47 projects can be categorized as 11 groups, namely, agriculture, water conservancy and inundation management, transportation and communication, urban and housing, recreation, culture and education, energy exploration, industry, environment protection, medical and health care, and regional development. Traditional evaluation of project performance is based on tabular information and statistical charts. This paper brings animation into its analysis functions and trying to link every project into its relevant data bases. Thus, animation and geographic information system (GIS) can provide new horizon to evaluation of project performance. Above all, at the implementation stage,

854

anyone can simply click mouse in a windowed environment to get all desirable information, even in a all Chinese environment.

METHOD

Traditional evaluation approaches were used to evaluate all major economic projects firstly. All output involving tables, statistics, charts, and recommendations then were configured into a Microsoft Windows environment. GIS was implemented to extracted associated spatial information from the existing data bases. Not every project has its own spatial data base. Overlay analysis and inquiry are two GIS functions that can be implemented right away. Animation is trying to depict statistics to its associate locations and also to display project progress changes in a sequence of time intervals.

A Windowed Environment

Mircosoft Windows provides "dynamic data exchange" function which enables two application programs to talk to each other by automatically exchange data. All tables, charts, project contents and management prescription can be reviewed by mouse clicks and a word processor is required. Figures 4 to 6 indicate the interface which was made by Visual BASIC. Interface on figures 4 to 6 can be changed into Chinese very easily once a Chinese version of Windows is used.

GIS Analysis

A commercial GIS package can be used in this regard. However, a self-developed GIS was used (Yang et. al. 1990). Reservoir construction and inundation management projects are suitable for overlay analysis and spatial inquiry. GIS analysis is confined to data base and its supporting functions.

Animation

3D Studio was used to do all animation works. An animation process is not so easy that only consists of several commands. It requires a little bit of training. If an animation file has been made, it is very easy to bring animation on a color monitor, say, no more than a mouse click.

MATERIALS

All 47 projects under strictly monitoring can be evaluated by above methods. This paper only discusses three reservoir construction projects and three inundation management projects in detail.

RESULTS AND DISCUSSION

Figure 4 indicates that all projects can be reviewed by its associated group. One moves mouse to the desired button and clicks it, then the more detail menu will show. If "Water Conservancy" was clicked, then figure 5 will display.

855

There are three reservoir construction buttons on figure 5. If "Nanhwa" was chosen, figure 6 comes up. Figure 6 shows all available information for the Nanhwa reservoir construction project. Table 1 will be obtained if "Progress - Table" button was clicked. Press "Progress - Chart" and figure 1 will show. Move cursor to "Budget - Expenditure" and click the cursor, then table 3 will display. "GIS Data Base" will bring the existing data base which is consisting of topography, land use, parcel map, road.

The "Progress - Animation" button will bring color animation on a color monitor. Every animation consists of at least 30 frames. Three monthly changes of ratios of expenditure to its approved budget of the Nanhwa project was created in the data base and can be shown on the color monitor by mouse clicks. Animation of ratios of expenditure to its approved budget for three reservoir construction projects for a given month was also made. Animation of monthly progress for the three reservoir construction projects was ready for display. Animation of monthly progress consists of 30 frames such that the first 10 frames show progress of January 1992, the second 10 frames show progress of February 1992, and the last 10 frames display progress of March 1992. An animation of monthly progress for the three inundation management projects indicates - 10.22%, +1.24%, and +0.07% respectively. For the Tali river project, there is a menu similar to figure 6. Parcel map of the Tali river can be shown by mouse clicks too.

Animation is a sequence of quick image display. A computer monitor display an image at a refresh rate of 1/30 second such that a 30-frame animation will show a very smooth "flying through".

A new project or new item can be added to figures 4 to 6 very easily and deleted too. Statistical charts can be automatically obtained and stored once an associate table was made. This is a function of "dynamic data exchange" provided in a windowed environment. GIS will bring spatial information on a monitor consisting of graphics, maps, images,, and attributes. Management prescription can be made more effectively because of spatial information of a given project can be obtained right away.

CONCLUSION

Major economic projects have profound impacts to the whole nation. Traditional approaches for the evaluation of project performance provide tabular information, statistical charts, and management prescriptions to ensure that every project on the right track. This paper brings animation as an improved method of viewing project performance changing in a sequence. GIS analysis has given new horizon to a project which is behind its schedule such that the government agency in charge can prescribe an effective measure. Animation and GIS is a good combination in solving evaluation of major economic projects. This paper is not only providing improved methods but also more user friendly environment. Further study should be addressed on topics trying to make animation an automatic process.

REFERENCES

Burger, Peter, Duncan Gillies. 1989. Interactive Computer Graphics. Addison-Wesley Publishing Company. 504 pages.

Energy Resource Laboratory, ITRI. 1990. Building Environmental Geology Data Base for the GIS of Li-Yu-Tan Reservoir Basin. ERL Report, No. ERL-79-R120. 46 pages.

Sanchez, Julio. 1990. Graphics Design and Animation on the IBM Microcomputers. Prentice Hall, Englewood Cliffs, N. J., U. S. A. 398 pages.

Taiwan Provincial Water Conservancy Bureau. 1990. A Pilot Study of Inundation Information Investigation Using Image Processing Techniques. Taiwan Provincial Water Conservancy Bureau project report. 91 pages.

Watt, Alan. 1989. Fundamentals of Three-Dimensional Computer Graphics. Addison-Wesley Publishing Company. 430 pages.

Yang, Lee-Chang, Meng-Hsi Chu, Mu-Lin Wu. 1990. A System Design for a PC-based 3-D Geographic Information System. In "Remote Sensing Science for the Nineties", Proceedings of IGARSS 1990, University of Maryland, U. S. A. pp. 2277.

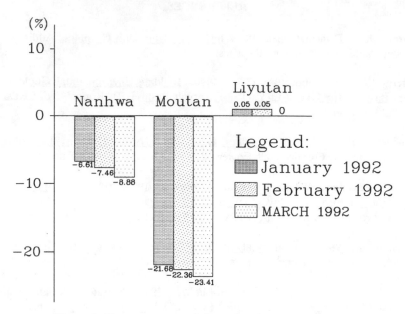

Figure 1. Progress report of the reservoir construction.

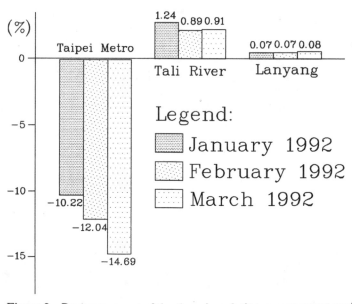

Figure 2. Progress report of the three inundation management projects.

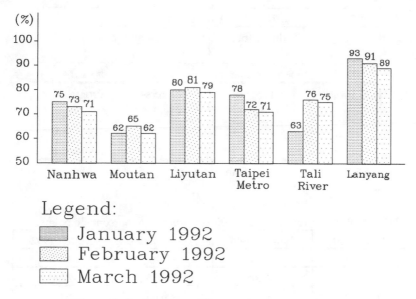

Figure 3. Ratio of expenditure to its approved budget.

PROGRESS REPORT BY GROUP		
AGRICULTURE	WATER CONSERVANCY	TRANSPORTATION
HOUSING	RECREATION	EDUCATION
ENERGY	INDUSTRY	ENVIRONMENT
MEDICAL CARE	REGIONAL DEVELOPMENT	

Figure 4. Inquiry menu of progress report by group.

PROGRESS REPORT		
GROUP TYPE:	WATER CONSERVANCY	
PROJECT:	RESERVOIR CONSTRUCTION	
Nanhwa	Moutan	Liyutan

Figure 5. Inquiry menu of progress report by project.

NANHWA RESERVOIR			
Administration:		Taiwan Provincial Government	
Agency in charge:		Taiwan Water Co.	
Project Content	Progress-Table	Progress-Chart	Progress-Animation
Budget-Expenditure		GIS Data Base	Discussion and Recommendation

Figure 6. All information available to a single project at Nanhwa reservoir.

Table 1. Progress report of reservoir construction in percentage points.

Reservoir	January 1992	February 1992	March 1992
Nanhwa	-6.61	-7.46	-8.88
Moutan	-21.68	-22.36	-23.41
Liyutan	+0.05	+0.05	0.00

Table 2. Progress report of the three inundation management projects in percentage points.

Project	January 1992	February 1992	March 1992
Taipei Metropolis	-10.22	-12.04	-14.69
Tali River	+1.24	+0.89	+0.91
Lanyang	+0.07	+0.07	+0.08

Table 3. Ratio of expenditure to its approved budget in percentage points.

Project	January 1992	February 1992	March 1992
Nanhwa	75	73	71
Moutan	62	65	62
Liyutan	80	81	79
Taipei Metropolis	78	72	71
Tali River	63	76	75
Lanyang	93	91	89

INCREASE GRAPHICAL DATA CONVERSION PRODUCTIVITY USING MULTI-TOLERANCE DATA CLEANING

Guangyu (Gary) Zhang
Xiaoqing (Rosina) Yei
MRF Systems Inc.
Suite 400, 604 - First Street S.W.
Calgary, Alberta, Canada T2P 1M7
Phone: (403) 265-3934
Fax: (403) 233-9188

ABSTRACT

Graphical data conversion is a process of converting graphical data from one format to a digital, more useful format. It is the bottle-neck of establishing a GIS. The major task of graphical data conversion is to correct errors in the input data such as over-shooting, under-shooting, duplicate points and lines, near-duplicate points and lines, etc. Almost every GIS vendor has a data cleaning package which uses a single tolerance in the cleaning process. This paper discusses a multi-tolerance data cleaning technique which preserves the accuracy of the input data and improves the productivity of graphical data conversion.

INTRODUCTION

Data Conversion can be defined as the process of converting geographic data from one format to a more useful one. It has been considered as the bottle-neck of establishing a GIS. In a data conversion process, we convert not only graphical data such as streets, parcels, and rivers, but also attribute data such as street names, river names and parcel owners. This paper will discuss techniques that will increase the productivity of converting graphical data.

In graphical data conversion, our input may be:

Case 1: paper maps,
Case 2: files containing graphical data which were scanned and vectorized,
Case 3: files containing raw data such as data captured using a total station or a stereoplotter,
Case 4: files from other digital mapping systems.

Paper maps can be digitized using a digitizing table or scanned and vectorized. The end result is a file of raw data to be further processed. Files from other digital mapping systems can also be treated as raw data if they have not been properly processed. So it is safe to assume that the input to editing/processing of the data conversion process is a set of raw data. The objectives of the editing/processing are to:

correct over-shooting,

861

correct under-shooting,
remove duplicate points and lines,
remove near-duplicate points and lines,
join singly-connected lines to form longer lines.

The following sections will discuss the need of multi-tolerance data cleaning, some test results, and some recommendations.

THE NEED OF MULTI-TOLERANCE DATA CLEANING

Many geographic features in different feature classes share common boundaries. The following are some examples:

A parcel and a street,
A forest stand and a river,
A park and a street,
A census tract and a street [Goodchild 1978].

Many government agencies collect, update, and disseminate certain commonly used geographic data [Steinitz et al 1976]. For many organizations, it is more cost-effective and timely to use data collected by the government agencies. It is likely that different data sets do not agree with each other. For example, forest boundaries may overlap with lakes. If these inconsistencies are not removed, the usefulness of the data sets will be significantly reduced. This situation requires that, in the editing/processing stage, the related feature classes should be processed together and not separately.

Different data sets were probably collected with different accuracy. For example, a parcel boundary may be more accurate than a school district. Almost every GIS vendor has a data cleaning package to break lines, arcs, etc. at intersections, remove over-shooting, under-shooting, and duplicate points and lines. Most, if not all, of the packages perform data cleaning based on a single tolerance which can usually be set by the user. Single-tolerance data cleaning has a few drawbacks:

1. Long manual editing time.

Assume that a forest company gets 1:20,000 digital topographic maps which have not been processed from a provincial/state government. The company has collected its forest stand maps at a scale of 1:45,000. Due to the difference in scale, the 1:20,000 maps are more accurate than the 1:45,000 forest stand maps. Assume that the typical over-shooting and under-shooting on the 1:20,000 maps and 1:45,00 0 maps are 2 meters and 4.5 meters respectively. If we use a single tolerance of 4.5 meters, points on the 1:20,000 maps which are within 4.5 meters of each other will be clustered together thus resulting in loss of accuracy. If a single tolerance of 2.0 meters is used, many of the over-shooting and under-shooting on the 1:45,000 maps will not be corrected automatically. They will have to be corrected manually and it is well known that manual editing time-consuming.

862

2. Random movements of points.

It is desirable to move points on the 1:45,000 maps to points on the 1:20,000 maps because the points on the 1:20,000 maps are more accurate. If a single tolerance is used, usually the movement of data is random, i.e., points on the 1:20,000 maps may move to points on the 1:45,000 maps.

Multi-tolerance data cleaning can be achieved using multi-tolerance fuzzy intersection and multi-tolerance clustering techniques [Zhang and Tulip 1990]. The accuracy of the intersection points between features of different feature classes (with different accuracies) can be rigorously estimated and stored in the GIS to assist users in their decision-making [Zhang 1991].

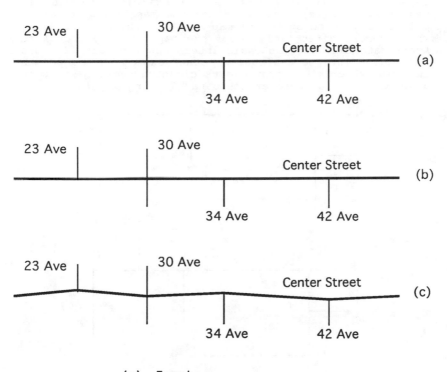

(a) Input;
(b) moving minor streets to major street;
(c) moving major street to minor streets.

Figure 1. Cleaning of a street network

Using multi-tolerance techniques, we can assign a tolerance of 2.0 meters to points on the 1:20,000 maps and 4.5 meters to points on the 1:45,000 maps. No point will be moved outside of its tolerance in the cleaning process to preserve the accuracy of input data. Points with larger tolerances will move to points with smaller tolerances. For example, if a point (A) on the forest stand boundary (from 1:45,000 maps) is closer than 4.5 meters to a point (B) on a lake boundary (from 1:20,000 maps), A will be moved to B. Typical over-shooting and under-shooting on all input layers will be corrected automatically. When input points have identical tolerances, the movement of points is performed in such a way that straight lines are kept straight - a desirable behavior for many GIS/LIS applications. For example, if you are editing a digitized street network, you will want the end points of minor streets to move toward the major street in order to keep the major street straight [Figure 1]. With multi-tolerance data cleaning, users have control over the movements of data. In mapping applications, for example, the user can assign the smallest tolerance for the map boundary so that points will be moved to the map boundary whenever possible. This process effectively keeps the map boundary unchanged. After data cleaning, no chain intersects with itself and no two chains intersect with each other except at end points. In summary, multi-tolerance data cleaning overcomes the two drawbacks of single-tolerance data cleaning.

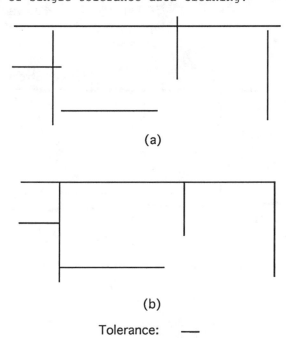

(a)

(b)

Tolerance: ───

Figure 2. Multi-tolerance cleaning of one layer

TESTING RESULTS

MRF Systems Inc. developed a new algorithm that improves the effectiveness and efficiency of the multi-tolerance fuzzy intersection and clustering technique [Zhang and Tulip 1990]. The new algorithm recognizes a line as a basic entity type and tries to cluster lines together whenever possible. As shown in the examples, this algorithm yields more desirable results than the purely point-based clustering technique.

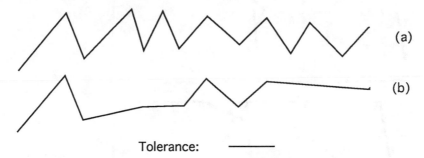

Tolerance: ————

Figure 3. Objective generalization using multi-tolerance cleaning

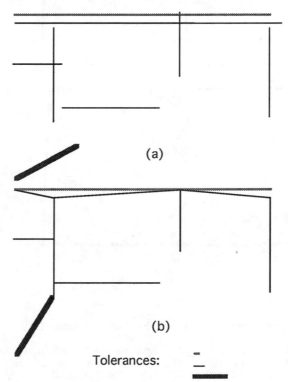

(a)

(b)

Tolerances:

(a). Input   (b). Output
Figure 4. Multi-tolerance cleaning of three layers

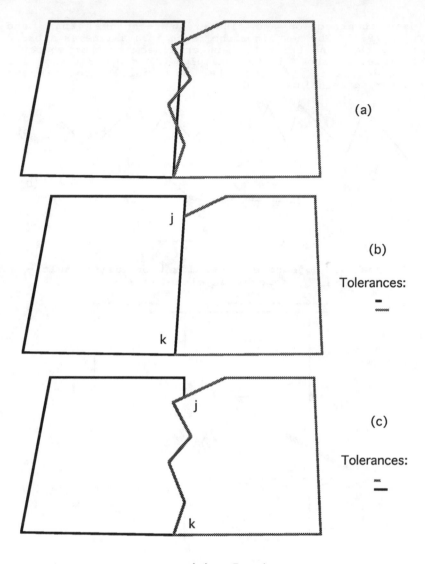

(a)

(b)

Tolerances:

(c)

Tolerances:

(a). Input
(b). Output 1
(c). Output 2

Figure 5. Line-based clustering

The new algorithm was implemented using MicroStation Development Language (MDL) and the released software package is called "MRF CLEAN". "MRF CLEAN" runs anywhere MicroStation runs and can be easily ported to other platforms since MDL uses the standard C programming language. The following testing results demonstrate the capabilities of "MRF CLEAN". They were obtained on a 486 PC using MicroStation as the graphics engine.

Figure 2 shows the behavior of "MRF CLEAN" when only one layer is cleaned. As mentioned earlier, "MRF CLEAN" tries to keep straight lines straight.

Figure 3 indicates that "MRF CLEAN" can be used to do an objective generalization. This capability removes excessive points generated by "stream" mode digitization. It can also be used to perform an automated scale change. For example, the federal government and the provincial government share the cost of mapping the Province at a scale of 1:20,000. However, the federal government needs the maps at a scale of 1:50,000. An objective generalization will cut down the editing time considerably. Note that the generalized line preserves the local characteristics of the input. It is transparent to users and yields satisfactory results. Initial testing indicates that it does not have the drawback the Douglas-Peucker algorithm has when generalizing polygons [Zhang 1990, Douglas and Peucker 1973].

In the cleaning process, a chain from a certain layer is generalized using the tolerance set for the layer. That is, chains from different layers are generalized based on different tolerances. This approach is more flexible and more appropriate than the one that applies generalization to all input chains using one uniform tolerance.

Figure 4 illustrates the advantages of using multi-tolerance data cleaning. Points with larger tolerances move to points with smaller tolerances and no point is moved more than its tolerance in clustering.

Figure 5 demonstrates the recognition of lines as entities in cleaning. Pure point-based clustering does not always yield such a satisfactory result. This ability connects line work nicely as long as the points are within the tolerances. No time-consuming snapping is necessary and operator productivity should improve considerably.

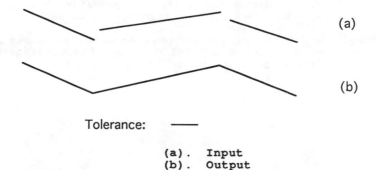

(a)

(b)

Tolerance: ——

(a). Input
(b). Output

Figure 6. Joining singly-connected lines to a longer one

"MRF CLEAN" joins singly-connected lines to form a longer line after clustering [Figure 6]. Figure 6 (a) shows three input lines while Figure 6 (b) shows the output where we have only one line.

DISCUSSION

A data cleaning package is one of the most important and most frequently used modules of a GIS. It is also the first defense against entering inconsistent data into a GIS [Zhang 1990]. The multi-tolerance cleaning technique not only minimizes the editing efforts and preserves the accuracy and quality of input data but also reduces the chance of introducing inconsistent data into the GIS since related layers can be processed together.

REFERENCES

Douglas, D.H. and T.K. Peucker (1973) "Algorithms for the Reduction of the Number of Points Required to Represent a Digitized Line or Its Caricature", The Canadian Cartographer, Vol. 10, No. 2, pp. 112-122

Goodchild, M.F., (1978), "Statistical Aspects of the Polygon Overlay Problem", First International Advanced Study Symposium on Topological Data Structures for Geographic Information Systems, Volume 6, Laboratory for Computer Graphics and Spatial Analysis, Graduate School of Design, Harvard University.

Steinitz, C.F., Parker, P. and Jordan, L., 1976. "Hand-drawn overlays: Their history and prospective uses. Landscape Architecture, pp. 444-455.

Zhang G. [1990] "Managing Consistency in a GIS", CISM Journal, VOL. 44, No. 4, pp. 417-423.

Zhang G. and J. Tulip (1990), An Algorithm for the Avoidance of Sliver Polygons and Clusters of Points in Spatial Overlay, Fourth International Symposium on Spatial Data Handling, July 23-27, 1990, Zurich, Switzerland.

Zhang, G. (1991) "Estimating the Accuracy of Areas in Simultaneous Multi-layer Spatial Overlay", Proceedings of EGIS'91, Brussels, Belgium. pp.

APPROACHING QUALITY GIS INTERFACE

Wang Zheng
Ph.D. Candidate

V.O.Shanholtz
Associate Prof./Director

Information Support Systems Laboratory
Department of Agricultural Engineering
Virginia Polytechnic Institute & State University
106A Faculty St., Blacksburg, VA 24060

ABSTRACT

A quality software must have software producibility, a goal of software engineering, and software usability, a goal of interface engineering. To the user, the user interface is the system. Despite significant accomplishments in human-computer interaction, molding important issues into a quality GIS user interface is still a somewhat elusive effort. This paper focuses on three critical issues for developing a quality GIS interface.

- Interface Development
- Key: User Centralness
- Interface Design Guidelines

INTRODUCTION

In the last decade, very active research programs have evolved in the area of human-computer interaction (HCI). Research and development have resulted in significantly improved techniques such as desk- top metaphors, direct inter-action, pull-down menus, asynchronous dialogue and graphical user interfaces. These techniques provide an opportunity to develop quality user interfaces, which not only must include aesthetically pleasing screens, but the interface must strongly emphasize software usability. Software usability is usually evaluated and/or measured by ease of learning and using, speed of user task performance, user error rate and subjective user satisfaction.

The area of human-computer interaction covers many disciplines. This paper does not deal with human factors, nor cognitive psychology; but by necessity it does discuss aspects considered critical for a quality GIS interface. In the first section, interface development is presented with principal focus on interface life cycle. The second section discusses user centralness, which has been significantly affected by current GIS data models. Interface design guidelines are covered in the last section.

INTERFACE DEVELOPMENT

A prevailing view of software engineering is that of a top-down, linear life cycle (Figure 1) (Boehm, 1988), which is being used for many software developments. Since this linear process somewhat isolates activities for requirements, design, implementation and testing, attempts to impose the classical "waterfall" paradigm on interface development are undoubtedly the cause of many poor interfaces. Based on empirical observations, a star life cycle for human-computer interface development was introduced by Hartson and Hix (1989), as shown in Figure 2 as a star-shaped network configuration with evaluation at the center. Unlike the conventional linear "waterfall" life cycle or its wheel-shaped variation with feedback for global iteration, this star life cycle is specifically for improving usability of the interface, and it revolves around continuous testing of the interface throughout the entire development process.

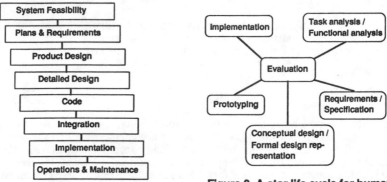

Figure 1. A linear "waterfall" life cycle

Figure 2. A star life cycle for human-computer interface development

Indeed, this star life cycle has given us a great idea for iterative refinement in developing user interface. However, it visually appears that interface development (focus of interface engineering) and software production (focus of software engineering) are separated. This impression comes up with the fact that some design activities such as specification, task/functional analysis and implementation are specified in the star life cycle as if they apply only to interface development. In fact, some of these activities are commonly shared processes, and the other (i.e. implementation) should be integrated through common efforts of both interface people and software people. To define a more applicable interface development management with more emphasis on GIS interface, a modification of the star life cycle is suggested (Figure 3), which is based on the following points.

- The processes, specification, task/functional analysis and implementation should be integrated with the similar processes specified in the life cycle of software production.
- Requirements and conceptual design are the processes that require interface developers and application developers (i.e. users) to work together. Thus products of these processes will contain user opinions/ideas/evaluation.
- To make the life cycle visually simple and easily managed, activities with similar operations, e.g. analysis, specification and requirements, which all produce documents, should be grouped together.
- Emphasizing effects of communication with GIS user during interface development, i.e. taking advantages that most of GIS users have basic skills of understanding of graphics, which is also an important characteristics of a GIS interface. Therefore, communication would be by using visual "language" or visual design as shown in Figure 3.

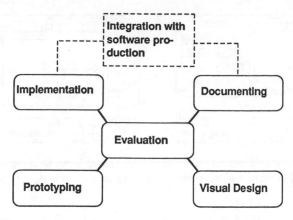

Figure 3. A star life cycle for GIS interface development

In the modified star life cycle, the documentation and implementation processes are connected with software production, i.e. some design activities in these two processes should be shared by both software development and interface development, and activities that are related to interface development are evaluated by the user. Activities for each process in the modified star life cycle are described as follows.

I. Documenting

- Goal analysis: objectives and short/long term goals.
- Task analysis: specify application tasks.
- Identify application objects, their properties and relationships.
- Specify both user's and software developer's perspectives of the objects.
- Functionality analysis: specify functionalities and technologies.
- User analysis: user characteristics and skills.
- Interface design guidelines.
- Design rules: detailed and specified design requirements based on the design guidelines.

II. Visual Design; To accomplish effective communication with GIS user in evaluation process, visual design would be the most desired representation technique. Two activities are suggested for the visual design.

- Station control design (SCD): SCD is a graphical representation of control flow within GIS interface. Station is used to group related tasks. For example, "Data Layer View" may be treated as a station, containing a group of tasks such as map display, attribute data display, zoom, query, etc. (refer to Figure 4). A station may contain several screens. A GIS interface may consist of different stations.
- Scenario design: developing a visual design/layout for the screens. It can be implemented with paper and pen or graphical edit programs, and should be deliverable, e.g. drawing a few representative screen sketches on transparencies or making slides. Figure 4 shows an example of two screens in "Data Layer View" station.

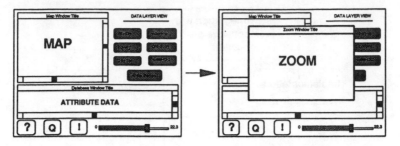

Figure 4. Scenario design on station of Data Layer View. Clicking "zoom" button in left screen, a zoom window pop-up as shown in right screen.

III. Prototyping; Using prototyping tools such as Open Look, OSF MOTIF, Windows, or HyperCard to produce an executable version of a partially specified interface (vertical prototyping) or a full-featured interface with less depth in functionality (horizontal prototyping), which allows the user to "take it for a spin", encourages early user participation and involvement. This technique allows early observation of user behaviors, and supports iterative refinement in response to user feedback.

IV. Implementation; This is an integrated process with software production. The coding for the interface component may be either evolutionary maturation or revolutionary maturation. Iterative refinement is still required in this stage because many computational components for realistic tasks can not be tested with the interface prototype. It is virtually impossible for an interface developer to provide realistic or appropriate messages in response to computational conditions.

V. Evaluation

- Subjective evaluation; obtain user opinions on design documents, visual design, prototype, or final product.
- Objective evaluation; activities include developing experiment, i.e. preparing facilities, sites, monitoring system, etc., observing user performing specific tasks, collecting experimental data (e.g. user error rate and speed of performance), analyzing the data, drawing conclusions and redesigning.

KEY: USER CENTRALNESS

"User-friendly interface" has been widely used as a "buzz word". If an interface is not user centralness, it is impossible to achieve user-friendliness. User centralness contains two facets: (1) the objects that the system should be structured around; and (2) techniques for developing user-centered dialogues. This section deals with the first issue while the second issue will be discussed in a later section on design guidelines.

To model and/or simulate real-world problems, geographers subjectively break the real-world entities into many different data layers (e.g. soil, landuse, hydrography, etc). This abstraction is established based on the users' perspective of data. However, to handle these data layers more efficiently in computer systems, GIS software developers further abstract different data types such as raster, vector and text-attribute from the data layers. This abstraction is based on the software builder's perspective of data; it is considered as very "low level" knowledge to the users, and should not be "seen" by the users. However, many current GISs do require their users to "touch" these "low level" data. This situation tends to require the user to have similar knowledge and skills as software developers as illustrated in Figure 5. For this situation, both GIS and its interface are structured around the "low level data", which is neither natural nor necessary. The users' focus should be on real-world problem solving rather than the "low level" data manipulation. This type GIS interface will unavoidably have the following limitations.

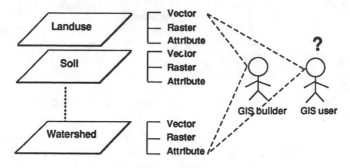

Figure 5. Perspective of GIS data.

- Poor quality; because it is system centered, not user centered. To manipulate different data types, the user must have a good understanding of data structures, naming, conversions, etc. This has required a very high price by the user, i.e. high "personware" cost.
- Existence of commonly shared data stored in each individual data type may result in inconsistency.
- Since the user is allowed to manipulate each individual data type, more human-computer interactions have to be performed. This is why so many GISs contain different subsystems to manage/manipulate vector data, raster data, and attribute database.
- More implementation efforts have to be made for the interactions. Research has shown that as much as 88% code is needed to support human-computer interactions in developing an interactive computer system. The larger the interface, the more code required.

As mentioned before, difficulty results when the GIS and its interface are structured around the "low level" data. Structuring around "low level" data is generally due to an incomplete definition of the data model. Data model is an integration of the two factors: (1) definitions for all data types, properties and their relations, and (2) strategies/methods for data storage and retrieval. If relationships or connections between user-defined objects (layers) and GIS builder-defined objects (data types) are not defined (e.g. something is missing between layers and data types as shown in Figure 5), the GIS data model is considered incomplete. If a data dictionary (DDIC) (Zheng, 1992) is created to "complete" the definitions for GIS data model, significant improvements can be accomplished. The DDIC functions as both spatial database administrator and a connector for data layers and different data types. It is because of the creation of a DDIC, the user's view point now can be "moved" as shown in Figure 6.

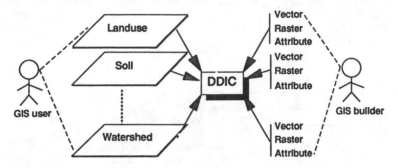

Figure 6. Improved perspective of GIS data.

Under this circumstance, the user is able to "see" the data layers directly, and the user interface now can be constructed around the data layers, which are abstracted directly by the users, and with which the users are more familiar. **The goal is to have the user manipulate data layers, and all necessary operations on the different data types performed in the background.** It is also because this type of GIS is structured around the "higher level" objects, code for human-computer interactions can be minimized and implementation becomes relatively simple.

GIS INTERFACE DESIGN GUIDELINES

Interface design guideline is an important composition of design documents, and they are usually generally worded. Good design guidelines not only establish development principles, but also help creativeness during the period of interface development. The GIS interface design guidelines discussed here are tailored based on Hartson's (1991) work for general human-computer interactions. As appropriate, a brief discussion is given relative to GIS interface to provide a perspective of scope of interface design guidelines.

1. **User Centralness**
   - Know user behavior and tasks; This is just as people can not develop deeper conversation without knowing each other very well. The most important characteristics of GIS interface is that many graphics such as maps, images and diagrams would be displayed, very often accompanied with various textual information to describe attributes or states of the spatial data. Thus, special attention should be paid to avoiding

874

cluttered desk syndrome. On the other hand, appropriate interaction style should be used such as direct interaction (e.g. moving cursor with mouse) for ease of navigation over the spatial data on screen.

- Use user-centered responses, not system-centered;

   a. Instructions: To conduct efficient dialogues between GIS and its user, familiar and understandable terminology and concepts should be used. For example, use generally accepted layer names instead of file names and use feet or meters as distance unit instead of pixels or cells. Effective instructions show user where to go and what is available.

   b. Feedback: Give instantly and informative feedback to user when system is performing a lengthy operation or computation.

   c. Error messages: Because error messages have great psychological impact on the user, use carefully-worded error messages, e.g. do not use violent, negative, or demeaning terms, use positive tone and constructive terms, be specific and concise.

- Account for human memory limitations; e.g.
  - Use visual cues to help user recognition, rather than recall.
  - Use slide bar to let user select model required parameter value.
  - Use menus or list box to help user select actions or items.
- Desk-top publishing (DTP); DTP is one of the best examples for keeping locus of control with user. Try every effort to make user feel in charge as if the user is working at his desk.

2. **Consistency**
   - Be conceptual consistent; (e.g. GIS displays maps, not pictures, therefore, include appropriate scientific and mathematic data such as latitude, longitude, scale, legend, etc. in map display).
   - Be consistent within a single interface; e.g.
     - All windows should have caption bars and titles.
     - On-line help information always appears on the bottom line.
     - Detail instructions are always displayed within a pop-up window.
   - Be consistent within entire GIS interface; e.g.
     - Push-buttons are used to activate tasks.
     - "Help" is called by clicking on "question mark" icon.
     - Dialogue is always conducted within a pop-up dialogue box.
   - For similar semantics, use similar syntax; e.g.
     - Use the same type of menu for the same level of operation.
     - Double clicking for execution and single clicking for selection.
     - Pressing Esc key always allows a return to the previous screen.

3. **Display Issues**
   - Use color conservatively; Almost every screen in GIS interface contains colorful map(s) or image(s). Design monochrome for other interface objects such as windows, buttons, menus, icons and texts.
   - Maintain display inertia; e.g.
     - Change as little as possible from screen to screen.
     - Keep unchanging objects in fixed position.
   - Do not pile up or clutter screen.
   - Use structured format for text.

4. **Other Guidelines**

a. Error messages (in detail).
b. Getting user's attention.
c. System Model.
d. Modality and reversible actions.
e. Designing for individual differences.
f. Windows.
g. Menus.
h. Boxes.
i. Form filling.
j. Command language.
k. Graphical user interface.

CONCLUSION

User interface should focus on software usability. To reach this goal, interface development management is extremely important. Using star life cycle suggests an iterative refinement based on user-based evaluation, which can have software interface eventually improved. A central theme is that the user interface is not an "add-on", which is created at the last minute when the "rest of the system" is completed. If the system can not communicate with its user, it matters little how well it can compute. In addition, problem with data model, as a primary problem affecting user centralness, should be resolved. This problem has been impacting on achieving quality GIS interface for many years.

REFERENCE

1. Boehm, B. 1988. A Spiral Model of Software Development and Enhancement. IEEE Computer, 21(3), 61-72.
2. Hartson, H.Rex and D. Hix. 1989. Human-Computer Interface Development Concepts and Systems for Its Management, ACM Computing Surveys, Vol.21, No.1, March 1989. 5-85.
3. Hartson, H.Rex. 1991. Human-computer Interaction. Lecture notes for CS 5714, Department of computer science, Virginia Tech, 1991.
4. Zheng, Wang and V.O.Shanholtz. 1992. Data Model for Spatial Information Systems. Paper No. 927002, ASAE St. Joseph, MI, 6 pp.

TOWARDS INTELLIGENT SPATIAL DECISION SUPPORT: INTEGRATING GEOGRAPHICAL INFORMATION SYSTEMS AND EXPERT SYSTEMS

Xuan Zhu
Department of Geography
The University of Edinburgh, Drummond Street
Edinburgh EH8 9XP, U.K.
(031) 6502532

Richard Healey
ESRC Regional Research Laboratory for Scotland
Department of Geography, The University of Edinburgh
Drummond Street, Edinburgh EH8 9XP, U.K.
(031) 6502534

ABSTRACT

Geographical information systems(GIS) have played an important role in the design and operation of spatial decision support systems. However, there is now some indication that current GISs do not adequately facilitate the decision process. The need for improvement of current GIS performance has promoted the integration of Expert Systems (ES) with GIS. The first part of this paper examines alternative approaches to the integration of GIS and ES, and discusses the advantages and disadvantages of each approach. In the second part, the structure of a knowledge-based spatial decision support system development environment is presented that integrates a GIS with an expert system tool, the latter being used to develop a domain-specific knowledge base, perform reasoning in a domain and guide users through the stages of GIS processing. The methodological approach in the development of the project is described.

INTRODUCTION

Environmental scientists are faced with an increasing number of resource and environmental problems. These complex spatial problems require resource managers, environmental planners and other decision-makers to process large volumes of spatial information. In turn, the effective utilization of information systems has become a significant factor which affects their decision-making efficiency.

Spatial decisions are often complex and require the exploration of many options(Franklin, 1979). Generally, the process of decision making involves a series of steps from definition of problems to analysis and determination of possible courses of action, to selection of a course of action, and finally evaluation of the choice (Simon, 1977). Throughout the phases of decision making, complex sets of multivariate information are combined and synthesized with multiple criteria to evaluate problems. Geographical information systems(GIS) are often implicitly designed to assist decision makers in this process.

Typically, a GIS system has following capabilities:

(1) extracting and describing spatial problems and their spatial relationships

877

(2) storing and managing large quantities of complex and heterogeneous spatial data

(3) structuring the available information using spatial or geographical models

(4) effectively manipulating and analysing spatial data, providing spatial data handling facilities from simple map overlay to complex spatial analyses and providing various display facilities.

The areas of application range from the monitoring and modelling of resource and environmental problems to the support of management and planning decisions.

With these capabilities, GIS systems can assist in many stages of the spatial decision making. They can be used to describe the past and present situation of resources and environment, identify problems when a decision is required, and generate possible solutions. More importantly, GISs provide the ability to ask "what if" questions in order to evaluate alternative scenarios(Teicholz and Berry, 1983) and examine the future. However, experience of the application of GIS techniques has highlighted a number of problems or limitations with existing systems. These problems mainly arise from the fact that the spatial decision making is based not only on spatial data analysis and mathematical modelling, but also on experience, preferences, intuition, judgement and the expertise of various human specialists(Gardels 1987). Many spatial problems are ill-structured(Densham, 1991). An ill-structured problem would not lend itself well to an algorithmic solution because there are many possibilities. Thus, much spatial decision making requires approaches that are more heuristic than algorithmic. The current GISs require problems to be quantified so that some mathematical techniques can be applied to them. Moreover,

(1) they provide few facilities for automatically detecting problems existing in resource and environmental management and planning, determining the causes of these problems, suggesting alternative solutions and explaining the analytical results.

(2) They implicitly assume that all information encoded is absolutely correct and precise. In essence, a natural phenomenon may not be uniquely definable. Exact data models and modelling techniques in GISs cause loss of information, inconsistency of phenomenon and inaccuracy in analysis (Berry, 1987; Burrough, 1986).

(3) Current GISs do not facilitate user interaction during the solution process (Frank et al, 1991). While GISs allow spatial databases to be integrated and manipulated through a set of procedures, a comprehensive spatial analysis from initial data entry through product generation requires extensive user interaction in the sequence of processing steps.

There is the potential for some of these problems and limitations to be overcome with expert system technology.

Expert systems(ES) perform decision-making tasks by reasoning using domain-specific rules that have been judged by an expert in his domain to be true. They are best suited for ill-structured problems. Typical tasks for expert systems involve automated interpretation, diagnosis, monitoring, and planning sequences of actions(Jackson, 1990). They put emphasis on developing and understanding nonnumerical methods for problem solving. An ES typically contains an inference engine and a knowledge base. The knowledge base is composed of the knowledge about solving specific problems. It exposes knowledge about some domain explicitly via symbolic data structures. The inference engine is a set of procedures which operate upon the knowledge base. Users supply facts or other related information to the ES and receive expert advice or expertise in response. The same knowledge encoded in the knowledge base can be simultaneously used for more than one purpose, such as solving a given problem, explaining the solutions produced by the system and offering advice about the problem. In addition, the ability of an ES can be extended either by expanding the knowledge base or by adding facilities to the inference engine(Duda and Reboh, 1984). The distinctive strength of ES can be summarized as (Jackson, 1990):

(1) handling imprecise data, incomplete and inexact knowledge
(2) exploiting knowledge "at the right time"
(3) explaining and justifying the reasoning that led to a conclusion
(4) changing or expanding knowledge relatively easily.

There is ample scope for applying ES technology in decision making processes. For example, we may use knowledge representation techniques to characterise decision-making domains, use heuristic methods to generate and evaluate decision options, apply inference and reasoning to explain and justify decisions, etc. However, expert system technology alone does not adequately support spatial decision making. It has the following limitations:

(1) Spatial decision-making requires large volumes of spatial data. These data mainly reside in GISs and not in ESs. ESs lack facilities for handling large-scale data sets (Stonebraker, 1986).

(2) Expert systems are concentrated on symbolic reasoning and do not provide good arithmetic capabilities (Jackson, 1990). Yet, arithmetic operations are required in spatial data handling. Mathematical models are often used in spatial analysis as part of a solution process.

(3) ESs are lacking in spatial data handling capabilities such as buffering and overlay which are unique and important to spatial analysis. The inference speed is too slow for many problems typically handled by a GIS(Webster, 1990).

(4) ESs do not provide facilities for representation of complex spatial problems and for output in a variety of spatial forms such as maps, images and other types.

The integration of GIS and ES may avoid some of the limitations and difficulties existing in each of them. The spatial decision process can be made more effective within such integrated systems. A conventional GIS is very suitable to well-structured spatial problem solving, while the integration of GIS and ES offers a best approach to solving ill-structured spatial problems. For example, if a suitable area for a proposed land use is to be chosen, a GIS can be employed to identify potential areas suitable for the proposed land use through spatial analysis techniques. However, the GIS can not propose a most suitable area. This can be done by the ES. Moreover, in such an integrated system, the ES can assist in various ways in the provision of an intelligent interface to enable users to use the GIS in the most effective way and guide users through the stages of analysis. The integration of GIS and ES is proposed as a step towards an improved spatial decision environment.

This paper first examines alternative approaches to the integration of the two components, and discusses the advantages and disadvantages of each approach. Then, the outline achitecture of a knowledge-based spatial decision support system development environment is presented that integrates a GIS(ARC/INFO) with an ES(CLIPS), and the methodological approach used in this project is described.

APPROACHES TO INTEGRATING GIS AND ES

Recently expert systems have begun to be applied to spatial decision support. Many efforts have been made to develop ES applications in the areas of natural resource management and environmental planning (for examples, see Davis and Clark, 1989; Robinson et al., 1987). However, most of the systems did not involve integration of ES with GIS. Some employed the spatial data handling capabilities of a GIS to perform spatial analysis but did not interface directly to it.

To date various approaches have been developed to integrate ES with information systems for business. These approaches can be summarized in Figure 1. They can be used to guide the integration of GIS with ES.

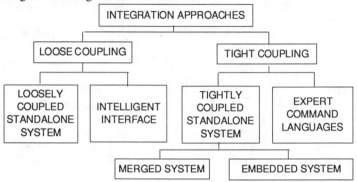

Figure 1   Strategies for Integration

LOOSE COUPLING APPROACH

In this approach, expert systems and GIS are "loosely" integrated by communication links. A communication channel exists between two components which transfers data from the GIS databases to the ES databases. This type of system is labelled a "loosely coupled standalone system".

Here, the GIS is used as a data source and a tool for displaying analytical results and other GIS processing, while the ES is used to develop a domain-specific knowledge base using rules, frames or logic which describe the criteria and relationships involved in the spatial analysis process and allow reasoning to be performed.

In this manner, the GIS creates data files by retrieving from the database or performing some GIS processing. These files describe different factors relevant to a specific application. They are written to some external files which can be read by the ES. Or, the GIS creates a data file which is a composite of several relevant coverages, then writes it to an external file. The ES carries out deductive reasoning on these input data and writes the analytical results to a prespecified file. It is then transferred back to the GIS for display or storage or further processing. Some systems use a specific program to activate the GIS, creation of a data file, data transportation, and execute the ES. ASPENEX (Morse, 1987) is an example of this variant.

Another method of loose coupling is to build an "intelligent interface" to a GIS, i.e. to incorporate expert knowledge about spatial modelling procedures and GIS operations into the knowledge base. This is a area which has attracted attention in the domain of integration of GIS and ES(Pearson et al., 1991; Usery et al.,

1988). As with a particular geographical application, the user is faced with a series of tasks: firstly, building the database relations and models; then, deciding modelling strategies; selecting appropriate data sets; choosing sequences of commands for analysis suited the data and the problems; and finally, displaying the results of the analyses or offering solutions to problems. These tasks involve knowledge about how to perform spatial modelling and how to run and use a GIS. Their implementation is often based on user's intuition and experience(Burrough, 1990). Although the command or macro languages provided by a commercial GIS can be used to create interfaces for non-specialists and inexperienced users, and can be used to perform spatial modelling automatically within a GIS environment, they can not be used to guide the user to proceed optimally through a set of operations to obtain the required answer. Just like traditional programming languages, they embed knowledge implicitly as procedures. They can only be used for one purpose and can not be easily changed or extended. So, it is appropriate to develop an intelligent front-end or interface to a GIS. Here, the knowledge base provides knowledge of how to use and run the GIS. We call the knowledge base a "tools knowledge base".

In this way, a command file is created and invoked by the ES according to the specific problem or situation. For instance, if the system is used to determine a set of land characteristics relevant to a particular crop, it simultaneously creates a command file which contains GIS commands to be performed for assessing these land characteristics. Then, the ES invokes and runs the command file to determine the land quality.

The loose coupling provides the ES with access to the data stored in the GIS or produced by the GIS. However, it does not provide the ES with the spatial data handling capabilities of the GIS. Application builders have to use different languages to create an application(King, 1990). In addition, the approach may not work if the data sets extracted from the GIS become too large to fit into the ES database. Other limitations are listed in Smith and Yiang(1991).

TIGHT COUPLING APPROACH

This approach can overcome some of shortcomings of loosely coupled systems. Tight coupling is to integrate ES with GIS using communication links in such a way that the GIS appears to the ES as an extension of its own facilities, or vice versa. Thus one component serves as a shell about the other. The system developed by this approach is called a "tight coupled standalone system". In relation to tight coupling, an ES can serve as a subsystem and be integrated with different subsystems of a GIS, which can be called a "merged system", or GIS subsystems can be extended to include deductive reasoning capabilities provided by an ES shell, which can be termed an "embedded system"(King, 1990).

In the former architecture, the system provides the user with a uniform language for implementation of each component, and reduces some of the inefficiencies inherent in the communication process, but the GIS still serves as a data source, and the ES does not provide the GIS with reasoning capabilities. In the latter, reasoning capabilities are added to different components of the GIS.

Similar to intelligent interfaces, tools expertise can be added to an ES in the tight coupling approach. That is to add ES reasoning capabilities to the "command" or "macro" languages of a GIS. This architecture is labelled as an "expert command language" approach. Few examples exist of integration in this way, even in applications for business. According to King's description(1990), expert command languages encompass elements of a standard ES shell. They can provide commands and an associated inference engine for defining and running sets of system operations needed to solve a problem.

INTEGRATION OF ARC/INFO AND CLIPS

A knowledge-based system development environment for spatial decision support is currently under development. It interfaces ARC/INFO with CLIPS through loose coupling. CLIPS (NASA, 1991) is a data-driven, rule-based expert system tool developed by NASA at the Artificial Intelligence Section of the Johnson Space Center. The rationale for employing loose coupling is that only minimal modifications to each system have to be made. Thus, the large-scale spatial decision support system development can be expedited through the loose coupling of GIS and ES, each representing the best available technology in a particular application area. Because the focus is on the integration rather than the individual programs, detailed examples of the individual operation of ARC/INFO and CLIPS will not be given in this paper.

THE ARCHITECTURE OF THE SYSTEM

In this system, ARC/INFO is used to acquire data describing the spatial entities in different forms and their associated attributes, build databases, and perform basic GIS operations such as overlay, proximity analysis, measurements, graphic and tabular display, etc. It prepares data through a series of processing steps for CLIPS reasoning and displays the reasoning results. CLIPS is used to develop a set of rules describing the criteria and relationships involved in decision-making, and to perform reasoning with them to infer the solution to the problem. In addition, CLIPS is used to develop a tools knowledge base to guide users through the whole analysis process. The communication medium between ARC/INFO and CLIPS is the file system. The architecture of the system is shown in Figure 2.

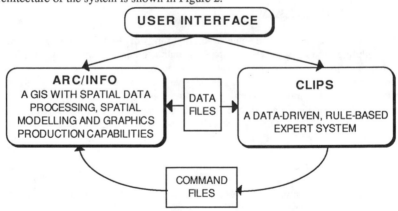

Figure 2   Integration of ARC/INFO and CLIPS

DATA EXCHANGE

One requirement for integration is that CLIPS and ARC/INFO be able to share a data base. ARC/INFO(ESRI, 1987) is considered as a spatial database with a set of spatial operators that manipulate and analyse geographical data. It consists of two main components: ARC and INFO. ARC is a set of software tools which handle the digital cartographic data for geographical features and perform various spatial data processing tasks, while INFO is a relational type database management system which stores and manages attribute data for geographical features. CLIPS needs INFO data for reasoning processes. The INFO data can be accessed either directly via the ARC/INFO programming language interface or via ASCII output files.

The INFO data file holds several predefined standard attributes about the geographical features. An INFO file is a feature attribute table. For each geographical feature, there is one record. For each record, there are several items. A record in an INFO table is analogous to a frame defined by the **deftemplate** construct in CLIPS. The item definitions can be used to define a frame with the items as slots. The values on the defined items are used to define specific instances of the frame. Figure 3 shows a sample dataset developed in INFO. A deftemplate may be defined for it as follows:

    (deftemplate land_feature (field recno (type INTEGER) (default 1))
               (field slope (type FLOAT) (default 0))
               (field soil_texture (type SYMBOL) (default LOAM))
               (field erosion (type SYMBOL) (default WEEK))
               (field soil_depth (type FLOAT) (default 0))
               (field flooding (type SYMBOL) (default NEVER)))

Figure 4 shows the fact dataset converted from this dataset, which is to be imported into CLIPS. The conversion is implemented automatically by a specific program. Once the fact data file is generated, CLIPS reads one line from the file each time for reasoning until the end of the file is reached.

$RECNO	SLOPE	SOIL_TEXTURE	EROSION	SOIL_DEPTH	FLOODING
1	1.5	LOAM	WEEK	85	OCCASIONAL
2	7.2	SANDY_LOAM	MEDIUM	60	OCCASIONAL
3	3.5	SANDY_CLAY	WEEK	30	NEVER
4	15.4	GRAVEL_CLAY	MEDIUM	20	NEVER

Figure 3   A Sample Dataset in ARC/INFO

(recno 1) (slope 1.5) (soil_texture LOAM) (erosion WEEK) (soil_depth 85) (flooding OCCASIONAL)
(recno 2) (slope 7.2) (soil_texture SANDY_LOAM) (erosion MEDIUM) (soil_depth 60) (flooding OCCASIONAL)
(recno 3) (slope 3.5) (soil_texture SANDY_CLAY) (erosion WEEK) (soil_depth 30) (flooding NEVER)
(recno 4) (slope 15.4) (soil_texture GRAVEL_CLAY) (erosion MEDIUM) (soil_depth 20) (flooding NEVER)

Figure 4   A Sample Dataset to be Imported into CLIPS

Assuming that the fact data file is named "land.dat", the file I/O provided by the CLIPS is illustrated in Figure 5.

In this example, a fact "(land_feature (recno 1) (slope 1.5) (soil_texture LOAM) (erosion WEEK) (soil_depth 85) (flooding OCCASIONAL))" is first asserted into the working memory in CLIPS for successive reasoning processes. Once the reasoning processes are finished, the results are written to a prespecified output file, and a second fact is read and reasoning begins again. This process continues until all fact data have been processed. Finally, the output file is converted into INFO data for display or further analysis.

```
(defrule open_file
    ?phase <- (phase open_file)
=>
    (retract ?phase)
    (open "land.dat" data)
    (assert (phase read_file)))

(defrule read_file
    ?phase <- (phase read_file)
=>
    (retract ?phase)
    (bind ?land0 (readline data))
    (if (eq ?land0 EOF)
        then (assert (phase close_file))
        else
            (bind ?land (str-cat "land_feature " ?land0))
            (str-assert ?land)
            (assert (phase land_classification)))))

(defrule close_file
    ?phase <- (phase close_file)
=>
    (retract ?phase)
    (close data))
```

Figure 5 Sample CLIPS rules for file I/O

INTERACTION BETWEEN ARC/INFO AND CLIPS

CLIPS was written in C, and designed especially to provide high portability and easy integration with existing or conventional software systems. ARC/INFO provides the ARC Macro Language(AML) through which a sequence of ARC/INFO commands can be organized to perform sophisticated spatial operations. The strategy we took is to have the two components communicate via command files.

CLIPS actually works as a top-level controller with a set of predefined rules, which are fired according to the state of the analysis. When GIS processing is needed, CLIPS generates a command file that contains ARC/INFO commands in AML. The **system** command of CLIPS is used to invoke the command file through a host language interface which in turn controls ARC/INFO and performs a series of GIS operations needed. When the operations have been executed, control is returned to CLIPS and the next rule is instantiated. The rule illustrated in Figure 6 is indicative of this approach. This rule is used to write a sequence of ARC commands to a command file "overlay.com" and activate this file using the **system** command to perform an overlay operation which overlays two coverages named SLOPE and SOIL_TEXTURE and creates a new coverage named SLOPETEX. The system design is being implemented on DEC VAX under VMS. "@overlay" is a DCL(Digital Command Language) command to execute the overlay.com file.

When INFO data is needed, CLIPS calls a program specially written to convert INFO data into a fact data file in the format described above and reads it into the working memory. The reasoning results are written to a output file and imported to INFO. ARCPLOT can be activated by CLIPS for displaying the results. All actions above are controlled by the rules in the tools knowledge base. No user interaction is required. The rules in the domain-specific knowledge base are used to perform reasoning on INFO data.

```
(defrule overlay
      ?phase  <- (phase overlay)
    =>
      (open "overlay.com" com_file)
      (printout com_file "$arc" crlf
                              "identify  SLOPE SOIL_TEXTURE SLOPETEX" crlf
                              "quit" crlf
                              "$exit"   crlf)

      (close com_file)
      (system "@overlay")
      (assert (phase read_SLOPETEX)))
```

Figure 6   A Sample CLIPS Rule for Activating ARC/INFO

CONCLUSIONS

GIS systems provide spatial decision makers with a means of integrating information to understand and address many spatial problems. Expert systems encode the expert knowledge, and use this knowledge, simulating human reasoning by heuristic and approximate methods, to solve problems. The integration of GIS and ES offers the capabilities for spatial modelling, large-scale spatial data management , as well as heuristic reasoning. The spatial decision making process involves not only the domain-specific knowledge of a particular geographical application, but also the knowledge of how to use analytical tools to perform the analyses required. The system described, which integrates ARC/INFO and CLIPS, provides the ability to capture both domain-specific knowledge and tools expertise so as to produce a flexible spatial decision support environment.

REFERENCES

Berry, J.K., 1987, Computer assisted map analysis: potentials and pitfalls. Photogrammetric Engineering and Remote Sensing, Vol.53, No.10.

Burrough, P.A., 1986, Principles of Geographical Information Systems for Land Resource Assessment. Oxford:Clarendon.

Burrough, P.A., 1990, Preface. Methods of spatial analysis in GIS. International Journal of Geographical Information Systems. Vol.4, No.3, pp.221-223.

Davis, J.R. and J.L. Clark, 1989, A selective bibliography of expert systems in natural resource management. AI Applications in Natural Resource Management, Moscow, Idaho.

Densham, P.J.; 1991, Spatial decision support systems. In: Maguire, D.J.; Goodchild, M.F. and Rhind, D.W.(eds.), Geographical Information Systems: Principles and Applications, Vol.1, Longman, London, pp.403-412.

Duda, R.O. and R. Reboh, 1984,   AI and decision making: the PROSPECTOR experience. In: Reitman, W.(edt.), Artificial Intelligence Applications for Business, Ablex Publishing Corporation, pp.111-147.

Environmental Systems Research Institute(ESRI), 1987, ARC/INFO Users Guide, Vol. 1 & Vol. 2, Redlands, California.

Frank, A.U., M.J. Egenhofer and W. Kuhn, 1991, A perspective on GIS technology in the Nineties. Photogrammetric Engineering and Remote Sensing, Vol. LVII, No.11, pp.1431-1436.

Franklin, J. F., 1979, Simulation modelling and resource management. Environmental Management, Vol.3, No.2.

Gardels, K., 1987, The expert geographic knowledge system: applying logic rules to geographic information. Auto Carto-8, Maryland, pp.520-529.

Jackson, P., 1990, Introduction to Expert Systems, Addison-Wesley Publishing Company.

King, D., 1990, Intelligent decision support: strategies for integrating decision support, database management, and expert system technologies. Expert Systems with Applications, Vol. pp.23-38.

Morse, B., 1987, Expert interface to a geographic information system. Auto Carto-8, Maryland, pp.535-541.

National Aeronautic and Space Administration (NASA), 1991, CLIPS Reference Manual 5.0, Software Technology Branch, Lyndon B. Johnson Space Center.

Pearson, E. J., G. Wadge and A. P. Wislocki, 1991, Mapping Natural Hazards with Spatial Modelling Systems. Proceedings of European Conference on Geographical Information Systems, Brussels, Vol.2, pp.847-855.

Robinson, V. B., A. U. Frank and H. A. Karimi, 1987, Expert systems for geographic information systems in resource management. AI Applications in Natural Resource Management, Vol.1, No.1, pp.47-57.

Simon, H. A., 1977, The new science of management decision. Englewood Cliffs, Prentice-Hall.

Smith, T.R. and Je Yiang, 1991, Knowledge-based approaches in GIS. In: Maguize, D.J., Goodchild, M.F. and Rhind, D.W.(eds.), Geographical Information Systems: Principles and Applications. Vol.1, Longman, London, pp.413-425.

Stonebraker, M., 1986, Triggers and inference in database systems. In: Brodie, M. and J. Mylopoulos (eds.), On Knowledge base management systems: Integrating artificial Intelligence and Database Technologies. New York: Springer-Verlag, pp.297-314.

Teicholz, E. and B. J. L. Berry, 1983, Computer graphics and environmental Planning, Prentice-Hall, Inc.

Usery, E.L., P. Altheide, R.R.P.Deister and D.J. Barr, 1988, Knowledge-based GIS techniques applied to geological engineering. Photogrammetric Engineering and Remote Sensing, Vol.54, No.11, pp.1623-1628.

Webster, C., 1990, Rule-based spatial search. International Journal of Geographical Information Systems, Vol.4, No.3, pp.241-259.

Zarko Sumić, Ph. D, Staff Engineer†*
†Puget Sound Power & Light Co.
One Bellevue Center Building
411 108th Avenue N.E., Bellevue, WA 98004-5515
(206) 462-3450

Erh-Chun Yeh, M.S.E.E, Graduate Student*
*Electric Energy Group,University of Washington
Department of EE, FT 10, Seattle, WA 98195
(206) 685-3603

A GIS-BASED EXPERT SYSTEM FOR RESIDENTIAL DISTRIBUTION DESIGN

1. INTRODUCTION

The design of the electrical supply system for new residential developments is an everyday task for electric utility engineers. Presently this task is carried out manually resulting in overdesigned, costly, and nonstandardized solutions. A complexity in automating plat design is imposed by the need to process spatial data such as circuit maps, records, and construction plans. The Intelligent Decision Support System for Automated Electrical Plat Design (IDSS for AEPD) is an engineering tool aimed at automating plat design developed jointly by Puget Sound Power & Light Co. (PSPL) and the University of Washington (UW) as an integral part of PSPL's GIS based Facilities Management System (FMS). The IDSS for AEPD combines the functionality of a geographically referenced database with the sophistication of Artificial Intelligence (AI) to deal with the complexity inherent in design problems.

The SEcondary Router (SER) is one of the Knowledge Sources (KS) used by AEPD. The task of SER is to emulate the design activity of human experts involved in the Underground Residential Distribution (URD) secondary cables routing and sizing. To perform this tasks, the problem solving environment must be capable of extracting geographically referenced information from digitized maps as well as mimicking spatial reasoning performed by human designers. A Geographic Information System (GIS) environment coupled with an Expert System (ES) shell is an appropriate combination for handling this type of engineering problem. In such an integrated environment GIS acts as a spatial reality factual source to provide an object-oriented data abstraction for rule based engineering reasoning performed by an expert system. To couple these two environments, a two way bridge between GIS and ES must be established. In addition to physical data transaction between these two software environments, the data bridge must be capable of restructuring two conceptually different data models used by them. SER is built using ARC/INFO (a GIS product from Environmental System Research Institute) and NEXPERT Object (a hybrid expert system shell from Neuron Data). NEXPERT Application Programming Interface (API) is used to embed NEXPERT Object into ARC/INFO to obtain user transparent expert system processing. ARC/INFO GIS provides the spatial manipulation capability to create the feasible tracings for cable routing, the database management system for graphic and thematic data, and cartographic display features to visualize results of different design stages. NEXPERT on the other hand, provides a knowledge based processing environment that can perform engineering reasoning on the spatial realty abstracted in a functional object-oriented way.

The paper presents results obtained in creating an integrated GIS-ES environment capable of emulating complex spatial problem-solving procedures, such as cable routing. After a brief description of the AEPD tool and its problem solving architecture, the paper will focus on SER development. The issues such as GIS-ES coupling, coverage to object data model restructuring, knowledge base design and spatial knowledge processing will be discussed. To prove appropriateness and efficiency of the developed SEcondary Router, the results of a testing performed on residential developments in Puget Sound Power & Light Co. service area will be given.

2. AUTOMATED ELECTRICAL PLAT DESIGN

In dealing with an underground residential distribution system design, two categories are identified: one is geographic layout design, the other is technical design. Geographic layout design starts from spatial constraints with other constraints relaxed. Therefore, it includes the subtasks of the URD design with spatial concerns, such as cable routing and device placement. Unlike the geographic layout design, the technical design takes care of the operational constraints. It deals mainly with the operation situations such as voltage drops, thermal limits of cables, system losses, and reliability analysis. However, neither of these approach alone can accomplish all the encountered tasks in the URD design. One subtask offsets others and all of them affect the total cost of the distribution system design. The final satisfactory design might efficiently come out after several interactive tunings performed by designers with the assistance of a smart computer tool. In the process of first satisfying the spatial constraints, the backbone of the distribution system is laid out. Fine tuning of this layout design is then achieved by relaxing spatial constraints with emphasis on the operational constraints, taking also the total cost into account. With this approach, the overall design procedure produces a spatially feasible design that satisfies technical performance and has acceptable cost.

The AEPD problem solving model is based on Intelligent Decision Support System Blackboard based architecture shown in Figure 1. As it is shown on this figure, the complete automated electrical plat design process can be divided into two modes:

- Batch mode consisting of preprocessing after which the initial set of geographically referenced data is posted on the blackboard. This information contains descriptions of relevant spatial relationships and possible device placement sites, routing corridors and street crossing options.
- Interactive mode during which the real design process, in symbiotic cooperation between human designer and knowledge sources, takes place. During this part each particular solution is evaluated from the economical and technical point of view to assist designer in determining the best design.

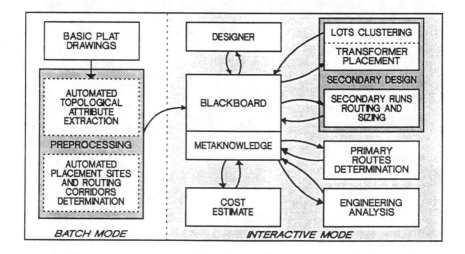

Figure 1. Blackboard Based AEPD Problem Solving Architecture

888

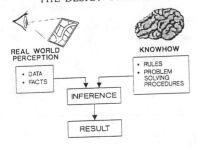

THE DESIGN PROCESS

REAL WORLD PERCEPTION
- DATA
- FACTS

KNOWHOW
- RULES
- PROBLEM SOLVING PROCEDURES

INFERENCE

RESULT

The electrical plat design process can be simply formalized as a composition of elementary operations, described with the following symbolic model:

what the designer sees on the map
+what the designer knows
= what the designer does

Successful repetition of these elementary operations moves the design process from the initial state to the goal state. As can be seen from Figure 2 and this description, visual perception is crucial for the electrical plat design. To computerize the plat design, spatial reasoning applied on top of data "visually" extracted from maps is required to implement rules like, "Place the street light at the street crossing first," or "Do not put the transformer at the end of the cul-de-sac."

Figure 2. Basic Elements of the Plat Design Process

After the basic engineering maps have been digitized, relevant information must be stored in the feature attribute tables of the appropriate coverages in the ARC/INFO database. During the AEPD preprocessing, required attributes can be entered manually or, what is much more efficient, can be extracted automatically from the spatial representation of the digitized maps, using ARC Macro Language (AML) programs. In preprocessing the plat, GIS is used as a substitute for the eyes of the artificial "blind" designer (expert system). This "blind" designer knows how to operate, but does not see the world it has to operate on. The ability of GIS to form the "real world" representation, without human intervention, enables automation of the design process.

In addition to enabling the computer to "see" the plat, the preprocessing part applies a set of spatial constraints to the plat design process. A set of AML programs is employed to identify relevant information such as street curbs, boundary lines between the lots, frontage corners, etc. As a result of the application of spatial rules, possible routing corridors and placement locations are automatically identified and stored as three coverages containing placement sites, routing corridors and street crossings. These three coverages CORRIDOR, CROSSING, and SITE, form the network of feasible cable routes and equipment placement sites. The application of spatial constraints used by designers within the GIS environment "prunes" the search space, and makes the overall search process manageable.

2.1. Blackboard

In the blackboard problem solving architecture, the blackboard is the communication medium through which one knowledge source communicate its findings to one another. The blackboard is the,

- Source of all data upon which a knowledge source operates.
- Destination for all conclusions from knowledge sources.

The blackboard in the AEPD tool is implemented in ARC/INFO proprietary Database Management System (DBMS) INFO. To support the rule based processing outside the ARC/INFO environment a special link is established between the expert system shell NEXPERT and INFO Blackboard. Similar links are established between the INFO blackboard and other numbercrunching procedures which operate outside ARC/INFO.

2.2. Metaknowledge and overall control

To guide the overall solution process and to respond to the designer's requests, a special knowledge source called metaknowledge is developed. The control strategy employed by the control knowledge source is request driven. The control knowledge source analyzes user requests, suggests the next step to be done in the problem solving process, and selects the domain knowledge source likely to provide the requested information. In addition, the metaknowledge acts as an agenda scheduler in suggesting which knowledge source to invoke and when it must be invoked to fulfill a particular need for any design scenario.

2.3. Transformer placer

The Transformer Placer knowledge source of the AEPD operates in ARC/INFO environment. To determine the best transformer location:

- A cluster of lots to be served with one transformer must be identified.
- Among the possible locations for transformer placement, defined during preprocessing, the site closest to the center of gravity of that cluster is chosen as the appropriate site.

The clustering process is crucial to the success of automated transformer placement and it later influences all activities of the plat design. Each cluster of lots starts to grow automatically from an identified origin. During this dynamic process some of the original clusters are merged, and some new cluster origins are created. New lots to be added to the cluster are identified based on their distance from the cluster origin and the existence of service paths through available corridors or street crossings.

After the lot clustering process is approved by the designer, transformer placement is performed using the minimization principle which states that an optimal location for the transformer is the center of gravity of the cluster. The user can accept this suggestion or relocate the transformer to explore different design alternatives.

3. SER: EXPERT GIS SECONDARY CABLE ROUTER

Secondary cable routing and sizing is one of the main design processes that has to be performed during the Underground Residential Distribution (URD) design. The task of the SER knowledge source is to emulate the design activity of human experts involved in the URD secondary cables routing and sizing. It is built using ARC/INFO and NEXPERT Object. These two software environments are coupled for spatial reasoning and knowledge processing.

The following sections will discuss intrinsic of GIS and expert systems as well as design and implementation issues of creating SER an Expert GIS environment for distribution cable routing and sizing.

3.1. Geographic Information System

Geographic Information System integrates database management, computer graphics, and spatial modeling into a software environment for managing geographic features. Instead of storing maps in the conventional graphical sense, a GIS stores data from which a user can create the desired view, drawn to suit a particular purpose. Thus GIS is an analysis tool allowing the user to identify spatial relationships between map features. The two basic types of information a GIS deals with are information describing the location and shape of geographic features and their spatial relationship to the other features, and descriptive (thematic) information about these features. Integration of these two types of data, and the ability to maintain the spatial relationship between the map features, creates a variety of ways for advanced, automated map data processing.

ARC/INFO, the GIS employed in SER uses a vector representation of spatial reality. In this GIS model, geographic reality is represented by nodes (points), lines (arcs) or regions (polygon). In URD design examples of these objects are transformers represented as points, secondary cables represented as arcs and residential lots represented as polygons. A point is recorded as a single (x,y) location in the coordinate system, lines are recorded as a series of ordered (x,y) pairs, and polygons are recorded as line segments that enclose an area. Topology, defined as a mathematical procedure for explicitly defining spatial relationships, is used to allow spatial analysis of geographic features. Thematic data, used for the geographic feature attributes description, are stored as a table of records. These feature attribute tables in ARC/INFO GIS are managed by INFO, a relational database management system.

Complete map information is often logically organized into sets of layers or themes of information. Each layer in ARC/INFO is called a coverage and has features with one common attribute, such as lines representing all roads in one area, or polygons representing lot division of residential plat development. A coverage is the basic unit of map information stored in vector representation. It consists of topologically linked geographic features and their associated descriptive data stored as an automated map. Information contained in various layers can be analyzed separately or combined to create new coverages which assist the user in decision making process. The ARC/INFO was originally developed for use in an interactive mode to support the user in identifying relations among the geographic features. However, commands, functions, and macros, provided within the fourth generation language (4GL), called ARC Macro Language (AML) enable fairly sophisticated spatial analysis to be performed in batch processing mode. The ARC/INFO ability to perform spatial analysis in the batch processing mode is used to automate data preparation and to generate the attributes required for the automated electrical plat design.

3.2. Expert System

Traditional algorithmic methods are intended to address well-defined problems that can be expressed by sufficiently accurate mathematical models. A solution methodology employed by the human designer in electrical plat design involves many heuristics and "rules of thumb." As such this methodology is not capable of formulation in an algorithmic form. This is why expert systems, an AI technology, have been employed as a problem solving paradigm for the secondary cable routing.

An expert system is an intelligent computer program which uses knowledge and inference procedures to solve problems difficult enough to require human expertise for their solution. The knowledge necessary to perform at such a level together with the inference procedures used can be thought of as a model of an expert in a specialized field or domain of interest. When using expert systems the reasoning of an expert is modelled with the formalism of rules and an appropriate inferencing mechanism used to draw valid conclusions. For the computational model of problem solving it is required to have a representation of the world in relation to the reasoning. This means that information about "things" must not only be available but also must be formalized so that it can be efficiently used during a problem solving activity.

The inference strategy employed by the ES can be forward chaining, backward chaining, or the combination of both. The forward chaining inferencing control strategy is intrinsic to the planning and configuring problems, while the backward chaining is well suited for diagnostic problems.

To mimic expert reasoning in cable routing domain SER uses a commercially available expert system shell NEXPERT Object. NEXPERT is a hybrid expert system shell with an inference mechanism called Agenda and an object oriented real world representation. Rules used in NEXPERT have a symmetric form. This means that for the knowledge processing depending on the inference procedure taking place, the same rule may be used for a backward chaining or a forward chaining. This particular ability is based on augmented rule format. In this case besides the Left Hand Side (LHS) and Right Hand Side (RHS), the rules in an augmented format have in addition a Hypothesis part.

The rule evaluation, while trying to find the status of the conditions on the LHS of the rule, leads to backward chaining as the default mode of the knowledge processing. In addition, the rule that has to be evaluated next can be determined in the forward chaining mode by the RHS action, or by the gates or context mechanism. The rules linked at the runtime by the backward chaining or forward chaining make the so called knowledge island. On the other hand, a context link as a rule scheduling mechanism is defined by the knowledge engineer to connect different knowledge islands that will not be automatically linked during the runtime.

891

Contrary to the conventional programming where the order of the processing is determined by a stack status in a last-in first-out (LIFO) mode, the AGENDA, an ordered stack of hypotheses in NEXPERT that are waiting to be evaluated, is dynamically modified according to the "conclusion" that is reached in the particular moment of knowledge processing. The AGENDA mechanism of rules scheduling allows the implementation of a complex inferencing control procedure.

An object-oriented data model is used in NEXPERT Object to represent the entities of the problem domain. The reason for adopting the object-oriented representations is that objects are closer to the expert perspective in terms of describing the real world. The objects are the basic entities in a domain, and the properties are used to describe the objects. A set of objects sharing some similar properties can form a class. The advantage of using the object oriented data model is not only that objects are closer to the human perception of the real world, but such a model enables use of inheritance, pattern-matching and interpretation mechanism.

3.3. Spatial Versus Object Oriented Data Model

The places where the data required for the rule based processing physically reside are feature attribute tables of different coverages that make AEPD spatial model. These data are used or created during the preprocessing, clustering or transformer placing procedures in ARC/INFO. The data stored in these tables represent facts that are input values for the rule based reasoning process.

In NEXPERT knowledge represented by rules and the set of objects that represent the world upon which the reasoning takes places is stored in several files, called knowledge base. Thus knowledge and facts are kept separately in the knowledge base and the external (GIS) database respectively. During the runtime facts from the INFO database must be mapped into the objects' properties, and the result of the reasoning must be written back to the INFO database. To enable this process a bridge allowing a two-way communication between the GIS database and the expert system shell must exist.

The ARC/INFO data structure organized into coverages as a set of related themes or layers with topological relationship among the geographic features, is appropriate to support procedural spatial analysis mostly encountered in GIS tools. On the other hand expert system reasoning often require different real world abstraction in the form of objects that will represent functional behavior of the entities in the problem domain. The knowledge acquisition procedure, performed as the initial part of the AEPD project, showed that objects upon which rules are operating, were dominantly structured from functional rather than from spatial perspective. This means, to perform engineering reasoning encountered in the secondary routing problem domain, spatial data model existing in GIS needs to be restructured to a functional object oriented data model.

In the secondary routing process the basic goal is *within each cluster find the best route to connect each customer (each lot) to the transformer*. Functionally the basic entities perceived by a human designer are customers. Therefore a customer is a basic object used in the NEXPERT object oriented data model. Most properties for the object CUSTOMER are extracted from the spatial AEPD model by mapping each customer to the appropriate parcel from the LOT coverage. Additional properties that describe parcel relationship to the possible trenching routes, adjacency to the possible placement site, and connectivity among them, are obtained by the process of vertical integration, overlaying CORRIDOR, CROSSING, and SITE coverages, and relating their feature attribute tables. Such a process results in encapsulation of the GIS data into properties of the object CUSTOMER.

The procedure that is used to restructure data from a spatial GIS model to a functional object oriented model, called TRANSFER, was first prototyped using INFO 4GL but the processing time was to long for an interactive designing tool. In order to get better performance the data restructuring procedure TRANSFER was finally implemented in C programming language.

In addition to the problem of conceptual data model restructuring, coupling an expert system shell with a GIS requires physical data transaction between the two stand alone software environments. The intersystem communication is established by employing data agents. The main task for a data agent is to retrieve, decode, and then code data as needed. All data files are then sent to a data pool, where all information from different systems resides. In SER, TRANSFER procedure restructures the AEPD spatial data model and output the information as ASCII flat files to the data pool, while a data agent retrieves the ASCII files and then codes them as the NEXPERT proprietary spreadsheet form files of the NXPDB-type. The result of the expert system processing is exported out of the NEXPERT environment through the ASCII file that is related back to the GIS environment using the INFO Relational Database Interface (RDBI).

The data bridge between the GIS and the ES established by the procedures for the data model restructuring and the intersystem communication is depicted in Figure 3. This bridge enables the integration of these two systems.

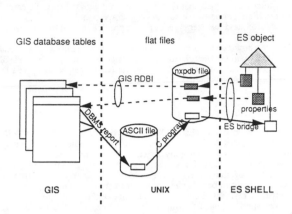

Figure 3. ARC/INFO-NEXPERT Object Intersystem Communication

3.4. Knowledge Processing

The spatial component is a crucial concern throughout the URD design process. Therefore, spatial reasoning plays a significant role in design activities. Most experiments about human thinking activity have shown that spatial reasoning performed by humans is using right-brain cognitive processes, which is very different from left-brain cognitive processes through which deductive or associative tasks such as language understanding and mathematical computations are performed. To emulate spatial reasoning is a challenge to computer applications, especially to automated layout or construction design.

In SER, most rules are dedicated to the spatial reasoning, therefore the main task of knowledge base design is to formulate spatial scenarios into rules in a case-based paradigm. All scenarios about the crossing and corridor selections, during the process of finding proper routes for cables, are abstracted from the spatial structure and then formulated as rules.

To accommodate different routing strategies used by human designers, rules in SER are divided into several knowledge bases. Those routing strategies provide more flexibility for SER in determining the routes for secondary cables. For example, the route for crossing a street can be either the shortest one or the most perpendicular one, and each of these strategies is handled by different knowledge base. However, during one cable routing session only one knowledge base is loaded and used.

Before the actual design of knowledge base started, the cognitive process of cable route determination was analyzed using the knowledge engineering method of verbal protocol recording. The findings of knowledge acquisition were that designers were dealing with clusters one by one, and within each cluster they were trying to connect customers one by one. These results were used to modularize the overall reasoning process according to the focus of attention at different stages. In this way, a skeleton of the reasoning process was conceptualized. The modularized reasoning process for adopts the group transaction method which is found to be the most efficient way from the data transaction overhead point of view. Two different ways can be used to form the dynamic objects during the runtime. One treats the whole coverage (the study plat) as one spatial problem domain, wheres the other sequentially focuses on clusters (the area served by one transformer) one by one. SER adopts latter approach since it is closer to the way how human designers perform the task. The advantages of this approach are: reduction in the complexity from exponential to linear, reduction in the failure rate and simplification in the redesign procedure. SER can identify the different cases by matching the spatial structure of object that is being processed with the episodic cases in the rule base. If the case can be classified as one of the scenarios, a corresponding routing method will be used. Otherwise, the case will fail, and the user is requested to provide the information required to complete routing process.

At the end of the knowledge processing, final conclusions are drawn. The conclusions refer to the results of routing and sizing cables. These results are stored back to the INFO database of their corresponding coverages by data agents. In this way, the results can be displayed in a cartographic form using ARC/INFO graphic capabilities.

4. TEST RESULTS

The AEPD is developed on DEC 5000 RISC workstation under ULTRIX 4.2 operating system. The present version uses ARC/INFO 6.0 and NEXPERT object 2.0B. During the design and testing phase AEPD and SER were intensively tested on more than fifty different residential layouts from the Puget Sound Power and Light Co. service area. These residential area layouts were extracted from the parcel coverage of the Pugets' FMS ARC/INFO library. The street layouts and lot arrangements varied in shape, number of streets, number of street intersections, number and shapes of cul-de-sacs, and number of residential customers. The size of the new developments ranged from as low as ten lots to over hundred and fifty lots. In all these cases AEPD tool was used to produce feasible trenching corridors, street crossings, and placement sites network during the preprocessing batch session. In an interactive mode AEPD knowledge sources were used to determine optimal lot clustering and transformer placement. These plats were then used to test the performance of the SER during the automated routing of secondary conductors. At the beginning of the testing process SER was able to provide correct solution in more than 85% of the test cases without user intervention, and with additional expansion of the rule bases the success rate in routing is now about 99%. In order to see how close SER mimic routing process of the human designers SER results were compared with cable routing results of human experts. The finding is that SER comes up with the results much of the same quality as those of human designers. One such comparison can be obtained from figures 6 and 7 that show designs of secondary distribution system provided by a human expert and AEPD tool respectively. Although the design process is highly subjective and even human designers are not producing the same layout of the distribution network for the same plat, the comparison of the manual design and automated design shows a close agreement. The main difference in routing between the two is the consequence of the rules in AEPD that does not allow placement of the devices (transformer or secondary junction boxes) on certain sites such as bottom of the cul-de-sac or street intersections.

The time required by human designers to produce manual design for a plat of this size (121 lot) is in the order of 16 hours (two working days). In addition to determine which lots will be served by one transformer, placing transformers and routing cables, most of the designers' time is spent manually measuring length of cables to assure proper sizes and to create bill of materials. For the sake of comparison to produce one design solution after preprocessing of the plat is done, designer using AEPD tool needs less then 3 minutes to perform lots clustering, transformer placement, and secondary cable routing. In addition to that, at the same time AEPD will perform cost estimation, run analytical study of the obtained electrical network to check violation of the operational constraints, and will provide user with bill of material. Since the time required to provide solution is so drastically reduced comparing with manual design, AEPD can be used as an intelligent decision support tool to provide distribution designers with an automated way to explore the impact of different design scenarios.

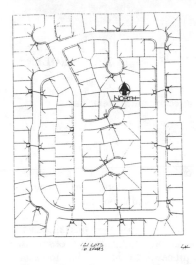

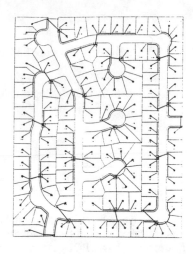

Figure 6. Manual Plat Design

Figure 7. Automated Plat Design

5. CONCLUSIONS

The successful results of testing the performance of AEPD tool and SER on number of different plats proved the feasibility of applying the AEPD concept and the SER structure in the automated URD design. Based on the concept of supporting and enhancing the designer's reasoning process and judgment, rather than only automating the design through prescribed computation, the AEPD tool is expected to be accepted by utility engineers. The AEPD tool provides an automated way to explore the impact of different design scenarios, where the nonexistence of such a capability in manual designs limits potential options designers can explore.

SER further proves the appropriateness of the integrated application of state-of-the-art technologies, namely Geographic Information System (GIS) and Artificial Intelligence (AI) in the area of electric distribution planning and design. They are uniquely qualified to structure the required information and to capture the designer's problem solving methodology. The tight cooperation between these two agents, one "being able to see", and the other "being able to think", provides the distribution system designers with a viable tool to support their decision makings. The concepts presented in this paper are generic and can be applied to a host of problems in the distribution planning and design area with spatial components. It is authors' hope that this work will establish a new standard for automated solutions to these problems.

6. ACKNOWLEDGEMENTS

The authors wish to acknowledge help of Ms. Helena Males-Sumic and Messrs Luis E. Gaggero, Xinguo Wei, and R. Atteri, all Puget Power Graduate Fellows, for their contribution during the course of this research project. Special thanks go to Prof. Mani Venkata from the University of Washington and Mr. Todd Pistorese, Puget Power FMS Manager, for their valuable input, and to Ms. Albie Merrill for her help in making this paper readable.

TOMORROW'S GOAL: EACH A ROLE IN AN INTEGRAL WHOLE!

John O. Behrens*
U.S. Advisory Commission on Intergovernmental Relations

ABSTRACT

Here is an eclectic look at what integration can mean for the geographic information systems of the near and far tomorrows, in terms of standards, infrastructure, information access and cost, data sharing and cooperation.

The paper's narrative has dissimilar but nevertheless compatible subjects, such as topological vector profiles, cadastral maps, Proposition 13, the U.S. Freedom of Information Law (FOIA), and reporting gross proceeds to the Internal Revenue Service. In addition, intergovernmental and private sector players gingerly probe the frontiers of GIS integration. One among them already has available on line data for real property in a majority of individual states.

INTRODUCTION

These 90's are heady times. We can enjoy technology of the twenty-first century even as we encounter remaining challenges of the twentieth. In our resolute quest for a more people friendly world, we welcome new departures like geographic information systems, noting their penetration into public and private applications. Indeed, their ubiquitous nature supports the notion that each has a relationship to each of the others, if only we can discover the involved complement. GIS technologies now radiate to every echelon of government, as well as to other activity centers, mundane or exotic.

Such is the orientation which conditions what follows. Particular aspects receive attention in the context of GIS technology. Included is a basic standard now emerging from the Department of Commerce, along with guidelines fashioned jointly by nominally disparate tax assessors and other information professionals. Property taxes, with their entire assessing and cadastral infrastructure, constitute information resources of singular worth, especially when linked with other topological data such as those of the Census Bureau's Topologically Integrated Geographic Encoding and Referencing System (TIGER). Similar alliances affect the accessibility spectrum that conditions the fortunes of those who disseminate or proscribe spatial information, with or without fee. One circumstance that can influence results is the extent to which data can be shared.

*The author, an attorney, is a Visiting Fellow at U.S. ACIR and President, Institute for Land Information. Views expressed are those of the author, and not necessarily those of any organization or any other person.

For a world inured to the "lack of data exchange standards" that has "inhibited the transfer of spatial data between ...producers and users,"* the Secretary of Commerce produced welcome news on July 29, 1992, when she approved the Spatial Data Transfer Standard (SDTS) as Federal Information Processing Standard (FIPS)

Publication 173. SDTS embodies specifications for "organization and structure of digital spatial data transfer, definitions of spatial features and attributes, and data transfer encoding. The SDTS promotes and facilitates the transfer of digital spatial data between dissimilar computer systems."** Publication 173 becomes effective February 15, 1993. Use of the FIPS involved is mandatory for all Federal agencies a year from that date, and is also available to state and local governments and to private sector and academic organizations.

The U.S. Geological Survey (USGS), as designated maintenance authority for SDTS, has the job of promoting its acceptance and coordinated use among Federal Agencies. As part of that assignment, USGS will develop "profiles," each of which is an explicitly defined subset of the standard, designed for use with specific data.*** The first profile is the "Vector Topological Profile." Others likely to result include a prototype raster profile, plus those for CAD/CAM and for graphics generally. USGA also will conduct workshops and training sessions, develop software, prepare attribute dictionaries, and bring into being a suitable data transfer processor.

Standards like SDTS are significant steps forward toward effecting a nationwide spatial data infrastructure. Even so, one GIS professional, noting the multi-year effort of Professor Harold Moellering of Ohio State University and more than 60 professionals to conduct experiments and report on SDTS, suggests that issues other than standards also require attention. The "other issues" include the following:****

1)   What about the chances for a national cadastre? Need U.S. citizens pay dearly for "an archaic and inefficient land recordation system (deed registry, uncoordinated surveying, title insurance) each time they buy or finance property? Federal agencies maintain that this is a local and private-sector matter, denying the fact that the feds own about one-third of the

---

*"The Spatial Data Transfer Standard: Status and Plans for Implementation" by Kathryn Neff, 5 The Iliad 2, Institute for Land Information, Winter/Summer 1992, pp. 4-5.

**"Technical Announcement of August 17, 1992, from U.S. Department of Interior, U.S. Geological Survey, National Mapping Division, Reston, VA.

***"Iliad, op. cit., p. 4.

****"See "TIGER/SDTS - Just Another Format?" by Donald F. Cooke, in GIS World, February 1992, pp. 96-99.

nation's land, and own more and more each time a bank fails and drops its mortgage portfolio into the government's lap."*

2) Who has the job to "maintain the street centerline component of the ...national spatial data infrastructure?" Best exemplar of the "centerline component" is TIGER, the Census Bureau's "direct-digitizing technical" success, whose originators do seem to like SDTS a lot, since it responds to the variety among databases by "defining a very general conceptual model of spatial data sets," specifying "modules" for encoding choices.**

3) Which spatial data infrastructure will administer "crucial environmental policies" like wetlands preservation? Is there a database of soils, vegetation, and hydrography that comes within a definition of "wetlands"?

4) Should USGS create all the 1:24,000, or 1:100,000 digital line graphs on a crash basis? Might a DLG series with a graduated accuracy series (rural to urban) be more useful?

The vital ingredient, awaited eagerly even as genuine advances like SDTS take hold, is a funded nationwide spatial data infrastructure, with cadastral, street-centerline, land use/land cover, and topological components.

Guidelines for Integration
Even as resolution of issues goes forward, efforts intensify among the disciplines to explore joint interests and develop guidelines for joint action. Such efforts reflect a natural, cooperative drive to implement GIS within multifunctional environments. One example is that of the International Association of Assessing Officers (IAAO) and the Urban and Regional Information Systems Association (URISA), both among the 35 organizations comprising the Institute for Land Information. IAAO membership consists largely of local tax assessors, while URISA has about 3,500, most of them data processing executives or staff specialists from local and other governments.

The two groups began in 1990 to work jointly on GIS guidelines for basic cadastral mapping systems, usually computerized, details to include content, design, preparation, materials, security, and maintenance. Because tax assessing and information processing are widely decentralized, with service delivery at the local level, administrative diversity abounds. GIS arises as a counter force, stimulating multilevel, multifunctional approaches. This has affected the product now near

---

*Ibid., p. 98.

**See "TIGER/SDTS: Standardizing an Innovation" by Beverly A. Davis, Jack R. George, and Robert W. Marx, paper prepared for SDTS Implementation and Testing Workshop, composed by Federal Geographic Data Committee, U.S. Geological Survey, and National Institute of Standards and Technology, Gaithersburg, MD, February 18-21, 1992, p. 4. See also TIGER/SDTS Prototype Files, Preliminary Description, presented for public comment August 10, 1991, revised July 1992. Prepared by U.S. Department of Commerce, Bureau of the Census, Geography Division, Robert W. Marx, Chief.

publication,* thoughclassic assessing functions of discovery, listing, and valuation condition fundamental content.

Thus the latter reflects attention to finding, accounting for, and valuing, as of the prescribed date, each parcel of taxable real property and (except in the nine states which exempt personal property) each account of personal property, in each assessing jurisdiction. Both mapping and valuation components of the assessing function benefit from the GIS infusion. With that, and with computer assisted mass appraisal (CAMA) already an ingredient in the high-tech valuing of the future, assessing sheds the "impossible job" attribute that had become its albatross before, during, and after World War II. Instead, GIS and CAMA impregnate the guidelines with practical, do-able tasks that can convert assessing in fact to what it had always been described as in glib statutes: the estimate, as of a prescribed valuation date, of the market value of each taxable property, based on its most recent sale, or that of comparable property, for existing use or for the property's highest and best use.

An effective mapping system, which GIS makes possible, means for each parcel, locating and uniquely identifying it, counting it, displaying it, and periodically correcting it as necessary. GIS means also that the jurisdiction's mapping system will possess geographic control (horizontal datum based on North American Datum of 1927 or 1983, vertical datum based on National Geodetic Vertical Datum of 1929, or North American Vertical Datum of 1988). There will also be base map, cadastral, and attribute data. Each parcel will have a parcel identification number (PIN) that is simple, unique, flexible, economical, and accessible.

CCI in the NSDI: Assessors, Their Records, and the Census Bureau

Without denigrating approval of SDTS as a FIPS, we note other progress as well, especially since it relates to data files from two well known sources, namely, local tax assessors cadastral records and census tracts.

What might be termed "cadastral-centerline" integration** (CCI) within a potential nationwide spatial data infrastructure (NSDI) is the linkage of these geographic systems that associate specific attribute information to particular locations. Steven French, who noted that tax assessors files contain a "gold mine" of information about the urban land market, such as land use, land value, and property transfer data, also noted that the land parcels may not be digitized. In an experiment he and associates digitized 336 assessor's map pages which comprised territory within 13 census tracts in Palo Alto (Santa Clara County), in California, then used a GIS overlay function to insert applicable tract number on each map page. It then became possible to link attribute data from the two systems, which exhibited a congruence

---

*Probable Title GIS Guidelines for Assessors. For details, contact International Association of Assessing Officers, 1313 East 60th Street, Chicago, IL 60637, Phone 312/947-2069.

**"Integration" here means reorganization of distinct data files in a way that produces congruence among them sufficiently accurate for use in the application involved.

index, calculated by French, of 98.8 percent.* French points out that such linkages could involve data from economic censuses the Census Bureau conducts at 5-year intervals, such as those for retail trade, wholesale trade, manufacturers, and service industries, each aggregating data by census tract.**

In addition to censuses French mentions, there is also the Census of Governments, conducted during each year ending in a "2" or "7". Each such census deals with and publishes findings on government operations, government finances, public employment, and taxable property values. For the last named study, the Census Bureau, generally for each fifth year beginning in 1957, has enumerated two samples. The first consists of recently sold parcels (randomly selected, usually from the public record of sales in local registers of deeds offices) and their assessed values (obtained from local assessment rolls). The second sample comprises individual parcel assessed values, enumerated by means of a stratified sample that yields selections either within an applicable "certainty" group of values, or on the basis of population of 50,000 or more as of the nearest applicable cutoff date.

The Census Bureau uses the first sample to conduct its nationwide quinquennial assessment-sales price ratio study, only one of its kind in the country. For each such effort the Census Bureau calculates de facto assessment sales price ratios, coefficients of intra-area dispersion (to measure uniformity achieved), and, effective tax rates (in each case, amount of property taxes expressed as a percentage of sale price). The second sample yields the data necessary for estimating assessed values and numbers of parcels, in total and for each of seven property uses most recently for the entire nation, each state, and individual jurisdictions down to 50,000 population in 1984. On two occasions since 1957, budget constraints abridged survey coverage. In 1972 available funds were sufficient for obtaining magnitude and composition results only for 28 large assessing jurisdictions. In 1987 budget constraints prevented the ratio study. For 1992 the Census Bureau will again produce nationwide assessment-sales price ratios, but not details on magnitude and composition of assessed values.

Property Values and Property Taxes
Approximately 13,500 tax assessors toil in the United States, most of them for local governments. They determine assessed values for about 118 million parcels of realty annually, for almost 66,000 among the 50 state and 83,126 local governments which have the power to levy property taxes. As shown in table 1, such taxes yielded $172 billion during 1991, an increase of 8.6 percent over the amount collected the previous year, and close to 170 percent more than the corresponding amount of $64.9 billion for 1978. It was June 8 during that year when California voters opted for Proposition 13. The 1991 yield constitutes 14.7 percent of all taxes of all types levied by all governments in the United States. This proportion, slightly

---

*"Reconciling Census and Tax Assessor Mapping Systems" by Steven P. French, in URISA 1992 Conference Proceedings, Urban and Regional Information Systems Association, Washington, DC, Volume IV, pp. 245-253.

**Other examples of linkage, and integration in general, come from Erma J. Thomas, Director of Cartography. Pinellas County Florida. They are detailed in URISA Proceedings for 1986, 1988, and 1992.

more than 12 percent in 1978 and close to 13 percent during the 80's, has been increasing since 1987.

The apparent resurgence follows years of proportionate decline in property taxation, as tables 2 and 3 indicate. The decline left property taxes robust enough to retain collective rank as the most prolific revenue source within local control. In fiscal 1990, local governments derived almost 26 percent of their "all sources" revenue from property taxes, which still account nationwide for three of every four local tax dollars. This does not happen in every jurisdiction, by any means. Property taxes are very local, varying substantially among jurisdictions. The ten largest cities illustrate the variety encountered (see table 4). In fiscal 1990, as a percentage of all taxes, the property taxes component varied from 19.5 percent in Philadelphia where a local income tax provided almost three of every five tax dollars, to 58.5 percent in Dallas. Twenty years earlier, for 1969-70, the range extended from 31.5 percent in Philadelphia to 69.3 percent in San Antonio. In fiscal 1980, Philadelphia was again lowest at 24.8 percent. At the top end was Houston which derived 62.2 percent of tax collections from property taxes.

The ten largest cities also underscore the growth of current charges as a revenue source, in the wake of Proposition 13. As table 4 indicates, the revenue share for current charges was higher in fiscal 1990 than what it was 20 years earlier for each city, though in Chicago, San Diego, and San Antonio it declined during the first decade. In four cities in fiscal 1990, Los Angeles, San Diego, Detroit, and Phoenix, current charges accounted for a greater revenue share than did property taxes.

Figures in table 5 show that the entire property tax base, including real and personal property, has increased from $280 billion in assessed value in 1956 to 6 trillion in 1989. Locally assessed realty has grown from assessed value totals of $210 billion in 1956 to $5.2 trillion in 1989. At market value the corresponding estimates are $700 billion in 1956 and $14.5 trillion in 1989.

Valuation remains the most important of the tax assessor's functions, especially so since technology, with the advent of computer assisted mass appraisal (CAMA) has made possible realistic annual use of the three approaches to value (comparison of sale, capitalized income, and depreciated replacement cost). In former days constraints of time forced assessors to rely on "the manual," a collection of costs per unit of area, with additions or deductions for departures from standard structure. Uniformity at market value remained a nominal standard, observed more in the breach than in the observance until the highest courts in three states, New Jersey in 1957, Kentucky in 1964, and New York in 1975, held in effect that assessors should follow the applicable law, setting assessments at market value if that is what the law required.* Legislative action in New Jersey and New York precluded any ascent to market value but did improve uniformity among specified groups.

Controverted assessed values have seldom come before the U.S. Supreme Court. One that did back in 1923 generated enduring support for the notion that uniformity

---

*The cases were Switz v. Middletown Township, 23 NJ 580, 130 A2d 15 (1957, New Jersey); Russman v. Luckett, 391 SW 2d 694 (1965, Kentucky); and Hellerstein v. Assessor of Town of Islip, 37 NY 2d 1, 332 NE 2d 279 (1975).

is the prime goal of tax assessors (though uniformity now has its own meanings, see *Nordlinger v. Hahn* below). In 1923, a plaintiff's property was assessed at 100 percent of true value to conform with state law, while other real estate locally was assessed at 55 percent of such value. The Court remanded the case, ordered reduction to the 55 percent of the assessed value on plaintiff's property, and said that "... when it is impossible to secure both the standards of true value, and the uniformity and equality required by law, the latter requirement is to be preferred as the just and ultimate purpose of the law."*

This year, on June 18, 1992, the U.S. Supreme Court found a "rational basis" for uniformity within a classification, when it affirmed the California Court of Appeal in *Nordlinger v. Hahn* (No. 90-1912), concluding that the radical Proposition 13 (formally Article XIIIA, added to the California constitution by 64.8 percent of the voters on June 8, 1978) did not violate the Equal Protection Clause of the U.S. Constitution. Proposition 13 limited realty taxes to 1 percent of "full cash value," the latter meaning assessed value as of base year 1975-76 initially, subject annually thereafter to decreases without limit and to increases of no more than 2 percent of base year value, subject as well to immediate revaluation at market value when a sale or other qualifying "change in ownership" occurs. Ms. Stephanie Nordlinger, a Los Angeles resident, bought a house there in November 1988 for $170,000, and as a result paid 5 times what her neighbors pay in property taxes. Her $1700 property tax bill was virtually identical to that for a Malibu beach front home worth $2.1 million (and not sold since 1975).

In a decision written by Justice Blackmun (Justice Stevens dissenting), the Court by a vote of 8 to 1, found in Proposition 13's "acquisition value," no jeopardy to fundamental interest and no categorizing according to any inherently suspect characteristic that would justify "heightened scrutiny" for possible violation of the Equal Protection Clause. The Court ruled out any violation of Ms. Nordlinger's right to travel, alleged to flow out of exemptions, from "change in ownership" reappraisal, that apply to: (1) transfers from parents to children of principal residence and from the first $1 million of value in other realty and (2) exchange of principal residence of any homeowner 55 years or older for home of equal or lesser value in same county. Eligibility for both hinges on California residency.

The Court said a "rational basis" exists for the "acquisition value" base year system. Classification based on time of purchase, opined the Court, legitimately discourages rapid turnover by permitting older owners to pay less taxes than new owners. An existing owner, has "reasonable reliance interests" (in a locked in assessed value) more worthy of protection than those of a new owner. Invoking like rationale, the Court said the exemptions foster neighborhood and family stability. Justice Stevens strongly dissented, noting that the "Squires" (his term), who are the 44 percent of homeowners with 1975-76 base year assessments, paid only 25 percent of total taxes paid by homeowners in 1989. Stevens called the two exemptions "medieval" privileges.

---

*Sioux City Bridge Co. v. Dakota County, 260 U.S. 441, 43 S Ct 190, 67 L Ed 340 (1923).

The Supreme Court insisted that its 1989 decision in *Allegheny Pittsburgh Coal Co. v. Webster County** is not applicable. In *Allegheny*, in a state (West Virginia) where the "Constitution and laws provide that all property of the kind held by petitioner shall be taxed at a rate uniform ...according to its estimated market value," the assessor used recent sales only for assessing the four coal companies, with no corresponding adjustments in the county. As a result, their properties were assessed 8 to 35 times higher than they should have been to be compatible with other assessments in the county. No inference was possible, from Webster County practice, that an "acquisition value" system was intended.

The U.S. Supreme Court has spoken, but the issue of whether "acquisition value" and "market value" are equally equitable bases for apportioning property tax burdens remains alive. *Nordlinger* simply added "time of acquisition" to "de facto level" and "market level" as classes within which states may require that values be uniform. A particular "value" figures in each: value at acquisition date, value at de facto level, and value at market level. Each may well have been a market value at some moment. Sales prices figure in all three, as transformation of assessing continues, often in GIS-CAMA environments.

To All, Some, or None -- At What Cost?

Precisely in such environments, now and in the future, the nature of information in assessors records assumes new importance. It is geographic, it is valuable, notably in its digital form, and most of it is public. What has this joinder of law and technology wrought? In job terms, it frequently means an added duty for the assessor, that of custodian of the most comprehensive database of property oriented, largely public information available.

What makes it public? Answer: its nature and genesis. It is communicated knowledge arising out of performance of a public function. Chances are this makes it publicly accessible, regardless of physical form. Accessibility is likely to be unfettered, in keeping with the idea behind the federal Freedom of Information Act (FOIA), passed in 1966, that the aim of such laws is "to ensure an informed citizenry" as something "vital to the functioning of a domestic society, needed to check against corruption and to hold the governors accountable to the governed..."**

The federal FOIA, applicable only to records of the 13 cabinet departments and 71 independent agencies in the executive branch of the Federal Government (excluding the 11 agencies in the executive office of the President), was most recently amended in 1987 when it acquired its first fee schedule spelling out the "full allowable direct costs" that fees may recover for the four categories of requesters

---

*488 U.S. 336 (1989).

**NLRB v. Robbins Tire and Rubber Company, 437 US 214 (1978).

903

specified (commercial use, educational and non-commercial scientific institution, requesters who are representative of the media, and all other requesters).[*]

In the United States each state has its own Freedom of Information (also called Open Records) law. They range from liberal (information arising from performance of a public function is public unless specifically designated otherwise) to conservative (information public if required by law to be kept). All still evolve, as new technology impinges on meaning of access, disclosure, cost, privacy, and value.

Access can relate to rights to see information, public or not. Under common law a person can see a public record (defined here as one required by law to be kept) if that person "has an interest in it" (is a litigant). A litigant can gain access also via civil or criminal "discovery," a pre-trial procedure enabling plaintiff to effect disclosure of defendant's information as a prerequisite for going ahead with trial. With respect to public information, access carries with it the right to disclose it unless an applicable law specifically prevents such action. Privacy in contrast, is a personal right to control collection, possession and release of information. Disclosure of information subject to privacy constraints cannot legally occur unless it is specifically authorized by applicable law.

Cost of public information is the sum total of public funds spent to acquire, develop, and maintain it. It may or may not be the same as value, given the influence of price level changes on individual information components, and given also any enhancement from synergistic performance of an information entity (a database, for example). Regardless of cost and any enhancing or other effect on value, accessibility of public information, as the crucial purpose respecting the public's right to know, means keeping fees to a minimum, or eliminating fees entirely when possible. This collides, in some views, with the notion that retaining some control while permitting limited access is valid where a costly data base is involved.

Subject to details in state statutes, fees for copies must be "reasonable" or limited to the cost of reproduction. In Kentucky anyone requesting a copy of all or part of a database or GIS must provide a certified statement containing details of the commercial purpose and use intended. Resulting contract with the public agency will show cost to be charged, based on cost to the agency in time, equipment, and personnel used to produce the database or system, and based also on value of commercial purpose for which the product will be used.[**] In states where statutes are silent regarding subsequent use, such use may be irrelevant to access.

---

[*]See Freedom of Information Act of 1986 (Public Law 99-570). For fee schedule, see Office of Management and Budget (OMB), FR Document 87-6951, filed 3-26-87, printed in Federal Register, Vol. 52, No. 59, March 27, 1987. Reprinted as Appendix E in Guidebook to the Freedom of Information and Privacy Acts compiled and edited by Justin D. Franklin and Robert F. Bouchard, 2nd edition, Clark, Boardman, Callaghan, 2 Corporate Drive, Cranbury, NJ 08512, February 1992.

[**]KRS Sec. 61 970.

Iowa has a provision* which states that "notwithstanding subsections 1 and 2 (establishing right to examine and copy public records), a government body which maintains a geographic computer data base is not required to permit access to or use of the data base by any person except upon terms and conditions acceptable to the governing body. The governing body shall establish reasonable rates and procedures for the retrieval of specified records, which are not confidential records, stored in the data base upon the request of any person."

This removal of a database from a requirement that access be unlimited is crucial to cost recovery and consistent with the idea some have that licensing agreements offer the best opportunity to retain control over a database and still permit access.

The accessibility spectrum extends from unfettered access to privacy. Preserving the latter means preventing any disclosure that the person named or affected by the information in question does not specifically authorize. Most of the states have either a separate privacy law or sections within other statutes that deal with privacy.**

Data Sharing, Sometimes Called Cooperation
In all appraisal, the primus inter pares among the three approaches to value is sales price at which a representative comparable property changed hands. With enough comparable sales, the market data approach is the way to go. In the so-called "real world," however, sales may be infrequent, not representative, not comparable. Thus any "mass infusion" of market data is a "plus." One such plus can take the form of affidavits completed to comply with a state or local transfer tax, a levy now existing in 36 states and the District of Columbia, on the amount paid for a sold property. In as many as 26 states affidavits containing sales price information are required under state law, and yet assessors are often left with insufficient sales.

The reporting environment changed, however, in 1986. As part of the Tax Reform Act of 1986, the party responsible for the closing is required to report the gross proceeds to the IRS under Section 6045(e) of the Code. Before TRA 86, IRS had required brokers to report "gross proceeds" for sales involving securities, commodities, regulated future contracts, and precious metals. Regarding realty sales, taxpayers were supposed to report (voluntarily) sale or exchanges on Form 2119 (gain or loss regarding sale of assets used in business), Form 4797 (gain or loss from involuntary conversion), and Form 6252, (computation of installment sale income). Such reporting was unsatisfactory. Compliance with Section 6045(e) means mandated reporting on Form 1099-S (see Exhibit 1) for all sales closing on January 1, 1991, or later, regardless of use of the sold property, subject to specified exceptions. It is expected that sales reported will eventually total 8 to 10 million annually. Actual and projected filings for the 90's are shown in table 6. Data sharing agreements through

---

*Iowa Code, Section 22.2, Subsection 3.

**For an excellent summary of state Open Records Laws, together with comprehensive analysis, see "A Survey of State Open Records Laws in Relation to Recovery of Data Base Development Costs: An Ends in Search of a Means" by Lori Peterson Dando, in URISA Proceedings 1992, Making Connections, Volume IV, pp. 85-110.

which IRS might make Forms 1099-S available to other governments (federal, state, or local) would seem possible under Section 6103 of the Code, as slightly amended.

Gaining access to sales and assessed value information can now occur in other ways. A legal referencing firm can now provide online access to individual assessed values, and frequently sales prices and other attribute data as well, within databases built up from public information in assessment and recording records in 30 states and the District of Columbia. A second firm makes available machine readable data on assessed values and sale prices for hundreds of assessing jurisdictions. Smaller firms do the same thing on a smaller numbers of assessing jurisdictions.

Outlook for the Future

Given the technology, it is tempting to conclude that the future will only slightly resemble the past. In important respects that is likely to be true. Indeed, our push-button, menu-driven world frequently has us wondering how efficacious such truth would be. Especially in such moments we can take comfort from the equally logical conclusion, Providentially rooted, that we humans were here first, to invent, test, refine, implement, and maintain the technology, and then repeat the process. The inference that began this "eclectic look" is unchallenged. In fulfilling tomorrow's goal, each will have a role in an integral whole.

Table 1 - Tax Revenue, All Levels of Governments, United States,

Years Indicated

Amount ($billions)

Taxes	1978	1979	1980	1981	1982	1983	1984	1985	1986	1987	1988	1989	1990	1991
All Taxes - Federal, State Local	na	539.9	590.3	671.1	669.4	689.9	756.4	821.4	863.6	962.5	1,014.5	1,100.0	1,151.1	1,172.0
All Federal Taxes	na	328.0	360.0	414.8	395.3	391.0	423.4	460.8	480.1	549.0	574.8	618.1	643.9	641.0
Income taxes	na	292.5	315.4	351.6	340.2	334.1	360.5	399.4	420.8	484.9	506.3	552.1	571.1	563.5
Individual	na	227.1	252.0	289.9	296.2	292.2	301.9	336.8	354.5	397.7	408.0	452.2	474.2	466.4
Corporate net income	na	65.4	63.4	61.7	44.0	41.9	58.6	62.6	66.3	87.2	98.3	99.9	96.9	97.1
Customs duties	na	7.7	7.4	8.7	8.7	9.2	12.0	12.3	13.7	15.6	16.6	16.5	16.8	16.3
Motor fuels	na	5.1	4.9	4.6	4.9	7.6	10.7	11.3	12.1	12.0	13.9	13.7	13.5	16.6
All State and Local Taxes	198.3	211.3	230.3	256.3	274.1	299.0	333.1	360.7	383.5	413.5	493.7	481.9	507.2	531.0
Income Taxes	46.6	51.8	57.4	63.4	66.3	73.3	85.4	91.5	98.1	109.9	114.7	126.7	129.2	133.5
Individual	35.5	38.8	43.6	48.9	52.8	59.3	68.8	77.6	76.7	86.9	90.0	101.5	106.1	110.4
Corporate net income	11.1	13.0	13.8	14.5	13.5	14.0	16.6	17.9	21.4	23.0	24.7	25.2	23.1	23.1
Property taxes	64.9	65.5	69.6	78.4	86.3	93.4	101.0	110.9	116.5	123.3	130.1	145.5	158.4	172.0
General sales tax	44.3	49.2	53.1	58.6	61.9	70.2	79.7	86.2	92.7	99.2	109.5	119.1	123.2	123.7
Motor fuels taxes	9.8	10.0	10.2	10.2	10.7	11.2	13.1	14.0	15.2	16.8	18.0	19.1	20.7	21.9
Tobacco product sales	3.8	3.8	4.0	4.1	4.2	4.2	4.4	4.5	4.8	4.9	5.0	5.5	6.0	6.2
Alcoholic beverage sales	2.5	2.6	2.7	2.9	2.9	3.0	3.2	3.3	3.3	3.4	3.4	3.4	3.6	3.7
Motor Vehicle; Ops. licenses	5.4	5.6	5.9	6.3	6.7	7.0	7.6	8.4	9.0	9.7	10.2	10.9	11.4	11.9
All other	21.0	22.9	27.4	32.6	34.6	36.1	39.2	42.7	43.9	46.3	48.8	51.1	54.8	58.0

Percent distribution

Taxes	1978	1979	1980	1981	1982	1983	1984	1985	1986	1987	1988	1989	1990	1991
All Taxes - Federal, State Local	100.0	100.0	100.0	100.0	100.0	100.0	100.0	100.0	100.0	100.0	100.0	100.0	100.0	100.0
All Federal Taxes	na	60.8	61.0	61.8	59.1	56.7	56.0	56.1	55.6	57.0	56.7	56.2	55.9	54.7
Income taxes	na	54.2	53.4	52.4	50.8	48.4	47.7	48.6	48.7	50.4	49.9	50.2	49.6	48.1
Individual	na	42.1	42.7	43.2	44.2	42.4	39.9	41.0	41.0	41.3	40.2	41.1	41.2	39.8
Corporate net income	na	12.1	10.7	9.2	6.6	6.1	7.7	7.6	7.7	9.1	9.7	9.1	8.4	8.3
Customs duties	na	1.5	1.3	1.3	1.3	1.1	1.6	1.5	1.6	1.6	1.6	1.1	1.5	1.4
Motor fuels	na	0.9	0.8	0.7	0.7	1.1	1.4	1.4	1.4	1.2	1.4	1.2	1.2	1.4
All State and Local Taxes	na	39.1	39.0	38.2	40.9	43.3	44.0	43.9	44.4	43.0	43.3	43.8	44.1	45.3
Income Taxes	na	9.6	9.7	9.4	9.9	10.6	11.3	11.1	11.4	11.4	11.3	11.5	11.2	11.4
Individual	na	7.2	7.4	7.3	7.9	8.6	9.1	9.0	8.9	8.9	8.9	9.2	9.2	9.4
Corporate net income	na	2.4	2.3	2.2	2.0	2.0	2.2	2.2	2.5	2.4	2.4	2.3	2.0	2.0
Property taxes	na	12.1	11.8	11.7	13.0	13.5	13.3	13.4	13.5	12.8	12.8	13.3	13.8	14.7
General sales tax	na	9.1	9.0	8.7	9.2	10.2	10.5	10.5	10.7	10.3	10.8	10.8	10.7	10.6
Motor fuels taxes	na	1.9	1.7	1.5	1.6	1.7	1.7	1.7	1.8	1.7	1.8	1.8	1.8	1.9
Alcoholic beverage sales	na	0.5	0.4	0.4	0.4	0.4	0.4	0.4	0.4	0.4	0.3	0.3	0.3	0.3
Motor Vehicle; Ops. licenses	na	1.0	1.0	0.9	1.0	1.0	1.0	1.0	1.0	1.0	1.0	1.0	1.0	1.0
All other	na	4.2	4.2	4.9	5.2	5.2	5.2	5.2	5.1	4.8	4.8	4.6	4.8	4.9

Source: Quarterly Summary of Federal, State, and Local Tax Revenue, GT-91-04, October-December 1991, U.S. Department of Commerce, Bureau of the Census, Washington, D.C., Issued July 1992, pp. 3, 4, 5. Plus earlier editions for periods shown.

Table 2. Revenue from Property Taxes, Years Indicated
(Dollar amounts in millions)

	State and Local Governments			State Governments			Local Governments		
	Property Tax Revenue	Percent of Revenue From All Sources	Total Tax Revenue	Property Tax Revenue	Percent of Revenue From All Sources	Total Tax Revenue	Property Tax Revenue	Percent of Revenue From All Sources	Total Tax Revenue
1989-90	155,613	15.1	31.0	5,848	0.9	1.9	149,765	25.8	74.5
1986-87	121,227	14.4	29.9	4,609	0.9	1.9	116,618	24.8	73.7
1981-82	81,918	15	30.8	3,113	1.0	1.9	78,805	25.2	76.0
1978-79	64,944	16	31.6	2,490	1.0	2.0	62,453	26.6	77.5
1977-78	66,422	17.9	34.3	2,364	1.1	2.1	64,058	29.9	79.7
1976-77	62,527	18.5	35.5	2,260	1.1	2.2	60,267	30.7	80.5
1971-72	42,133	22.2	38.7	1,257	1.1	2.1	40,876	36.1	83.5
1966-67	26,047	24.4	42.7	862	1.4	2.7	25,186	39.0	86.6
1961-62	19,056	27.4	45.9	640	1.7	3.1	18,416	42.6	87.9
1956-57	13,097	28.5	45.1	479	1.9	3.3	12,618	43.4	89.0

Sources: Taxable Property Values, Vol. 2 1987 Census of Governments, GC 87(2)-1, p. XIV; also Governmental
Finances: 1989-90 GF/90-5, table 6. Both from U.S. Bureau of the Census, Washington, DC.

Table 3. Percent Distribution, State and Local Government Revenue Structure, Years Indicated
(Dollar amounts in millions)

Item	State Governments					Local Governments					Exhibit: Revenue 1989-90
	1957	1966-67	1976-77	1986-87	1989-90	1957	1966-67	1976-77	1986-87	1989-90	
Revenue from all sources	100.0	100.0	100.0	100.0	100.0	100.0	100.0	100.0	100.0	100.0	580,193
Intergovernmental Revenue	15.9	23.4	23.8	19.8	20.0	26.2	31.2	39.2	33.3	32.9	190,723
From federal govt.	14.2	22.3	22.5	18.5	18.7	1.2	2.9	8.5	4.2	3.2	18,449
From State Govts.						25.0	28.3	30.7	29.1	29.7	172,274
From Local Govts.	1.7	1.1	1.3	1.3	1.3						
Taxes, All Types	58.9	52.3	49.4	47.8	47.5	49.9	44.8	38.1	33.7	34.7	201,130
Property Taxes	1.9	1.4	1.1	0.9	0.9	43.4	39.0	30.7	24.8	25.8	149,765
Income, Individual	6.3	8.0	12.5	14.7	15.2	0.7*	1.4 *	1.9*	1.6	1.6	9,563
Income, Corporate	4.0	3.7	4.5	4.0	3.4					0.3	1,815
Sales and Gross Receipts	34.2	30.4	25.6	23.2	23.3	3.5	3.0	4.2	5.2	5.3	30,815
Other	12.4	8.8	5.8	5.0	4.7	2.3	1.5	1.3	1.6	1.4	8,354
Current Charges	5.0	6.9	6.2	6.0	6.8	8.6	9.7	9.7	11.6	12.5	72,795
Misc. Gen. Revenue	2.8	2.7	3.6	7.4	7.6	3.5	4.1	4.2	8.6	8.2	47,674
Other Than Gen. Revenue	17.3	14.7	16.9	18.9	18.2	11.8	10.1	8.8	12.6	11.7	67,871

*Corporate income taxes combined with individual income taxes for local govts. in 1976-77 and earlier Governmental Finances.
Sources: Taxable Property Values, Vol. 2 1987 Census of Governments, page XV, table G. Also, Governmental Finances: 1989-90,
table 6. Both from U.S. Bureau of the Census, Washington, D.C.

908

Table 4. Selected Revenue or Percentage of Total General Revenue
Ten Largest Cities, Years Indicated

	All Taxes			Property Taxes			Current Charges			Intergovernmental Revenue		
	1969-70	1979-80	1989-90	1969-70	1979-80	1989-90	1969-70	1979-80	1989-90	1969-70	1979-80	1989-90
New York City	45.7	42.9	47.3	27.9	20.6	20.6	6.3	8.5	9.5	45.8	44.0	38.5
Los Angeles	58.1	45.2	49.4	31.3	16.0	18.0	12.9	15.9	22.3	17.2	27.3	12.5
Chicago	61.5	44.4	52.7	36.3	19.0	20.7	8.0	6.9	15.4	24.2	43.3	24.8
Houston	73.7	51.3	53.4	46.9	31.9	29.0	10.4	16.4	27.8	3.7	14.1	4.4
Philadelphia	66.3	53.3	61.4	20.9	13.2	12.0	9.3	9.6	9.8	14.4	31.7	23.6
San Diego	45.8	34.1	32.0	24.8	11.2	12.7	15.6	10.5	17.2	18.0	34.7	18.0
Detroit	54.3	30.8	34.4	30.6	15.1	12.0	12.3	14.3	14.8	27.4	49.9	44.4
Dallas	74.0	57.4	58.6	50.7	32.6	34.3	15.2	15.5	22.8	4.0	17.7	4.7
Phoenix	50.8	33.5	33.1	17.4	10.8	12.2	11.5	12.4	18.8	26.0	45.8	32.7
San Antonio	59.2	32.2	37.7	41.0	17.9	20.8	15.8	15.3	16.1	5.9	37.7	18.0

909

Table 5 - Summary, Gross Assessed Value and Indicated Market Value
In the United States, and Changes Therein,
Type of Property and Periods Indicated

Year or period	All taxable property (including State-assessed)		Locally assessed realty		
	Assessed Value		Assessed Value		Indicated
	Amount ($ billions)	Percent Change	Amount ($ billions)	Percent Change	Market Value ($ billions)
1956	280		210		700
1961	366		281		970
1956-61		30.5		33.8	
1966	499		393		1,277
1961-66		36.4		39.9	
1971	718		574		1,755
1966-71		43.8		46.1	
1976	1,229		993		3,202
1971-76		71.2		73.0	
1981	2,958		2,515		6,760
1976-81		140.7		153.2	
1986	4,818		4,105		10,250p
1981-86		62.3		63.0p	
1989	6,013		5,220		14,500p

p - personal estimate by author.

Source: *Taxable Property Values and Assessment-Sales Price Ratios*, vol. 2, 1982 Census of Governments. Adapted by author from Tables A and C, pp. X and XII, and from material on pages XXXIII and XXXIV. Also, *Taxable Property Values*, vol. 2, 1987 Census of Governments, table 3. Both from U.S. Department of Commerce, Bureau of the Census, Washington, D.C., issued in 1984 and 1989 respectively. 1989 data from unpublished memoranda of Governments Division, Taxation Branch, January 14, 1992.

Table 6 - *Selected Withholding and Information Forms*
Filed on Paper and Tape, Filing Years 1991-98
(Data in Thousands)

| | Form W-2 | Form 1099-S | | |
	Tape	Paper	Tape	Total
Actual				
1990	203,578	1097	2946	4043
Projected				
1991	198,471	907	2613	3520
2	192,261	1114	3322	4436
3	199,225	1114	3378	4492
4	203,359	1188	3633	4821
5	206,730	1250	3837	5087
6	208,602	1261	3881	5142
7	209,400	1240	3821	5061
8	208,619	1225	3773	4998

Source: **Projections: Information and Withholding Documents Calendar Years 1991-98, United States and Service Centers**, Dept. of the Treasury, Internal Revenue Service, Research Division, Document 6961 (Rev. 4-91), Catalog No. 63437X, Washington, DC 20224, 1991, Update, Tables 2 and 3.

EXHIBIT 1

FILER'S name, street address, city, state, and ZIP code		☐ VOID ☐ CORRECTED	OMB No. 1545-0997	Proceeds From Real Estate Transactions
			1992	
FILER'S Federal identification number	TRANSFEROR'S identification number	1 Date of closing (MMDDYY)	2 Gross proceeds $	Copy A For
TRANSFEROR'S name		3 Address or legal description (including city, state, and ZIP code)		Internal Revenue Service Center
Street address (including apt. no.)				File with Form 1096. For Paperwork Reduction Act Notice and instructions for completing this form, see
City, state, and ZIP code				
Account number (optional)		4 Check here if the transferor received or will receive property or services as part of the consideration ▶ ☐		Instructions for Forms 1099, 1098, 5498, and W-2G.

Form **1099-S**   Cat. No. 64292E   Department of the Treasury - Internal Revenue Service

Do NOT Cut or Separate Forms on This Page

911